Trends and Innovations in Urban E-Planning

Carlos Nunes Silva
University of Lisbon, Portugal

A volume in the Advances in Electronic Government, Digital Divide, and Regional Development (AEGDDRD) Book Series

IGI Global
PUBLISHER of TIMELY KNOWLEDGE

Published in the United States of America by
IGI Global
Information Science Reference (an imprint of IGI Global)
701 E. Chocolate Avenue
Hershey PA, USA 17033
Tel: 717-533-8845
Fax: 717-533-8661
E-mail: cust@igi-global.com
Web site: http://www.igi-global.com

Library of Congress Cataloging-in-Publication Data

Names: Silva, Carlos Nunes, editor.
Title: Trends and innovations in urban e-planning / Carlos Nunes Silva, editor.
Description: Hershey, PA : Information Science Reference, [2022] | Includes bibliographical references and index. | Summary: "The aim of this book is to provide an updated picture of some of the main trends and recent innovations in the field of urban e-planning through the critical perspectives of authors from different disciplinary backgrounds and geographies"-- Provided by publisher.
Identifiers: LCCN 2021051152 (print) | LCCN 2021051153 (ebook) | ISBN 9781799890904 (hardcover) | ISBN 9781799890911 (paperback) | ISBN 9781799890928 (ebook)
Subjects: LCSH: City planning--Information technology. | Land use, Urban--Information technology.
Classification: LCC HT166 .T744 2022 (print) | LCC HT166 (ebook) | DDC 307.1/2160285--dc23/eng/20211029
LC record available at https://lccn.loc.gov/2021051152
LC ebook record available at https://lccn.loc.gov/2021051153

This book is published in the IGI Global book series Advances in Electronic Government, Digital Divide, and Regional Development (AEGDDRD) (ISSN: 2326-9103; eISSN: 2326-9111)

British Cataloguing in Publication Data
A Cataloguing in Publication record for this book is available from the British Library.

For electronic access to this publication, please contact: eresources@igi-global.com.

Advances in Electronic Government, Digital Divide, and Regional Development (AEGDDRD) Book Series

Zaigham Mahmood
University of Derby, UK & North West University, South Africa

ISSN:2326-9103
EISSN:2326-9111

Mission

The successful use of digital technologies (including social media and mobile technologies) to provide public services and foster economic development has become an objective for governments around the world. The development towards electronic government (or e-government) not only affects the efficiency and effectiveness of public services, but also has the potential to transform the nature of government interactions with its citizens. Current research and practice on the adoption of electronic/digital government and the implementation in organizations around the world aims to emphasize the extensiveness of this growing field.

The Advances in Electronic Government, Digital Divide & Regional Development (AEGDDRD) book series aims to publish authored, edited and case books encompassing the current and innovative research and practice discussing all aspects of electronic government development, implementation and adoption as well the effective use of the emerging technologies (including social media and mobile technologies) for a more effective electronic governance (or e-governance).

Coverage

- ICT Infrastructure and Adoption for E-Government Provision
- Social Media, Web 2.0, and Mobile Technologies in E-Government
- E-Government in Developing Countries and Technology Adoption
- E-Citizenship, Inclusive Government, Connected Government
- Frameworks and Methodologies for E-Government Development
- Adoption of Innovation with Respect to E-Government
- Urban Development, Urban Economy
- E-Governance and Use of Technology for Effective Government
- ICT within Government and Public Sectors
- Knowledge Divide, Digital Divide

IGI Global is currently accepting manuscripts for publication within this series. To submit a proposal for a volume in this series, please contact our Acquisition Editors at Acquisitions@igi-global.com or visit: http://www.igi-global.com/publish/.

The Advances in Electronic Government, Digital Divide, and Regional Development (AEGDDRD) Book Series (ISSN 2326-9103) is published by IGI Global, 701 E. Chocolate Avenue, Hershey, PA 17033-1240, USA, www.igi-global.com. This series is composed of titles available for purchase individually; each title is edited to be contextually exclusive from any other title within the series. For pricing and ordering information please visit http://www.igi-global.com/book-series/advances-electronic-government-digital-divide/37153. Postmaster: Send all address changes to above address.

Titles in this Series

For a list of additional titles in this series, please visit: www.igi-global.com/book-series

Planning and Designing Smart Cities in Developing Nations
Saleem Gregory Zoughbi (Independent Researcher, Palestine)
Information Science Reference • © 2022 • 282pp • H/C (ISBN: 9781668435090) • US $215.00

Multidimensional Approach to Local Development and Poverty Causes, Consequences, and Challenges Post COVID-19
João Conrado de Amorim Carvalho (Centro de Ensino Superior Dom Bosco, Brazil) Francisco Espasandín Bustelo (Universidad de Sevilla, Spain) and Emmanuel M. C. B. Sabino (Centro de Ensino, Formação e Pesquisa. Brazil)
Information Science Reference • © 2022 • 353pp • H/C (ISBN: 9781799889250) • US $195.00

Blockchain Technologies and Applications for Digital Governance
Pradeep Nijalingappa (Bapuji Institute of Engineering and Technology, India) and Mangesh Manikrao Ghonge (Sandip Foundation's Institute of Technology and Research Centre, India)
Information Science Reference • © 2022 • 266pp • H/C (ISBN: 9781799884934) • US $225.00

Advancements in Global Cyber Security Laws and Regulations
Shahid M. Shahidullah (Hampton University, USA) Zina McGee (Hampton University, USA) and Carla Miller Coates (North Carolina Agricultural and Technical State University, United States)
Information Science Reference • © 2022 • 305pp • H/C (ISBN: 9781799845850) • US $195.00

Handbook of Research on Nurturing Industrial Economy for Africa's Development
Frederick Muyia Nafukho (Texas A&M University, USA) and Alexander Boniface Makulilo (University of Dodoma, Tanzania)
Information Science Reference • © 2021 • 510pp • H/C (ISBN: 9781799864714) • US $245.00

Global Perspectives on the Impact of Mass Media on Electoral Processes
Stella Amara Aririguzoh (Covenant University, Nigeria)
Information Science Reference • © 2021 • 187pp • H/C (ISBN: 9781799848202) • US $195.00

Impact of Political Socialization on the Support for Democratic Principles Emerging Research and Opportunities
Allison Clark Ellis (University of South Carolina Upstate, USA)
Information Science Reference • © 2021 • 204pp • H/C (ISBN: 9781799842910) • US $175.00

IGI Global
PUBLISHER of TIMELY KNOWLEDGE

701 East Chocolate Avenue, Hershey, PA 17033, USA
Tel: 717-533-8845 x100 • Fax: 717-533-8661
E-Mail: cust@igi-global.com • www.igi-global.com

Table of Contents

Detailed Table of Contents

E-government in Europe has changed significantly in the years since 2017. This chapter begins by briefly summarizing the nature and impacts of e-government. It then offers a broad overview based on the continent's internet penetration rates and the United Nations' e-government readiness and e-participation scores. Third, it points to national variations in e-government implementation, noting successes and failures, and stresses the significant geographic unevenness among and even within countries. It discusses these patterns in light of the continent's digital divide and varying governance structures. The fourth part addresses technical and social changes in the period 2017-2021, including cloud computing, mobile e-government, trans-national e-governance, and the covid pandemic. Throughout, the chapter emphasizes the temporal and spatial variability in e-government practices as a cautionary tale against simplistic interpretations that eschew local contexts.

A growing interest in participatory planning practices in local governments in Poland resulted in the development of new methods supporting citizens' engagement in decision-making processes. The latest solutions in the public participation methods catalog are based on geoweb tools. One of such tools is the geo-questionnaire developed within the project: Geoportal supporting public participation in spatial planning (GEOPLAN) – an R&D project conducted by the interdisciplinary team of researchers from Adam Mickiewicz University in Poznań supported by IT specialists and urban planning practitioners. The chapter presents the features of PPGIS tools, the development of PPGIS in Poland, the evaluation of the pilot implementations of geo-questionnaires in decision-making processes, and the future of PPGIS in the light of new post-pandemic reality.

Civic technology models propose that urban environmental sensing and analysis can inform community-based decision making, but technology-driven planning and design is still subject to systemic challenges including a lack of community trust and local government capacities. This chapter examines the problem of integrating urban sensing into engaged planning and design. The authors assess the outcomes of two planning and design efforts in Virginia, USA utilizing the community-centered urban sensing (CCUS) and engagement platform. Developed by planners, urban designers, landscape architects, and data scientists at the University of Virginia, CCUS addresses the need for actionable information on the urban environment through community-engaged urban data collection and analysis. The cases provide evidence for the importance of engagement around sensing itself. Based on these findings, the authors propose a conceptualization of urban sensing applications that accounts for a range of local contexts from high capacity and trust to dysfunction and mistrust.

Based on a case study from Rennes, the chapter presents how a group of urban public actors re-uses methods and technology from citizen sciences to raise the urban air quality issue in the public debate. The project offers to a group of inhabitants to follow air quality training and proceed PM2.5µm measurements. The authors question the impact of the ongoing hybridisation between citizen science and urban public action on participants' commitment. They present how the use of PM2.5-sensors during 11 weeks led to a disengagement phenomenon, even if we observe a strong participation to workshops. Finally, the authors present a spatial analysis of the measurements produced and analyse how the inhabitants use the sensors in their daily lives. These results come from an interdisciplinary methodology using observations, interviews, and data analyses.

Slum upgrading comprehensive plans and urban regularization plans are two planning tools used to promote integrated interventions in Brazilian slums and illegal settlements. Aiming at urban improvements as well as land regularization and community development, these plans have been, however, criticized for being too technical, time consuming, expensive, and top-down oriented, lacking sufficient participation to achieve community consensus on priorities, under severe budget restrictions in complex and fast changing realities. This chapter discusses the results of recent experiences in Belo Horizonte, Brazil using Geodesign framework and geovisualization strategies for collaborative planning in two illegal

Telemedicine solutions help improve the delivery of and access to healthcare services. The recent pandemic has given a burst to the use of telemedicine services, allowing to overcome or set aside, at least temporarily, some of the difficulties which usually hamper the implementation of innovative services. It has also spurred a new attention to the issues of access to healthcare services, and to the role telemedicine services could play in overcoming or reducing the physical distance between patients and healthcare premises. This study makes a preliminary step to sharpen the topics for the Piedmont region. Data retrieved from the Regional Health Information System are used to profile the health mobility pattern in the regional local areas, through a set of spatial performance indicators. A few hypotheses of the possible impact of telemedicine on health mobility are formulated and their impact is investigated in an exploratory analysis of the regional health flows.

The situation over the COVID-19 pandemic has affected and continues to affect the mobility of the population in several countries. Traditional approaches in transport planning have had to be replaced by new methods using digital technologies and e-planning methods. The COVID-19 pandemic brought the opportunity to use such a situation to enhance feedback from the citizen's perspective. This chapter describes the approach of population involvement through e-planning in the field of sustainable mobility. The case study itself is made based on data from the town of Banská Bystrica, where the process of preparing the strategic document Sustainable Urban Mobility Plan began last year. It can be stated that perhaps also thanks to the COVID-19 pandemic, the population was involved to a much greater extent than in other periods.

Ubiquitous digitalization in our technology-mediated cities has recently pushed the computation from virtual worlds into the cybernetic sensor-guided cites. AI-driven architecture and planning assumedly make cities efficient, sustainable, and "smart" in a linear, mechanical manner. However, cities are complex adaptive, self-organizing systems evolving through transitions. They are hard to control using any planning tool founded on a rational comprehensive model. Computational, generative methods are established tools in the urban design and planning providing novel perspective for smart city planning. Moreover, the paradigm of complex systems and (universal) self-organization enable crossing of ideas in biological systems, cities, and digital systems. This generates intellectual progress in both philosophical and methods development for e-planning tools. Here, the author introduces an e-planning method applying self-organization of information, capable of reflecting the complex urban dynamics. She concludes by illustrating possible futures in self-organizing technology and planners' roles in it.

Preface

The ubiquitous and growing digitalization of all spheres of Local and Urban Governance continuously transforms how Urban Planning is practiced, and raises, at the same time, new and complex ethical, social, and political challenges to the planning process, as Silva (2010, 2013), Anttiroiko (2012), Klosterman (2012), Williamson and McFarland (2012), among others[1], identified and anticipated, over a decade ago. In fact, one of the most sticking features of Urban e-Planning is its continuous change, pushed by groundbreaking innovations in information and communication technologies, applied to the governance of cities and other human settlements, as revealed by the studies published in the International Journal of E-Planning Research during its first decade of existence (Silva, 2021a).

The book 'Trends and Innovations in Urban E-Planning' gives continuity to the collective effort of mapping the technical and social innovations taking place in the field of Urban e-Planning, a project carried out in the context of the 'Urban e-Planning Research Network' (UEPNET), in close association with the International Journal of E-Planning Research (IJEPR). This volume adds new empirical evidence and new insights to the growing library on Urban E-Planning, which already includes the books 'Citizen-Responsive Urban E-Planning: Recent Developments and Critical Perspectives' (Silva, 2020), 'New Approaches, Methods, and Tools in Urban e-Planning' (Silva, 2018) and 'Emerging Issues, Challenges, and Opportunities in Urban E-Planning' (Silva, 2015).

The book examines and discusses some of these shifts through a series of empirical case-studies, located in different regions of the World, and through the discussion of the impacts of some of these innovations on basic planning concepts. It does this through interdisciplinary approaches and methodologies, using observations, interviews, and other forms of data collection and data analysis. The aim is to explore and discuss, through this collection of studies, as the three previous books did, trends and innovations in urban e-planning, in the case of this fourth volume, from e-government infrastructure, to Geo-web tools, autonomous computing, or mobile e-government, among others, which somehow frame new and emerging urban planning paradigms. Although not the main focus of this collection of essays, the impact of the COVID-19 Pandemic in Urban e-Planning is also considered in two of the studies included in the book[2].

The book organizes the chapters into five main groups or sections, each dealing with one of the trends identified: the first section, with one chapter, deals with the increasing digitization of urban government; the second, with three chapters, deals with innovations in citizen e-participation; the third, with two chapters, addresses collaborative e-planning through Geodesign; the fourth, with just one chapter, explores new e-planning tools in response to climate change; the fifth, with two chapters, deal with digital innovations for sustainable mobility; and the sixth and final section, with one chapter, offers a theoretical discussion on self-organization in smart urban planning.

The first of these trends explored in the book is the rapid transformation of public administration into increasingly advanced and more mature forms of e-government, which offers the possibility for the development of modes of e-planning increasingly more sophisticated. This growing maturity in e-government, and its highly unequal geographical distribution, are examined in chapter 1 'E-Government in Europe: Changes and Developments, 2017-2021', in which Barney Warf offers an informed account of the national variations in e-government development in Europe since 2017. The differences among these countries in the level of internet penetration and e-government maturity, and within each of them, do suggest the existence of significant differences in what respects the conditions for the development of Urban e-Planning. The digital divide and the different governance cultures, two dimensions responsible for the spatial variability of Urban e-Planning, among countries and within each of them, are also examined in the chapter.

This trend towards increasingly more mature modes of e-government, in developed countries at least, as the case of Europe examined in Chapter 1 illustrates well, despite the enormous spatial variability, among them and within each of them, associated with changes in the way the role of citizens in the public policy process is perceived in advanced democracies, opened new possibilities for the engagement of multiple actors in the urban planning decision-making process. In the case of urban planning this is taking place through methods and Geo-web tools that facilitate the e-participation of citizens and other stakeholders in the planning process. It is the case of the Internet tools based on GIS. This trend and innovations are explored and examined in the next chapter, 'PPGIS Potential in Local Governance and Spatial Planning: Lessons From Poland' by Edyta Bąkowska-Waldmann and Tomasz Kaczmarek. The chapter discusses one of these innovative tools for citizen e-participation in the planning process, the geo-questionnaire developed within the project GEOPLAN, a Geoportal supporting public participation in spatial planning, through the presentation of the results of six pilot studies of e-consultation performed in Poland. Edyta Bąkowska-Waldmann and Tomasz Kaczmarek discuss the main characteristics of PPGIS tools, assess the possibilities of their use in spatial planning, and explore legal and organizational challenges related to the implementation of e-consultation processes in urban planning.

Related with this second trend—new possibilities for the engagement of multiple actors in the urban planning decision-making process—primarily associated with the potential opened by the extensive use of Geo-web tools, is the trend towards Community-Centred Urban Sensing or, in other words, community-engaged urban data collection and data analysis. This is addressed in Chapter 3, 'Evaluating Urban Sensing Within Engaged Planning and Design', in which Andrew Mondschein and Mona El Khafif examine the problem of integrating urban sensing into engaged planning. The chapter starts from the following research question, 'To what extent enhanced urban data and analysis can augment resident engagement in planning and design, rather than hinder it, even when current urban planning and design practices are dysfunctional?' and addresses it through the analysis of the outcomes of two planning and design initiatives in Virginia, USA, suggesting the need for smart and engaged urban planning to take account of urban dysfunctions, in order to encompass modes and methods that work well even when local urban systems are dysfunctional.

New and innovative modes of citizen engagement in urban planning, in this case through the use of Citizen Science methods and tools, is also the issue addressed by Gwendoline l'Her, Myriam Servières, and Daniel Siret in the following chapter, 'Citizen Commitment in City-Organized Air Quality Monitoring'. Based on a case study in Rennes, France, the authors present and discuss how a group of urban public actors re-used methods and technology from Citizen Science, acting as Citizen-Sensors, to raise the urban air quality issue in the public debate. A group of local citizens received training on air quality measurement before engaging in the project. The findings presented led the authors to question the real and practical impact of hybridisation between citizen science and urban public action on participants' commitment, as some level of disengagement has been noted, a pattern that needs further exploration and analysis in other settings before firm conclusions can be taken. These findings, as well as some of those in the previous chapters, do suggest the need to be careful and critical about simplistic views of the real impact and social consequences of these innovations in urban e-planning.

This trend towards the introduction and implementation of new modes of citizen engagement in the urban planning process is also dealt with in the following two chapters, both centred on the use of the Geodesign framework. In Chapter 5, 'Creating Collaborative Environments for the Development of Slum Upgrading and Illegal Settlement Regularization Plans in Belo Horizonte, Brazil', Rogério Palhares Zschaber de Araújo, Ana Clara Mourão Moura, and Thaisa Daniele Apóstolo Nogueira develop a methodology, based on the Geodesign framework, that seeks to create a collaborative environment and to enhance the participation of citizens and other stakeholders in the context of illegal settlement regularization plans. The findings, in this Brazilian case study, shows a participatory process that proved to be more effective and inclusive for slum upgrading and land regularization than the conventional planning process, which has been largely criticized for being too technical, top-down oriented, marked by lack of sufficient citizen participation, time consuming, and expensive. In other words, conventional urban planning in contexts such as the one examined in this case study proved to be, time and time again, unable to achieve community consensus on priorities, and to face complex problems and fast changing realities, in particular under severe budget restrictions. Also in this case, groundbreaking innovations in planning tools, the use of the Geodesign framework, made a difference, a positive one, in the outcome of the urban planning process.

The analysis of this trend—use of the Geodesign framework for engaging citizens in urban planning—continues in Chapter 6, 'Bridging the Gap Between Strategic Environmental Assessment and Planning: A Geodesign Perspective Revised', in which Michele Campagna, Elisabetta Anna Di Cesare, Andrea Matta, and Matteo Serra explore and discuss the use of the Geodesign approach in a specific case study in Italy, and how it contributed to overcome some of the constraints imposed by regulations aiming at environmental protection and risk reduction. The chapter deals with the role Geodesign can play in the current practice of Urban Planning and on that of Strategic Environmental Assessment, by taking advantage of the Spatial Data Infrastructures in the EU, stressing the innovation associated with it but also the shortcomings and pitfalls.

Another critical trend increasingly observed in urban e-planning is related with the innovations introduced in response to the challenges of climate change. Nano Langenheim and Marcus White in Chapter 7, 'Responding to Future Climatic Conditions of Heat and Flood: A Design Decision Support System for Street Trees', address the escalating intensity and duration of heat and flood events, one of the most challenging issues in contemporary cities, and in futures ones as well. This requires an increase of the green infrastructure, which in turn calls for greater emphasis on the design of green infrastructures capable to respond to climate change challenges. Nano Langenheim and Marcus White demonstrate, in the chapter, the potential of a rapid precinct scaled design approach for street tree planting choices, based on a street tree design decision support system (DDSS), which enables simultaneous assessment of visual impact and environmental performance factors, including shade quantification and flood responsiveness. This decision support system allows the integration of aspects related to climate change alongside community aesthetic preferences, offering at the same time the possibility of designing more resilient and comfortable urban spaces. The DDSS presented by Nano Langenheim and Marcus White draws from advances in computational botany, entertainment industries, light engineering, and hydrology modelling, and was tested at both street and precinct scale using two case studies.

The last two chapters deal both with mobility issues and illustrate both how extreme events, such as the COVID-19 pandemic, can enhance and speed up the adoption of innovative approaches and tools in urban e-planning. In the first of these chapters, 'Health Mobility, Telemedicine, and Local Areas: Exploring the Relationships in the Piedmont Region', Sylvie Occelli and Marco Dalmasso examine how telemedicine can improve the delivery of and access to health care services, a key issue in health policy but also in spatial planning, highlighting how an extreme event such as the COVID-19 pandemic helped to overcome some of the resistances and difficulties which usually slow down the implementation of innovations in the public policy sector. The data collection and data analysis procedures applied in this study are good illustrations of what is being done in urban e-planning through the use of new Information Systems made available by an increasingly digitized public administration. In Chapter 9, 'COVID-19 and Citizen Engagement in the Sustainable Mobility Planning: A Case Study From City of Banska Bystrica', Marian Gogola, Stanislav Kubaľák, and Mikulas Cerny explore the impact of the COVID-19 pandemic on the mobility of the population and examine how traditional approaches in transport planning had to be replaced by new ones based on innovative digital technologies, a move facilitated, as in the case examined in the previous chapter, by the pandemic, an extreme event that required new answers, which among other aspects needed a better and quicker feedback from the citizens. The chapter describes these changes in the city of Banská Bystrica, Slovakia, in the context of the elaboration of the Sustainable Urban Mobility Plan, a strategic spatial plan, in whose preparation were successfully applied e-planning digital tools.

In the last chapter, 'Learning From Self-Organizing Universe: Towards Evolutionary E-Planning', Jenni Partanen offers a critical discussion of how ubiquitous digitalization of our cities has pushed computation from virtual worlds into the cybernetic sensor-guided cites, making them 'smart' in a linear, mechanical manner, although, as Jenni Partanen argues, cities are complex adaptive, self-organizing systems, which evolve through transitions, turning them hard to control through planning tools founded on a rational comprehensive model, as is currently still the case in most cities in the developed and in the developing countries as well. In alternative, new computational methods seem to provide the means for the development of new modes of smart urban planning, as is the case of the e-Planning method that applies self-organization of information to reflect the complex urban dynamics, presented and discussed in this chapter, which raises new philosophical and methodological issues in the field of urban planning, questioning from a post-positivist perspective the role of planners in the urban planning process itself.

The cases examined and discussed in the book illustrate some of the trends identified in recent years in the way Urban e-Planning is developing. If, as in years before, the evidence suggests a potential for a better and more inclusive urban planning, or urban governance more generally, the cases examined also point out for critical issues that require prudence in the conclusions, calling for further research, as it can be pernicious for those with less power in the urban arena. As before, the findings in these essays suggest the existence of a risk of giving over emphasis to technology-oriented approaches in detriment of more citizen-centric approaches, with all the consequences that might entail.

In conclusion, the book, *Trends and Innovations in Urban E-Planning*, developed in the framework of the Urban e-Planning Research Network (UEPNET), offers new light on the use of digital tools in Urban e-Planning. It is of interest for scholars, researchers, students, planners, and policy makers, in particular those working in the field of Urban and Regional Planning, Urban Informatics, Environmental Sciences, and Geography.

Carlos Nunes Silva
University of Lisbon, Portugal

REFERENCES

Anttiroiko, A.-V. (2012). Urban Planning 2.0. *International Journal of E-Planning Research*, *1*(1), 16–30. doi:10.4018/ijepr.2012010103

Klosterman, R. E. (2012). E-Planning: Retrospect and Prospect. *International Journal of E-Planning Research*, *1*(1), 1–4. doi:10.4018/ijepr.2012010101

Silva, C. N. (2010). *Handbook of Research on E-Planning. ICTs for Urban Development and Monitoring*. IGI Global. doi:10.4018/978-1-61520-929-3

Silva, C. N. (2013). *Citizen E-Participation in Urban Governance, Crowdsourcing and Collaborative Creativity*. IGI Global. doi:10.4018/978-1-4666-4169-3

Silva, C. N. (2015). *Emerging Issues, Challenges, and Opportunities in Urban E-Planning*. IGI Global. doi:10.4018/978-1-4666-8150-7

Silva, C. N. (2018). *New Approaches, Methods, and Tools in Urban e-Planning*. IGI Global. doi:10.4018/978-1-5225-5999-3

Silva, C. N. (2020). *Citizen-Responsive Urban E-Planning: Recent Developments and Critical Perspectives*. IGI Global. doi:10.4018/978-1-7998-4018-3

Silva, C. N. (2021). Urban E-Planning and the COVID-19 Pandemic: Public Health Response and Transformative Recovery. *International Journal of E-Planning Research*, *10*(2), vi–xiv.

Silva, C. N. (2021a). A Decade of Urban E-Planning. *International Journal of E-Planning Research*, *10*(3), v–xii.

Williamson, W., & McFarland, P. (2012). Investigating the Role of Electronic Planning within Planning Reform. *International Journal of E-Planning Research*, *1*(2), 65–78. doi:10.4018/ijepr.2012040104

ENDNOTES

1 See the four issues of the first volume (2012) of the International Journal of E-Planning Research (IJEPR), which provides an empirically grounded view of the state of Urban e-Planning a decade ago, how it developed until then, and what were the prospects for its future, as seen by leading experts in the field, at that turning point in the history of Urban Planning.

2 For a developed approach of the impact of the Covid 19 pandemic in urban e-planning and its response to it see the IJEPR, Vol. 10 (2), a special issue on Covid-19 and Urban e-Planning, and Silva (2021).

Acknowledgment

I would like to thank each of the contributors to this volume, for their work and commitment to this project. I am in debt to all colleagues that assisted me in the review process of the original manuscripts. Their useful comments helped to improve the quality of the final manuscripts. I also thank the publishing team at IGI Global.

Carlos Nunes Silva

Editor

Chapter 1
E-Government in Europe:
Changes and Developments, 2017-2021

Barney Warf
University of Kansas, USA

ABSTRACT

E-government in Europe has changed significantly in the years since 2017. This chapter begins by briefly summarizing the nature and impacts of e-government. It then offers a broad overview based on the continent's internet penetration rates and the United Nations' e-government readiness and e-participation scores. Third, it points to national variations in e-government implementation, noting successes and failures, and stresses the significant geographic unevenness among and even within countries. It discusses these patterns in light of the continent's digital divide and varying governance structures. The fourth part addresses technical and social changes in the period 2017-2021, including cloud computing, mobile e-government, trans-national e-governance, and the covid pandemic. Throughout, the chapter emphasizes the temporal and spatial variability in e-government practices as a cautionary tale against simplistic interpretations that eschew local contexts.

E-GOVERNMENT IN EUROPE: CHANGES AND DEVELOPMENTS, 2017-2021

Among the innumerable transformations unleashed by the internet, the growth of electronic government, or e-government, ranks among the most significant. Definitions of e-government vary (Yildiz 2007), but all revolve around the use of the internet to deliver information and services online, streamline administrative procedures, and solicit citizen input and participation. The expansion of Web 2.0, which allows users to interact with websites rather than just passively receive information, has expedited this process considerably. The topic has been extensively scrutinized by scholars (Rocheleau 2007; Dwivedi et al. 2012; Taipale 2013).

Among the world's major geographic regions, Europe has done relatively well in the adoption and implementation of e-government. According to United Nations data on e-government implementation, European countries routinely score among the best in the world in this regard, although some East Asian states (e.g., Japan, South Korea, Taiwan) also rank highly. This observation in and of itself is not

DOI: 10.4018/978-1-7998-9090-4.ch001

surprising: the region is wealthy, with high internet penetration rates, populations that often possess the requisite technical skills, and states that view e-government as a way to improve efficiency, cut costs, and facilitate the delivery of public services. European e-government is also profoundly shaped by the European Parliament, the European Commission, the European Court), whose recommendations are binding on all EU countries. But the status of e-government is highly uneven across the continent, and fluctuates constantly, a reflection of changes in public policies, the public's willingness to use the internet to access public services, and technological changes (e.g., 5G broadband and the growth of the mobile internet), among other things.

Warf (2018) conducted a survey of e-government in Europe that portrayed its uneven status in 2016. It revealed the uneven distribution in the extent of e-government services across the continent, the importance of the digital divide among and within countries, and offered numerous examples of its implementation. A geographic perspective serves as a corrective to common but erroneous "one-size-fits-all" approaches to this issue, emphasizing the importance of local political contexts in the understanding of how e-government unfolds over time and space. The adoption of e-government reflects, among other things, the public's trust in the state, the prevalence of essential digital skills, the effectiveness of political leadership, and how well designed government web pages are, or are not. In short, e-government includes varying practices that reflect varying political, cultural, and institutional contexts (see Bannister and Connolly 2012). Like cyberspace more broadly, e-government is wrapped up with geography. This means that it cannot be understood independently of local and national contexts, including demographic structures, standards of living, educational levels, and political formations.

Since this survey appeared, several changes have unfolded over the European e-government landscape. This essay surveys the changing status of European e-government in 2020. It opens with a brief summary of the types and impacts of e-government. Second, it offers an overview of e-government on the continent, focusing on rising internet penetration rates and two widely used measures of e-government implementation, the e-government development index and the e-government participation score, which reflect the degree of sophistication of e-government and the public's ability and willingness to utilize public services online, respectively. The third part offers some recent examples of e-government implementation in different states to illustrate the enormous variability in its nature and success. Fourth, it focuses on changes in the region's e-government over the last four years and the forces likely to reshape in the near future. The conclusion briefly draws these lines of thought together.

E-GOVERNMENT IN PERSPECTIVE

e-Government includes a wide variety of practices in which residents and firms interact digitally with the state. Common examples include: on-line registration of births, marriages, and deaths; applications for permits and licenses; corporate registrations; payments of fees, dues, tickets, taxes, and utility bills; electronic proposals, bids, and contracts; applications for public services and programs; downloading forms; emergency and disaster alert systems. internet teaching; telehealth; videos of public meetings; and access to online census data, maps, and other types of public records. Some, such as Estonia and France, have toyed with internet voting.

There is a fecund and informative literature on e-government, so there is no need to recapitulate its features in detail here. Research on this issue has revealed to basic types, including government-to-business (G2B), government-to-government (G2G), and government-to-citizens (G2C) (Fountain 2001).

A variety of levels of sophistication can be found, ranging from simple webpages that provide information to two-way digital portals that allow for interaction and citizen feedback. Yet others point to different models of e-government ranging from the managerial to the participatory (Chadwick and May 2003).

e-Government is widely heralded for the potential benefits it offers. Administrations often hope it will generate cost savings. By making information readily available, it reduces uncertainty, improves decision making, and decreases inconvenience. Some claim it reduces corruption, elevates accountability, and improves the efficiency of public bureaucracies (Anderson 2009). Others note it can raise transparency by making public data and records more available. Trips to government offices and waiting times are reduced. When e-government elevates the productivity of the state, public approval of administrations tends to rise. Others note the utility of telemedicine (e.g., teleconsultations with physicians), distance learning programs, and digital access to library resources, providing access to people who may not otherwise enjoy it, notably in remote areas. Others voice concerns that e-government may encourage the hacking of government databases, invasions of privacy, and that it may exacerbate the digital divide between information haves and have-nots (Belanger and Lemuria 2008).

Social and spatial discrepancies in internet access are a major barrier to the successful adoption and implementation of e-government (Yigitcanlar and Baum 2006). Fountain (2001, 48) notes that "An increasingly digital government favors those with access to a computer and the Internet and the skills to use these sophisticated tools competently." Urban-rural differences within many states remain profound. In many countries in which internet penetration rates have risen, the digital divide has been reduced in significance. However, the issue has become more complex over time. Rather than a simple dichotomy of access/non-access, digital divides increasingly reflect a continuum of skills and modes of information delivery. For example, broadband access, which is essential for loading graphics-intensive webpages, has become more important (Riddlesden and Singleton 2014). In accentuating the differences between those skilled at internet use and those who are not, e-government may actually increase the digital divide (Dugdale et al. 2005).

EMPIRICAL CONTOURS OF EUROPEAN E-GOVERNMENT IN 2020

Three dimensions of e-government are particularly important in understanding its uneven geographies: internet penetration rates, the quality of government digital services, and the degree to which the public embraces this mode of information and service delivery.

Among the most important forces affecting e-government is the growth of internet penetration rates, which now average above 82% in the continent (Fig. 1). In Western and Northern Europe, which had much higher penetration rates in 2016 than did Eastern Europe, the internet has become ubiquitous, with penetration rates typically above 95 percent. The greatest growth in penetration rates has occurred in Eastern Europe, where rates were typically much lower; even there, today penetration rates are often above 85%. Ukraine, for example, exhibited a remarkable jump from 43% in 2016 to 93 percent in 2020. The Balkans remains the region with the lowest rates of internet use. The spread of the internet, including mobile access, has greatly improved the ability of many Europeans to access e-government services. Thus Fernandez-i-Marín (2011) found that e-government adoption rates were much higher in countries where internet access was abundant. The rapid growth of internet penetration rates in Eastern Europe implies that the digital divide among countries in the region is waning, although important divisions remain within countries, particularly between rural and urban areas. Nonetheless, the digital divide continues to remain significant in determining access to European e-government services (Botrić and Božić 2020). Age and gender remain important dimensions of this phenomenon, particularly in the lower income, less developed countries in Eastern Europe (Seljan et al. 2020).

Figure 1. Internet penetration rates, 2020
Source: author, using data from internetworldstats.com.

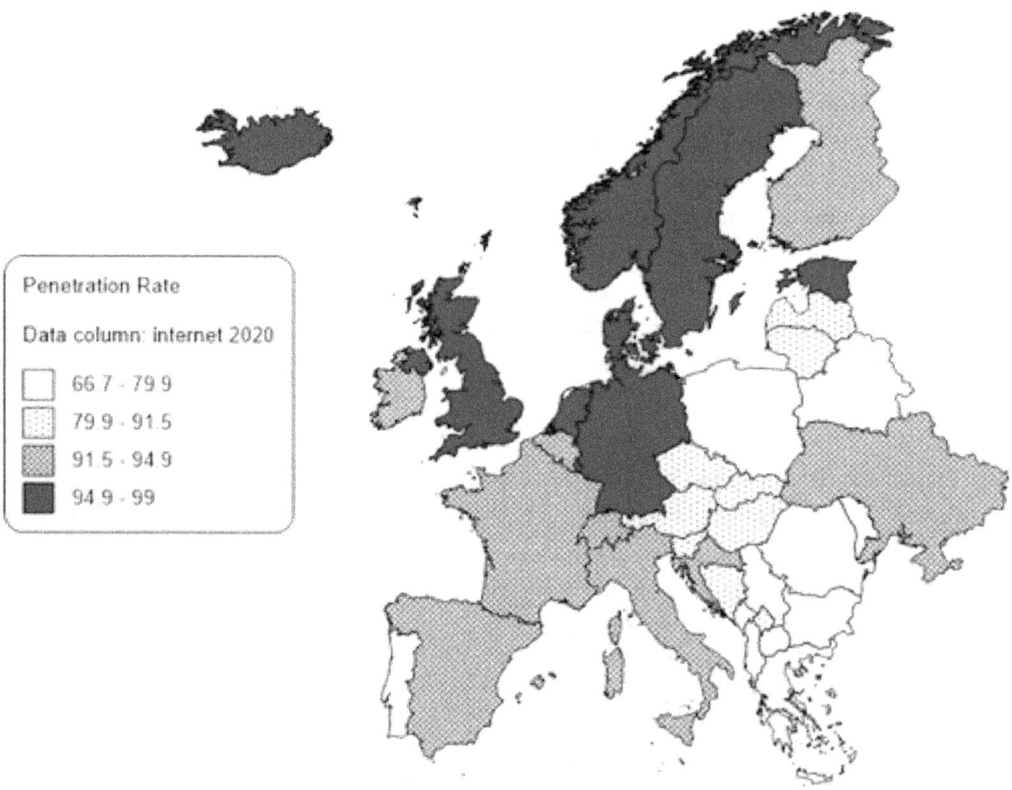

A widely used measure of e-government implementation is the United Nation's e-government development index (EGDI), published bi-annually. It ranges from zero to 1.0 It is a composite measure that consists of three components: the telecommunications infrastructure index, a human capital index, and an online services index. It combines assessments of countries' digital infrastructure, internet affordability, human capital resources, public webpage designs, and the scope and effectiveness of government services delivered online. The index has its critics (e.g., Whitmore 2012), who point to margins of error in the data used to calculate it.

This index varies widely across the European continent (Figure 2), reflecting different rates of e-government adoption (Yera et al. 2020). Broadly, the pattern reflects the long-standing division between East and West European countries (Bovaird 2017). The highest ranked countries include Britain, the Scandinavian region, and Estonia; an intermediate tier is composed of Germany, Poland, France, and Spain; a third rung consists of Italy, Portugal, Greece, and several Central and Eastern European countries; near the bottom are relatively poor Eastern European states such as Slovakia, Serbia, Bosnia, Albania, Romania, and Ukraine. Surprisingly, several countries witnessed declines in their e-government readiness between 2016 and 2020, including Montenegro, Italy, Lithuania, and Spain. This likely reflects several causes, including measurement errors, budget cuts in public internet services, including those to reduce the digital divide, and the failure of complacent bureaucracies to update their webpages in a timely manner. As Stratu-Strelet et al. (2021) argue, this distribution reflects the priorities of political leaders as much as it does the availability of internet technologies such as servers and routers. They also argue that governments' willingness to accept user input is as important in the dissemination of e-government as the technical infrastructure.

Figure 2. e-Government development index (EGDI), 2020
Source: author, using United Nations data.

Another popular measure of e-government is the e-participation score, which reflects the degree to which populations participate in e-government programs, the interactive properties of government webpages, and programs to enhance access in low-income and rural communities. As with e-government readiness, e-participation varies unevenly across Europe (Figure 3). Although most countries that do well on this score also have high EGDI indices (e.g., Britain, Finland), others have lower EGDI scores but high e-participation ones (e.g., Italy, Spain), indicating that the population is well ahead of the state in its receptivity to digital governance. Typically, European governments that put more resources into e-government witness higher levels of participation and satisfaction (Pérez-Morote et al. 2020). The geography of digital skills is thus an important determinant of e-government participation (Rodriguez-Hevia et al. 2020). Such cases indicate governments that have put insufficient effort into moving public services online. The lowest tier of e-participation scores includes Iceland, Ireland, Portugal, and most of Eastern Europe. Oddly, Switzerland belongs to this group as well. Countries with low e-participation scores but higher EGDI indices are indicative of governments that have pushed e-government at a more rapid pace than their populations are ready to accept. Equally important, the quality of e-government is critical to citizens' satisfaction and encouraging their use (Ma and Zheng 2018, 2019).

Figure 3. e-Participation scores, 2020
Source: author, using United Nations data.

NATIONAL VARIATIONS E-GOVERNMENT IMPLEMENTATION 2017-2021

The diversity of European e-government is evident in the varying strategies states utilize, the varying amounts of resources they dedicate to it, their relative success in implementation, and the uneven outcomes and impacts of such initiatives. Some, such as France, Estonia, and Denmark, use highly centralized approaches; others, such as Germany, Switzerland, and the Netherlands, are resolutely decentralized in their focus, including data infrastructures (Bharosa et al. 2020).

Estonia has arguably established the most sophisticated e-government system in Europe (Anthes 2015). Starting in the 1990s, its e-Estonia project has made great gains (Heath 2019). Roughly 99% of government services are available online. The country's system is highly centralized and standardized to allow different agencies to share sign-in information and databases. The country is a pioneer in internet voting, and introduced it in national elections in 2003; half the population uses internet voting for European Parliament elections. The system is very good at ensuring voter identity and securing privacy. In 2014 it opened its e-government to foreigners, allowing entrepreneurs to establish an e-residency in order to start companies there. However, Estonia's e-government has been criticized on the grounds that it is used to promote a particular ethno-nationalist vision of the country and its citizens (Björklund 2016).

Scandinavian countries continue to expand and deepen their already formidable e-government systems. Denmark has a state-of-the-art G2B e-government portal that allows for corporate registration, payments of VAT, customs declarations, and environmental permits. Similarly, Sweden has portals for the Swedish Employment Agency, the Swedish Companies Registration Office, the Swedish Tax Agency, and the Swedish Agency for Growth (Williams et al. 2018b). Finland's centralized e-government portal, Suomi.fi, started as part of the KaPa program that ran from 2014 to 2017. It reflects the input from multiple stakeholders, including the media, the European Union, and third party software integrators (Päivärinta et al. 2019). The country moved aggressively to open its National Land Survey, including GIS software that allows users to conduct searches (Ahonen 2018).

The United Kingdom continues to enjoy its status as one of the world's e-government success stories. Its centralized portal gov.uk sees roughly 14 million visitors per month. The state has trained more than 30,000 civil servants in digital skills. It has made steady progress in increasing the number of online services (Government of the United Kingdom 2017). It offers online identity verification services through gov.uk Verify, receives millions of payments daily through gov.uk Pay, and fields millions of inquiries annually through gov.uk Notify.

Other countries have also enjoyed less success. Despite having the largest economy in Europe, Germany is something of a laggard in e-government. The government released the National E-Government Strategy in 2015 with the aim of coordinating e-government across offices and regions, but the country has not lived up to its expectations (Hofmann et al. 2020). As Ogonek and Becker (2018, 2256) point out, "the federal structure especially in Germany is said to bring along a mass of different IT systems that need to become integrated and interoperable." Germany also has used e-government to facilitate the integration of refugees (Sideridis et al. 2017). France has sought to deliver government services online through the website Service-public.fr, the official website of the French administration, with mixed success; one notable accomplishment, however, is the LegiFrance portal (www.legifrance.gouv.fr), which provides legal assistance. In Spain, the National Action Plan of 2017-2019 sought stronger cybersecurity measures for the protection of digital data and rights (e.g., privacy). Oddly, Switzerland, a country with considerable resources, lags behind much of Europe in e-government implementation (Pardal 2020). Although its EGDI score is high, its e-participation rate is low. This status likely reflects bureaucratic inertia and lack of systematic leadership. As Mettler (2018, 184-185) notes, "Switzerland's decentralized form of governance significantly increases the complexity of the country's digital transformation and ultimately negatively influences the pace at which emerging technologies are introduced."

In Eastern Europe, E-government participation varies among the post-communist countries of the region (Szopiński and Staniewski 2017), including propensities to email government officials and use of public administration websites. Even the more limited group of the four Visegrad Group countries (Czech Republic, Hungary, Poland and Slovakia) exhibit uneven rates of progress. Whereas Poland has enjoyed successes, the Czech Republic's e-government status may even have deteriorated (Majerova 2019). Czech corporations often complain that the government has not digitized a sufficient number of services, has confusing web portals, and suffer from lax data security standards (Strouhal et al. 2020). Even Ukraine, which generally scores poorly in e-government rankings, has made progress in the hopes of democratizing rule there (Hapanovych 2019). In 2019, it created a Ministry of Digital Transformation to advance this agenda. Its government website, www.kmu.gov.ua, offers 120 services online (Shkarlet et al. 2020).

The Balkans, a region that has long lagged in this domain, has exhibited tentative signs of progress. Albania, for example, established the E-Albania website that offers roughly 1,344 services online and has move many public financial operations online (Veliu and Kocibellinj 2018; Shehu and Xhina 2019). In a classic example of technological leapfrogging, the country's government has also moved quickly into big data (Elezaj et al. 2018). Greece has made a concerted effort to offer government agricultural services online (Bournaris 2020). North Macedonia has a public e-learning portal called Moodle. Croatian e-government remains poorly developed, in part because it faces significant obstacles in perceived quality, especially among the young (Zoroja et al. 2020).

Differences in e-government implementation reveal themselves not only among countries but within them. The quality of e-government in the south of Italy lags behind the north (Alfonso and Pariso 2019). In Spain, citizen use of e-government reflects the country's digital divide (Robles et al. 2021). Typically, urban areas fare best. Some use regional e-government platforms, such as Aquitaine Numérique and Auvergne Pro in France, OpenToscana and ioPartecipo+ in Italy, and Talk London in the UK. It may well be the case that e-government not only reflects existing regional inequalities, but also amplifies them.

Across Europe, cities have taken the lead in implementing e-government. Many, in the quest to be designated "smart cities," have tried to shift from e-governance to smart governance, a umbrella term that emphasizes transparency, citizen participation, and collaborative decision making (Bernardo 2017). Thus, Spain has used e-government to increase the transparency of municipal governments, including budgets and financial information (López-López et al. 2018). Czech cities have experimented with crowdsourcing as an e-government strategy (Haltofová 2020), allowing citizen input via their smartphones to supplement existing databases. German cities, for example, continue to exhibit considerable variation in the quality of their e-government portals (Distel and Becker 2018). In sum, e-government is typically tailored to meet the needs of different countries. The agendas and competence of government officials vary enormously among and often within states. Coalitions of stakeholders vie to shape the design and function of e-government services. Thus, the geographic variability in European e-government reflects the fact that it is a deeply political process as much as it is a technological one.

RECENT CHANGES IN EUROPEAN E-GOVERNMENT

e-Government in Europe is never in stasis, and its geographies are always in flux. A variety of studies have revealed longitudinal changes that reflect new technologies, the diffusion of digital skills, and new applications, notably for the business community (Faura 2020; Mulder and Snijders 2020).

Cloud computing has become increasingly common in European e-government and figures as a key objective of the European Commission (Zwattendorfer et al. 2013; Almarabeh et al. 2016). Often called G-cloud, or government cloud, it figures prominently in many national strategic IT plans for its scalability, ability to generate technological and administrative cost savings, pool resources, energy efficiency, and storage, backup, and recovery possibilities. The Estonian government has taken the lead in this domain, using it to facilitate procurement by public sector bodies. The Irish government's public service reform plans put cloud computing front and center with Cloud4Gov, a public-private partnership. Across Europe, smart city initiatives invariably include cloud computing. However, cloud computing also creates challenges concerning privacy, lack of user control, and data leakage.

Similarly, open government data (OGD) has grown enormously in popularity. OGD refers to digital stockpiles of government files, databases, contact data, maps, historical documents, and so forth that are made publicly available in machine-readable formats. Peer-to-peer open data frameworks offer the possibility of enhanced civic participation. Nordic European countries, particularly Sweden and Finland, have been particularly successful in this regard (Kassen 2020).

Mobile e-government, or m-government, is another technology that has had important implications. Although it is still in its infancy, it has the potential to change the nature and use of e-government significantly. The ubiquity of smartphones has considerable potential to alleviate the digital divide. Especially among the young, most of whom are digital natives who grew up with digital technologies, use of a smartphone to access websites and services is commonplace, even taken for granted. Among the elderly, however, this mode of access tends to be much less popular. Given Europe's demographic situation, with low birth rates and a rapidly growing elderly population, m-government for the elderly is a top priority (Molnár and Kő 2020). Examples include having larger fonts on web pages and an emphasis on mobile health services, including teleconsulting. Following the Japanese model, some local governments are experimenting with tracking devices for the impaired elderly that trigger alarms if no movement is detected for extended periods.

Closely associated with m-government is the growing use of social media; indeed, for many people, e-government and social media have become inextricably fused (Magro 2012). Many governments use email, Facebook, and Twitter to inform their citizens of various updates and changes, and some use to prepare for emergencies (Dwivedi et al. 2017). Many users are more comfortable with social media than with centralized government websites, which may be clumsy and infrequently updated, and the integration of e-government services with social media improves citizen satisfaction and participation.

Another important recent change has been the internationalization of e-government services. Encouraging cross-border linkages in this way has been a rising priority of the European Union (Kalvet et al. 2018). From 2017 to 2019, the European Commission launched the "The Once-Only Principle Project" to advance cross-border e-services, typically with an eye on the business community (G2B e-government). The plan focused on linking national registries to facilitate data interoperability, facilitate business registrations, and reduce the duplication of tasks. The growing popularity of secure, ready-made platforms for e-health, e-signatures, and e-immigration (sometimes called eIDAS, or electronic IDentification, Authentication and trust Services) has led many countries to adopt shared standards and compatible information technology infrastructures (Protopappas et al. 2020). As a result, cross-border cooperation is at an all-time high. In a study of the internationalization of e-government in the Baltic region, Williams et al. (2018a) note that many states sought to form digital single markets with services aimed at business (G2B e-government), including cross-border document authentication and delivery; linguistic differences and online identity authentication were the major barriers. Even e-justice has become internationalized (Velicogna 2018).

The covid-19 pandemic of 2020 illustrated the strengths and weaknesses of e-government. As Gabryelczyk (2020) argues, the contagion accelerated the digitization of European government services. The European Commission (2021) quickly implemented a new digital strategy to encourage this move. Fears of contagion led large numbers to stay home, and many civil service workers could work remotely. Lockdowns encouraged many people to use digital services out of necessity. Digital contact tracing and apps became widespread. Many states used social media to update their citizens about the number of cases, their growth, and fatalities, as well as preventative measures to slow the virus's spread. Hodžić et al. (2021, 166) offer several examples:

Estonia has developed a community engagement application that lets local governments distribute information and guidelines to prevent the spread of the coronavirus. In Sofia, Bulgaria, the city administration initiated an application through which citizens can report on the COVID-19 social distancing orders to support the work of police officers. In Barcelona, Spain, a new economic recovery project called "Barcelona Never Stops" has been established.

All over Europe, with court buildings closed and hearings suspended, the judicial system moved largely online. Croatia developed an online system during the pandemic, e-Gradani, to monitor people living in isolation, grant travel permissions, and launched a medical digital assistant (chatbox) named Andrija, after a pioneer in social medicine (Bušelić et al. 2020).

A host of emerging technologies and developments will continue to reshape e-government in Europe in the future. Besides cloud computing, other technological changes include the fifth generation (5G) internet, blockchains and cryptocurrencies, artificial intelligence, machine learning, neural networks, the Internet of Things, and big data, to name a few (Alexopoulos et al. 2019). These are often used for forecasting models; to analyze social media data; text mining; biosecurity and immigration control; health, transportation, and agricultural planning; and the management of energy and information flows.

Other changes are not technical in nature, including the increasing popularity of electronic signatures, e-authentication, and the diffusion of best practices. So are programs to encourage trust and transparency. Public-private partnerships (PPPs) have become increasingly common, including in the development of national electronic identification cards (Medagalia et al. 2017). Denmark has been at the forefront of this type of strategy. The growing popularity of PPPs raises the specter of semi-public e-government, with concerns about the influence of corporations.

CONCLUSION

European e-government consists of a diverse set of practices and policies with uneven impacts over the region. Varying agendas and priorities, internet penetration rates, and governance structures (e.g., centralized v. decentralized) contribute to this variability. So do the geographies of citizen trust and technical skills. Thus, e-government encompasses different levels of sophistication and has uneven impacts. Such comments caution us to reject one-size-fits-all models that fail to account for geographic variability. Moreover, e-government is constantly changing, often rapidly so, as new technologies and organizational forms arise.

Across the European continent, the northern sections exhibit the most developed and effective forms of e-government, including Britain, Scandinavia, and Estonia. While some countries in Eastern and Southern Europe have made strides, typically budgets and technical capabilities in such regions are not as high. The geographies of the European digital divide, which, while diminishing, are still important, are an important part of this uneven map. E-government thus mirrors – and may reinforce – longstanding regional differences in wealth, education, and power across Europe. It may also do so within countries, where wealthier urban areas tend to pull ahead of poorer, rural ones.

Finally, e-government in the continent is being rapidly and continuously transformed by a host of technological and social changes. Cloud computing, m-government, social media, smart cities, and open government data are especially important. The growth of transnational e-governance and responses to emergencies such as the covid pandemic have also contributed to its evolution. These comments serve to indicate that the trajectory of e-government in Europe is path-dependent and open-ended, continually evolving, and forever coming into being.

ACKNOWLEDGMENT

The chapter builds upon an article published in the International Journal of E-Planning Research (IJEPR).

REFERENCES

Ahonen, P. (2018). Learning from opening data in the context of e-governance: Finland, with special reference to government location data. In L. Muñoz & M. Bolivar (Eds.), *International e-government development* (pp. 271–290). Palgrave MacMillan. doi:10.1007/978-3-319-63284-1_11

Alexopoulos, C., Lachana, Z., Androutsopoulou, A., Diamantopoulou, V., Charalabidis, Y., & Loutsaris, M. (2019). How machine learning is changing e-government. In *Proceedings of the 12th international conference on theory and practice of electronic governance* (pp. 354-363). Berlin: Springer. 10.1145/3326365.3326412

Almarabeh, T., Majdalawi, Y., & Mohammad, H. (2016). Cloud computing of e-government. *Communications and Network*, *8*(01), 1–8. doi:10.4236/cn.2016.81001

Andersen, T. (2009). E-Government as an anti-corruption strategy. *Information Economics and Policy*, *21*(3), 201–210. doi:10.1016/j.infoecopol.2008.11.003

Anthes, G. (2015). Estonia: A model for e-government. *Communications of the ACM*, *58*(6), 18–20. doi:10.1145/2754951

Bannister, R., & Connolly, R. (2012). Forward to the past: Lessons for the future of e-government from the story so far. *Information Polity*, *17*(3,4), 211–226. doi:10.3233/IP-2012-000282

Belanger, F., & Lemuria, C. (2008). Trust and risk in e-government adoption. *The Journal of Strategic Information Systems*, *17*(2), 165–176. doi:10.1016/j.jsis.2007.12.002

Bernardo, M. (2019). Smart city governance: from e-government to smart governance. In *Smart cities and smart spaces: Concepts, methodologies, tools, and applications* (pp. 196–232). IGI Global. doi:10.4018/978-1-5225-7030-1.ch009

Bharosa, N., Lips, S., & Draheim, D. (2020). Making e-government work: learning from the Netherlands and Estonia. In *Proceedings of the international conference on electronic participation* (pp. 41-53). Cham: Springer. 10.1007/978-3-030-58141-1_4

Björklund, F. (2016). E-government and moral citizenship: The case of Estonia. *Citizenship Studies*, *20*(6-7), 914–931. doi:10.1080/13621025.2016.1213222

Botrić, V., & Božić, L. (2020). The digital divide and E-government in European economies. *Economic Research-Ekonomska Istraživanja*, *33*, 1–20.

Bournaris, T. (2020). Evaluation of e-government web portals: The case of agricultural e-government services in Greece. *Agronomy (Basel)*, *10*(7), 932–951. doi:10.3390/agronomy10070932

Bovaird, T. (2017). Improving public services through e-government and citizen engagement: Comparing emerging lessons in east and west European countries. In E. Loffler & M. Vintar (Eds.), *Improving the quality of East and West European public services* (pp. 252–256). Routledge. doi:10.4324/9781351155922-20

Bušelić, V., Slamić, M., & Kovačević, A. (2020). Complex digital citizen services in coronavirus age. In *Central European conference on information and intelligent systems 2020* (pp. 89-96). Faculty of Organization and Informatics Varazdin.

Chadwick, A., & May, C. (2003). Interaction between states and citizens in the age of the internet: 'e-Government' in the United States, Britain, and the European Union. *Governance: An International Journal of Policy, Administration and Institutions*, *16*(2), 271–300. doi:10.1111/1468-0491.00216

Distel, B., & Becker, J. (2018). A long and winding road? Analyzing e-government website maturity in Germany. In *Tagungsband Multikonferenz Wirtschaftsinformatik* (pp. 621-632). https://mkwi2018.leuphana.de/wp-content/uploads/MKWI_103.pdf

Dugdale, A., Daly, A., Papandrea, F., & Maley, M. (2005). Accessing e-government: Challenges for citizens and organizations. *International Review of Administrative Sciences*, *71*(1), 109–118. doi:10.1177/0020852305051687

Dwivedi, Y., Rana, N., Tajvidi, M., Lal, B., Sahu, G., & Gupta, A. (2017). Exploring the role of social media in e-government: an analysis of emerging literature. In *Proceedings of the 10th international conference on theory and practice of electronic governance* (pp. 97-106). New York: Association for Computing Machinery. 10.1145/3047273.3047374

Dwivedi, Y., Weerakoddy, V., & Janssen, M. (2012). Moving towards maturity: Challenges to successful e-government implementation and diffusion. *ACM SIGMIS Database*, *42*(4), 11–22. doi:10.1145/2096140.2096142

Elezaj, O., Tole, D., & Baci, N. (2018). Big data in e-government environments: Albania as a case study. *Academic Journal of Interdisciplinary Studies*, *7*(2), 117–124. doi:10.2478/ajis-2018-0052

European Commission. (2021). *Digital Solutions During the Pandemic*. https://ec.europa.eu/info/live-work-travel-eu/coronavirus-response/digital-solutions-during-pandemic_en

Faura, J. (2020). Longitudinal study of the development of e-government in Europe. In How can digital technologies build a more integrated Europe? (pp. 75-90). Nomos Verlagsgesellschaft mbH.

Fernandez-i-Marín, X. (2011). The impact of e-government promotion in Europe. Internet dependence and critical mass. *Policy and Internet*, *3*(4), 1–29. doi:10.2202/1944-2866.1093

Fountain, J. (2001). The virtual state: Transforming American government? *National Civic Review*, *90*(3), 241–251. doi:10.1002/ncr.90305

Gabryelczyk, R. (2020). Has COVID-19 accelerated digital transformation? Initial lessons learned for public administrations. *Information Systems Management*, *37*(4), 303–309. doi:10.1080/10580530.2020.1820633

Government of the United Kingdom. (2017). *Digital Government – Maintaining the UK Government as a World Leader in Serving its Citizens Online*. https://www.gov.uk/government/publications/uk-digital-strategy/6-digital-government-maintaining-the-uk-government-as-a-world-leader-in-serving-its-citizens-online

Haltofová, B. (2020). Critical success factors of geocrowdsourcing use in e-government: A case study from the Czech Republic. *Urban Research & Practice*, *13*(4), 434–451. doi:10.1080/17535069.2019.1586990

Hapanovych, Y. (2019). Prospects for the implementation of the European experience of e-democracy and e-government in Ukraine. *Rocznik Administracji Publicznej*, *5*(5), 115–124. doi:10.4467/24497800RAP.19.007.11470

Heath, N. (2019). How Estonia became an e-government powerhouse. *TechRepublic*. https://www.techrepublic.com/article/how-estonia-became-an-e-government-powerhouse/

Hodžić, S., Ravšelj, D., & Alibegović, D. (2021). e-Government effectiveness and efficiency in EU-28 and COVID-19. *Central European Public Administration Review*, *19*(1), 159–180. doi:10.17573/cepar.2021.1.07

Hofmann, S., Madsen, C., & Distel, D. (2020). Developing an analytical framework for analyzing and comparing national e-government strategies. In *International conference on electronic government* (pp. 15-28). Cham: Springer. 10.1007/978-3-030-57599-1_2

Internet World Stats. (2021). https://www.internetworldstats.com/

Kalvet, T., Toots, M., van Veenstra, A., & Krimmer, R. (2018). Cross-border e-government services in Europe: expected benefits, barriers and drivers of the once-only principle. In A. Kankanhalli, A. Ojo, & D. Soares (eds.) *Proceedings of the 11th International Conference on Theory and Practice of Electronic Governance* (pp. 69-72). New York: Association for Computing Machinery. 10.1145/3209415.3209458

Kassen, M. (2020). Open data and its peers: Understanding promising harbingers from Nordic Europe. *Aslib Journal of Information Management*, *72*(5), 765–785. doi:10.1108/AJIM-12-2019-0364

López-Lópeza, V., Iglesias-Antelo, S., Vázquez-Sanmartín, A., Connolly, R., & Bannister, F. (2018). e-Government, transparency & reputation: An empirical study of Spanish local government. *Information Systems Management*, *35*(4), 1–18. doi:10.1080/10580530.2018.1503792

Ma, L., & Zheng, Y. (2018). Does e-government performance actually boost citizen use? Evidence from European countries. *Public Management Review*, *20*(10), 1513–1532. doi:10.1080/14719037.2017.1412117

Ma, L., & Zheng, Y. (2019). National e-government performance and citizen satisfaction: A multilevel analysis across European countries. *International Review of Administrative Sciences*, *85*(3), 506–526. doi:10.1177/0020852317703691

Magro, M. (2012). A review of social media use in e-government. *Administrative Sciences*, *2*(2), 148–161. doi:10.3390/admsci2020148

Majerova, I. (2019). e-Government as a smart solution for public administration: A case of Visegrad group. *Balkans Journal of Emerging Trends in Social Sciences*, *2*(2), 115–123. doi:10.31410/Balkans.JETSS.2019.2.2.115-123

Marino, A., & Pariso, P. (2019). E-government and its impact on national economic development: a case study concerning Southern Italy. In *Proceedings of the 2019 3rd International Conference on E-commerce, E-Business and E-Government* (pp. 1-4). New York: ACM. 10.1145/3340017.3342242

Medaglia, R., Hedman, J., & Eaton, B. (2017). It takes two to tango: Power dependence in the governance of public-private e-government infrastructures. *Proceedings of the 38th international conference on information systems.* https://core.ac.uk/download/pdf/132283312.pdf

Mettler, T. (2018). The road to digital and smart government in Switzerland. In A. Ladner, N. Soguel, Y. Emery, S. Weerts, & S. Nahrath (Eds.), *Swiss public administration: Making the state work successfully* (pp. 175–186). Springer.

Molnár, T., & Kő, A. (2020). Exploring usability and acceptance factors of m-government systems for elderly. In *Proceedings of the international conference on electronic government and the information systems perspective* (pp. 121-134). Springer. 10.1007/978-3-030-58957-8_9

Mulder, E., & Snijders, D. (2020). Playing the telephone game in a multilevel polity: On the implementation of e-government services for business in the EU. *Government Information Quarterly*, *37*, 101526. doi:10.1016/j.giq.2020.101526

Ogonek, N., & Becker, J. (2018). Can we Learn from Down Under How to Rise Up in E-Government? A Comparative Analysis of the Public Sector Competences in the German and Australian Higher Education Systems. *Proceedings of the 51st Hawaii international conference on system sciences.* https://scholarspace.manoa.hawaii.edu/bitstream/10125/50171/1/paper0284.pdf

Päivärinta, T., Smolander, K., & Yli-Huumo, J. (2019). Towards Stakeholder Governance on Large E-Government Platforms-A Case of Suomi.fi. *10th Scandinavian Conference on Information Systems, 11,* 1-12. https://aisel.aisnet.org/scis2019/11

Pardal, I. (2020). E-government in Europe: On the backwardness of the Swiss government. *Futuribles*, *437*(4), 89–96. doi:10.3917/futur.437.0089

Pérez-Morote, R., Pontones-Rosa, C., & Núñez-Chicharro, M. (2020). The effects of e-government evaluation, trust and the digital divide in the levels of e-government use in European countries. *Technological Forecasting and Social Change*, *154*, 119973. doi:10.1016/j.techfore.2020.119973

Protopappas, L., Sideridis, A., & Yialouris, C. (2020). Implementation issues of cross border e-government systems and services. In Proceedings of the 9th international conference on information and communications technologies in agriculture, food and the environment. Thessalonika. http://ceur-ws.org/Vol-2761/HAICTA_2020_paper25.pdf

Riddlesden, D., & Singleton, A. (2014). Broadband speed equity: A new digital divide? *Applied Geography*, *52*, 25–33. doi:10.1016/j.apgeog.2014.04.008

Robles, J., Torres-Albero, C., & Villarino, G. (2021). Inequalities in digital welfare take-up: Lessons from e-government in Spain. *Policy Studies*, *42*(3), 1–16. doi:10.1080/01442872.2021.1929916

Rocheleau, B. (2007). Whither e-government? *Public Administration Review*, *67*(3), 584–588. doi:10.1111/j.1540-6210.2007.00741_1.x

Rodriguez-Hevía, L., Navío-Marco, J., & Ruiz-Gómez, L. (2020). Citizens' involvement in e-government in the European Union: The rising importance of the digital skills. *Sustainability*, *12*(17), 6807. doi:10.3390u12176807

Seljan, S., Miloloza, I., & Pejicbach, M. (2020). E-government in European countries: Gender and ageing digital divide. *Interdisciplinary Management Research*, *16*, 1563–1584.

Shehu, J., & Xhina, E. (2019). Modeling an ontology for public e-government services in Albania. *International Academic Conference on Research in Engineering and Technology*. https://www.dpublication.com/wp-content/uploads/2019/09/620-R.pdf

Shkarlet, S., Oliychenko, I., Dubyna, M., Ditkovska, M., & Zhovtok, V. (2020). Comparative analysis of best practices in e-Government implementation and use of this experience by developing countries. *Administratie si Management Public*, *34*, 118–136. doi:10.24818/amp/2020.34-07

Sideridis, A., Protopappas, L., Tsiafoulis, S., & Pimenidis, E. (2017). Smart cross-border e-gov systems: An application to refugee mobility. *International Journal of Electronic Governance*, *9*(3-4), 246–266. doi:10.1504/IJEG.2017.088219

Stratu-Strelet, D., Gil-Gómez, H., Oltra-Badenes, R., & Oltra-Gutierrez, J. (2021). Critical factors in the institutionalization of e-participation in e-government in Europe: Technology or leadership? *Technological Forecasting and Social Change*, *164*, 120489. doi:10.1016/j.techfore.2020.120489

Strouhal, J., Horák, J., Bokšová, J., Bokša, M., & Pavlica, K. (2020). Corporate perspective on development of Czech e-government services. In *Proceedings of the 13th International Scientific Conference* (pp. 114-125). https://fes.upce.cz/sites/default/files/public/rast3126/public_administration_2020_155249.pdf#page=115

Szopiński, T., & Staniewski, M. (2017). Manifestations of e-government usage in post-communist European countries. *Internet Research*, *27*(2), 199–210. doi:10.1108/IntR-01-2015-0011

Taipale, S. (2013). The use of e-government services and the Internet: The role of socio-demographic, economic and geographical predictors. *Telecommunications Policy*, *47*(4-5), 413–422. doi:10.1016/j.telpol.2012.05.005

Velicogna, M. (2018). e-Justice in Europe: From national experiences to EU cross-border service provision. In International E-Government Development (pp. 39-72). Cham: Palgrave Macmillan.

Veliu, S., & Kocibellinj, M. (2018). The implementation of e-government in Albania. *Ditet e Studimeve Shqiptare*, 87-192. https://uet.edu.al/new/wp-content/uploads/2021/06/Proceedings-Book-1.pdf#page=187

Warf, B. (2018). Geographies of e-government in Europe. *International Journal of E-Planning Research, 7*(4), 61–77. doi:10.4018/IJEPR.2018100104

Williams, I., Falch, M., & Tadayoni, R. (2018a). Institutional legitimacy and digital public cross-border service delivery between Denmark/Sweden and Denmark/Germany. *Journal of NBICT, 1*, 1–30.

Williams, I., Falch, M., & Tadayoni, R. (2018b). Internationalization of e-government services. In *2018 11th CMI international conference: Prospects and challenges towards developing a digital economy within the EU* (pp. 19-31). IEEE. 10.1109/PCTDDE.2018.8624828

Yera, A., Arbelaitz, O., Jauregui, O., & Muguerza, J. (2020). Characterization of e-government adoption in Europe. *PLoS One, 15*(4). doi:10.1371/journal.pone.0231585 PMID:32302326

Yigitcanlar, T., & Baum, S. (2006). E-government and the digital divide. In M. Khosrow-Pour (Ed.), *Encyclopedia of e-commerce, e-government, and mobile commerce* (pp. 353–358). IGI Global. doi:10.4018/978-1-59140-799-7.ch058

Yildiz, M. (2007). E-government research: Reviewing the literature, limitations and ways forward. *Government Information Quarterly, 24*(3), 646–665. doi:10.1016/j.giq.2007.01.002

Zoroja, J., Strugar, I., & Jaković, B. (2020). E-government in Croatia: Perceived quality and obstacles among younger users. *International Journal of E-Services and Mobile Applications, 12*(3), 85–96. doi:10.4018/IJESMA.2020070105

Zwattendorfer, Stranacher, Tauber, & Reichstädter. (2013). Cloud computing in e-government across europe. In *International conference on electronic government and the information systems perspective* (pp. 181-195). Springer.

Chapter 2
PPGIS Potential in Local Governance and Spatial Planning:
Lessons From Poland

Edyta Bąkowska-Waldmann
Adam Mickiewicz University, Poznań, Poland

Tomasz Kaczmarek
Adam Mickiewicz University, Poznań, Poland

ABSTRACT

A growing interest in participatory planning practices in local governments in Poland resulted in the development of new methods supporting citizens' engagement in decision-making processes. The latest solutions in the public participation methods catalog are based on geoweb tools. One of such tools is the geo-questionnaire developed within the project: Geoportal supporting public participation in spatial planning (GEOPLAN) – an R&D project conducted by the interdisciplinary team of researchers from Adam Mickiewicz University in Poznań supported by IT specialists and urban planning practitioners. The chapter presents the features of PPGIS tools, the development of PPGIS in Poland, the evaluation of the pilot implementations of geo-questionnaires in decision-making processes, and the future of PPGIS in the light of new post-pandemic reality.

INTRODUCTION

At least two decades have passed since the change in the paradigm of management in public policy and administration in Europe and transition from the Weberian or neo-Weberian model of governing to the model of public governance (Rhodes 2000; Kettl 2002). The essence of this transformation is the cooperation between public and commercial spheres, participation of stakeholders, cooperation of government institutions, local governments and non-governmental organizations, decentralization

DOI: 10.4018/978-1-7998-9090-4.ch002

of decision-making, and replacement of the hierarchic axis of dependencies by the horizontal links, including networks (Sullivan & Skelcher 2002; Armitage et al. 2007). The changes have been driven by the intention to eliminate the negative features of bureaucracy from the managerial process. It also was reflected in the reduction of social conflicts by inviting many entities to decision-making. A new and based on a larger platform model of inter-sector and partner management is implemented in smaller managerial structures, includin the local level of government (Geddes 2006). The decision-making at this level should involve different groups of actors. Nonetheless, in Europe the top-down approach to spatial development with minimum involvement of residents is still strong (Horelli & Wallin 2013). Thus, it is very important to share the information between the stakeholders representing different needs and using the data for different purposes (Kahila & Kytta 2009).

Changes in the paradigm of local governance towards cooperation and consensus of interested stakeholders also include the area of spatial planning (McCall & Dunn 2012). According to Haughton et al. (2010), the premises for governance in spatial planning include increase in the number of spatial conflicts, stakeholders and the complexity of the process of spatial planning together with the development of civil society. These authors also point to the hermeticity of spatial planning systems and inflexible legal solutions, rigidity of administrative structures and formation of new type of flow-spaces or soft spaces (see also: Mikuła 2019).

Haughton et al. (2010) have introduced the term of 'new spatial planning' that is based on strong mechanisms of large groups' consultations with a greater emphasis on solving problems of social integration within the spatial strategies. The above authors treat the spatial planning as a continuous process that does not develop according to one convention only. With reference to the earlier forms of spatial planning the new spatial planning approach is characterized by the following features (op. cit.):

1. Devolution of ideas and approaches to spatial planning from the European and international level to the national, regional and local levels.
2. Departure from the sector (silos) approach to spatial planning to the integrated planning that combines the functional-space, economic, social and ecological elements.
3. Assuming the determining role of coordination and negotiations in the process of spatial planning even though the process gets longer and more complex.
4. Increasing significance of evaluation and monitoring of the effects of spatial planning decisions and their execution.
5. Increasing the role of spatial planning procedures and instruments leading to deformalization, decentralization and socialization of the process.

Recently, also in Poland the trend towards the increasing participation of stakeholders in decision-making has been promoted. The notions such as co-management, participative management, cyberdemocracy, deliberative democracy, interactive democracy etc. have acquired some meaning (Jankowski et al. 2017b). On the one hand, it is a consequence of development of civil society, while on the other, it is a result of systematic adaptation of new methods of management that have been worked out in western Europe. Commonly, in democratic societies public participation in planning and local decision-making is legally regulated (Siemiński 2014). This includes the right to vote, organize referenda, and to create citizen advisory bodies. According to one of the first Polish definitions of public participation in spatial planning, given by Wiench (1991), it is a legal action of individuals and groups of citizens to influence the determination of spatial development plan. Public participation in Poland is a relatively young phenomenon, mostly identified with the process of spatial planning, regulated by the Act of 27 March 2003 on Planning and Spatial Development. According to the Act, the participation of the public in spatial planning is realized in two steps. The first step begins with the approval of the preparation of a local spatial development plan (local plan) and a study of spatial planning conditions and directions and includes the collection of suggestions (concerning the project of a local plan). The second step is implemented at the advanced phase of a decision-making process, when the first draft of the plan is ready and involves submitting comments about the project and participation in a public discussion.

The subsequent sections of this chapter are devoted to presentation of Public Participation Geographic Information Systems tool (geo-questionnaire) and examples of its use for public consultations. The results presented were collected by the authors during the pilot implementations of the geo-questionnaire in Poznań and Łódź agglomerations carried out in 2015-2017 in cooperation with local authorities. The pilot studies were performed within the project "Geoportal supporting public participation in spatial planning" ("GEOPLAN") conducted at Adam Mickiewicz University, Poznań, in 2015–2018. The aim of the project was to create two Internet applications (tools) – geo-questionnaire and geo-discussion – used for public consultations, together with the methodology of their use, based on implementation in the planning practice on the local level (cities, communes and counties). The project was financed by the National Center for Research and Development (NCBR), supporting the creation of modern solutions and technologies that increase innovation, and thus the competitiveness of the Polish economy.

AIM AND METHODS OF STUDY

The purpose of this chapter is to present experience in the development and implementation of PPGIS in Poland, with focus on recent and unexpected challenges for its future posed by the COVID-19 pandemic. The conducted research investigated the transition observed between 2014 and 2020 in the number and the legal form of entities interested in PPGIS implementation, thus showing the evolution of actors responsible for the popularization of such solutions and the diffusion of innovation. In-depth analysis of the national (Polish) specificity of PPGIS development potential was presented on the example of six public consultation processes conducted in Poznań and Łódź agglomerations in cooperation with local authorities.

In order to analyze the number of PPGIS implementations in Poland, use was made of a query method and the research was conducted among the entities providing public consultation services for local government units. To this end, what was made first was the identification of the entities (private companies, NGOs, research units) implementing and developing such solutions, which were then asked to impart information about the tools and the processes in which they were used in subsequent years. On the other hand, for the purpose of an in-depth analysis of PPGIS implementation processes, use was made of the elements of the Action Research method aimed at solving practical problems of a given organization (i.e., local government), which allows changes in the functioning of this organization and the creation of conditions conducive to learning them (Greenwood & Levin 2007; Chrostowski & Jemielniak 2011).

The Action Research approach was used to propose a model procedure for the implementation of PPGIS tools in spatial planning. It consisted of four stages: planning, acting, observing, and reflecting (Figure 1). The first stage - planning - included identification of the scope and key partners, making decisions about the aim of public consultations, preparing geo-questionnaires, and other organizational tasks (formal-administrative and technical) of geo-questionnaire implementation. After completing all the tasks at this stage, the acting phase followed, including testing of the geo-questionnaire, its implementation and conducting public consultations along with the promotion necessary to engage local communities. In the third stage - observing - coinciding with the duration of public consultations, monitoring of current results was carried out. It ended with interviews conducted with partners of public consultations, analysis of results and preparation of reports. Reflecting was the last stage. It included evaluation and summary of results as well as revision of the entire process, which resulted in the recommendations for changes in subsequent implementations. The research contained only some elements of the AR, others were difficult to implement (e.g., the lack of repetitive cycles due to the characteristic of planning procedures and their long duration).

Figure 1. Action Research approach on the example of geo-questionnaire pilot implementations
Source: Own elaboration.

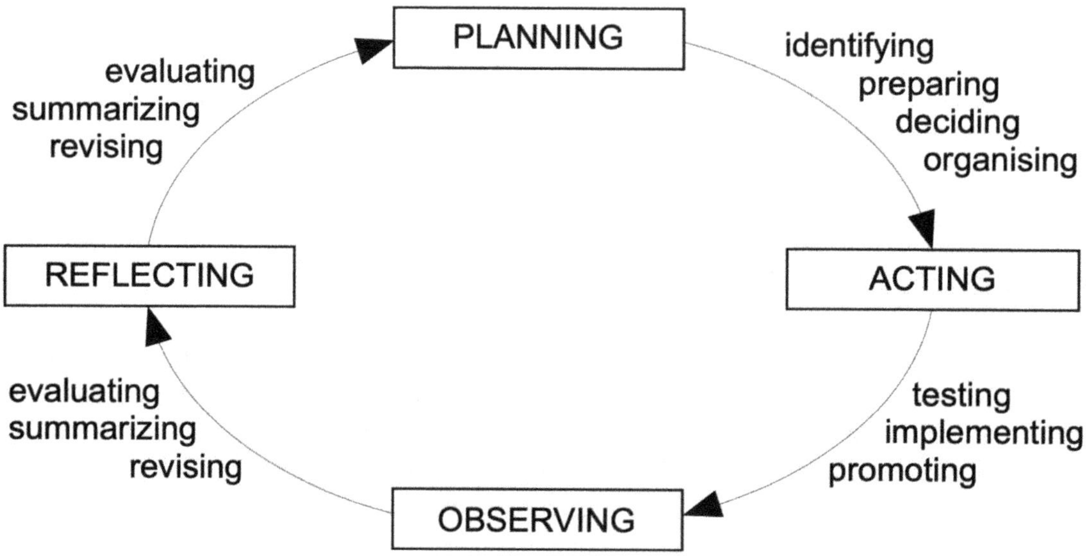

The authors carrying out the study assumed that the social reality is best understood by the researchers engaged in it (Glaser & Strauss 1967). The traditional functionalistic approach in which the community is analyzed with the use of the earlier prepared model was rejected. Following the assumptions of the grounded theory (op. cit.), collecting new information on each case, the authors made new observations that were used to develop the procedures of public consultations at a local level. Thanks to the engagement in preparation of the geo-questionnaires, analysis and reporting of the results, the authors have become participants in the process of consultations, supporting the urban planners and local authorities in organization of e-consultations. The adopted approach to the research process also included a case study analysis dedicated to the development of knowledge in the field of social sciences (Flyvbjerg 2018). According to the methodology proposed by Yin (2003), the case study method consists of a problem definition, data collection, data analysis, and reporting. Therefore, it has an important role at the stages of observation and conclusion in the Action Research method. In this research, the case study method was used to identify effective and ineffective mechanisms related to the implementation of PPGIS on specific examples, which in turn allowed creating a model procedure for the implementation of PPGIS in decision-making processes in spatial planning.

The results presented in the chapter are based on six cases (pilot implementations of the geo-questionnaire) from Poznań and Łódź agglomerations conducted in 2015–2017 by researchers engaged in the implementation of the "GEOPLAN" project (including authors). Studied processes concerned the procedure of the preparation of local spatial development plans (documents regulated by the Act on Spatial Planning and Development) and the modernization of public spaces – inner-city markets and squares. The detailed characteristics of the studied cases are presented later in the chapter. The consultations were carried out in cooperation with local authorities, office-workers and urban planners. The procedure covered: (1) identification of the subject of consultations, (2) preparing the contents and configuration of the geo-questionnaire tool addressed to residents, (3) organization of campaigns promoting public consultations, (4) analysis of results and preparation of reports to be used for further stages of urban planning. The authors not only participated in the organization of e-consultations, but also prepared reports on their results and conducted interviews with organizers of public consultations (local authorities, urban planners) and surveys among residents to gather opinions on the use of those PPGIS tools. The evaluation of public consultations was performed based on the level of public participation frequency, participants' socio-demographic features and place of residency, as well as respondents' opinions on that method of public consultations, and in-depth interviews with the representatives of local authorities and urban planners engaged in the processes. The research strategy which adopted in-depth case study analysis, seemed to be the most effective in a situation in which there was almost no experience in the use of Internet tools in public consultations in Poland.

In this study, the research questions were as follows:

1. Who are the main actors responsible for PPGIS development and application in Poland?
2. Which areas of spatial planning can be addressed in the e-consultation procedure?
3. What are the strong and weak points of the use of geo-questionnaire in public consultations?
4. Which measures should be taken into account to ensure effective implementation of geo-questionnaire in decision-making in spatial planning at a local level?
5. How useful is the Action Research approach in solving the practical problems of participative urban planning in local government?

The following paragraphs of the chapter present the typical features of PPGIS tools, development of PPGIS in Poland, and the evaluation of the pilot implementations of geo-questionnaires in decision-making processes in Poznań and Łódź agglomerations. The final paragraphs are devoted to the analysis of the PPGIS future in the face of a new post-pandemic reality.

TYPICAL FEATURES OF PPGIS TOOLS

Since the 1970s the rational model of spatial planning has been criticized as technocratic, imposed by key authorities, and not related to local social groups. At the same time methods of communication that would prompt increased involvement of the public in the decision-making have been searched for. Of key role in development of such methods was the appearance of the PPGIS in the middle of 1990s and then their implementation in the public participation processes to support development of structures of democratic society (Kahila & Kytta 2009). Schroeder (1996) has defined PPGIS in a broad sense as "different approaches to the use and availability of GIS tools for decision-making on spatial planning to all subjects interested in the decision-making process". PPGIS permit wider and more reliable (as based on the interactive map) communication with local inhabitants than the other techniques of public consultations (meetings, debates) (Bugs et al. 2010). PPGIS operating online offers the possibility of participation in consultations at any time and from any place (Kingston 2011; Carver et al. 2001), in contrast to the traditional methods which require the presence at a specific place at a certain time. The PPGIS advantage to the planners is the possibility of getting information on the preferences and experience of the local residents, their participation in spatial planning that aims at elimination or minimization of conflicts at subsequent stages of planning. Moreover, approval of residents legitimizes publicly accepted decisions (Sieber 2006, Brown & Raymond 2014; Jankowski et al. 2016; Bąkowska et al. 2017). The most important advantage of PPGIS tools is the possibility of collecting large amount of spatial information and its wide consultation. In these tools, the GIS can be used to characterize a local space based on the knowledge and opinions of local inhabitants and promote deeper and more expanded forms of public participation than the traditional forms (Bugs et al. 2010). The use of PPGIS permits collecting information on the behavior, attitudes, and preferences of residents on the spatial development on the scale of housing estates, commune, municipal and on larger areas such as agglomerations, functional areas, etc. The information can supplement the specialist knowledge of architects and be an important source of data useful for the process of spatial planning. The consultations applied at proper stages of spatial planning can limit the conflicts among the participants and lead to publicly accepted solutions (Sieber 2006; Brown & Raymond 2014; Jankowski et al. 2016).

Recently, the interest in the tools of Internet communication based on two-dimensional and three-dimensional maps in the context of specific physical locations has much increased. Their implementation in spatial planning has been mostly observed in the USA and countries of western Europe, e.g., in Finland, UK, Italy, Austria (see: Carver et al. 2001; Atzmanstorfer & Blaschke 2013; Kahila-Tani et al. 2015; De Filippi et al. 2017). In Finland, since 2003 the SoftGIS approach based on PPGIS has been used. Through this procedure local citizens are included in the process of spatial planning by providing the so-called soft information on the everyday used pathways, attitude to specific places or buildings and activities taking place in given sites, that are introduced into the GIS hard database on the natural environment (Manzo 2003; Rantanen & Kahila 2008; Kahila and Kyttä 2009).

PPGIS DEVELOPMENT IN POLAND

The analysis of PPGIS development in Poland included a query of the public consultation processes carried out at local level from 2014 to the mid-2020, in which such solutions were applied. The analysis of the implementations that took place after the second half of 2020 was abandoned due to the sudden appearance of numerous quasi-PPGIS applications and their questionable quality and functionality. This rapid development of new tools was a consequence of the COVID-19 pandemic restrictions abandoning personal contacts (also according to functioning of local administration). When analyzing the development of interest in the use of PPGIS by local governments, both the area of their application and entities that directly supported local governments in this process were taken into account.

According to the information presented in Fig. 2, before 2016 academic institutions were the only actors responsible for the implementation of PPGIS at local level. Although the development of tools was carried out in cooperation with IT developers, representatives of the scientific community were responsible for the practical application of the methods in public consultation processes. In 2016, a gradual commercialization of the process and cooperation of private entities (SME) with local government units regarding the application of these solutions was noticed. This process reached its peak in 2018, when private entities were responsible for over 75% of implementations for local governments throughout the country. At that time, the role of academia remained at a similar level – until 2019, since when no public consultation process supported by scientific units has been recorded. In 2017, NGOs became another actor supporting local governments in implementing PPGIS, and which, thanks to the obtained subsidies, became the primary entity stimulating the use of PPGIS by local governments. The statistics presented in Fig. 2 show only the use of PPGIS tools for decision-making processes carried out in local government units. It should be taken into account that the role of academic centers in the dissemination of tools such as a geo-questionnaire is still of great importance, but currently the primary purpose of the use of PPGIS at universities is scientific research, the results of which can then be used by local governments. On the other hand, the role of private entities is twofold. First, these are IT companies developing software which then can be used by NGOs, academic units and other private companies. Secondly, they are consulting companies that use tools (software) distributed by IT companies and on its basis provide services for local offices.

Figure 2. Implementation of PPGIS tools in decision-making processes in Poland from 2014 to the mid-2020 by entities
Source: Own elaboration based on the data collected from entities implementing PPGIS tools in Poland.

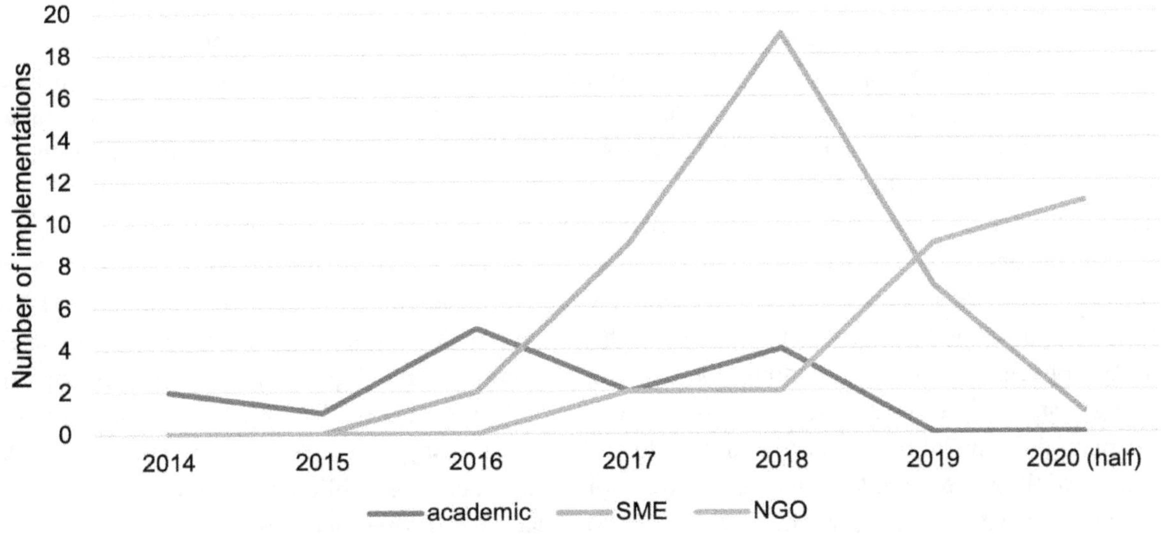

After the experimental phase, in which academic researchers played the main role, and the transitional phase, when IT companies, apart from creating software, provided also consulting and analytical services for local governments, NGOs took the dominant role in the dissemination of PPGIS. However, this support depends on the grants obtained by those entities, which usually provide support for one-time use of PPGIS in a given unit. It seems that, ultimately, the most important actor in the dissemination of

PPGIS will be consulting companies and other companies supporting local governments in conducting public consultation processes in spatial planning, such as private urban planning studios, which, however, will use software developed by IT companies. For local government units in Poland, there is currently no clear interest in the purchase (one-off or in the form of packages or subscription) of software and its independent use by office workers. IT companies, on the other hand, are not interested in providing analytical and consulting services. From the office workers' point of view, it is particularly important to obtain a full service from one supplier, i.e., receiving a finished product – a software (an application), its configuration, implementation, and analysis of the results. Such a situation is generated not only by the burden on administrative staff, but also by their potential lack of competences to use such tools and data analysis. This range of services may be provided by consulting companies. However, because of the need to purchase applications and support from IT developers (business-to-business services), costs for local government units may be higher.

GEO-QUESTIONNAIRE PILOT IMPLEMENTATIONS IN POZNAŃ AND ŁÓDŹ AGGLOMERATIONS

The development of the geo-questionnaire was initiated at Adam Mickiewicz University in Poznań within the project entitled "An Experimental Study of Public Participation in Planning Decision Making Using Web-based Geographic Information System" realized in the years 2012-2015. Geo-questionnaire used in presented research is based on traditional form of questionnaire supplemented with questions related to the aspects identifiable in space. The methodology of using the tool results from the previous development and knowledge in the field of PPGIS (Jankowski et al. 2016; Jankowski et al. 2017a). Geo-questionnaire permits the respondents to indicate, characterize and - if needed - evaluate the elements of space.

Fig. 3 presents an exemplary page of the geo-questionnaire used in public consultations on local spatial development plans in Rokietnica, (Tab. 1, P2). The questions evaluate the representative character of the group of respondents who are asked to mark their dwelling place on the map so that it contained information about the spatial distribution of respondents.

Through the interactive map, the respondent can use three tools for drawing objects: points (Fig. 4a), lines (Fig. 4b) and polygons (Fig. 4c) to mark the sites relevant to a given question, e.g., localization of dwelling place, preferred localization of greenery. Moreover, residents can be asked to respond to additional questions related to a given object, e.g., on the frequency of visiting a given site, by marking sites, pathways, or regions. Geo-questionnaire permits supplementing typical questions with the information referring to space. Therefore, besides the characteristics of the respondent it is possible to get information on his/hers/their spatial behavior, e.g., daily life paths, features of given sites and preferences as to the directions of their use. Thanks to the use of this tool in the process of public consultations it is possible to collect information in the form of spatial databases ready for further analyses.

The abilities needed to fill in the geo-questionnaire are similar to those needed to use Internet websites e.g., respondents are expected to be able to fill in and send Internet forms and work with maps available through Internet browser (move towards a given point or away from it, drawing points, lines or areas) (Jankowski et al. 2016). Geo-questionnaires can also include questions not related to spatial planning like e.g., those referring to demographic characteristic of a respondent.

Figure 3. An exemplary page of the geo-questionnaire

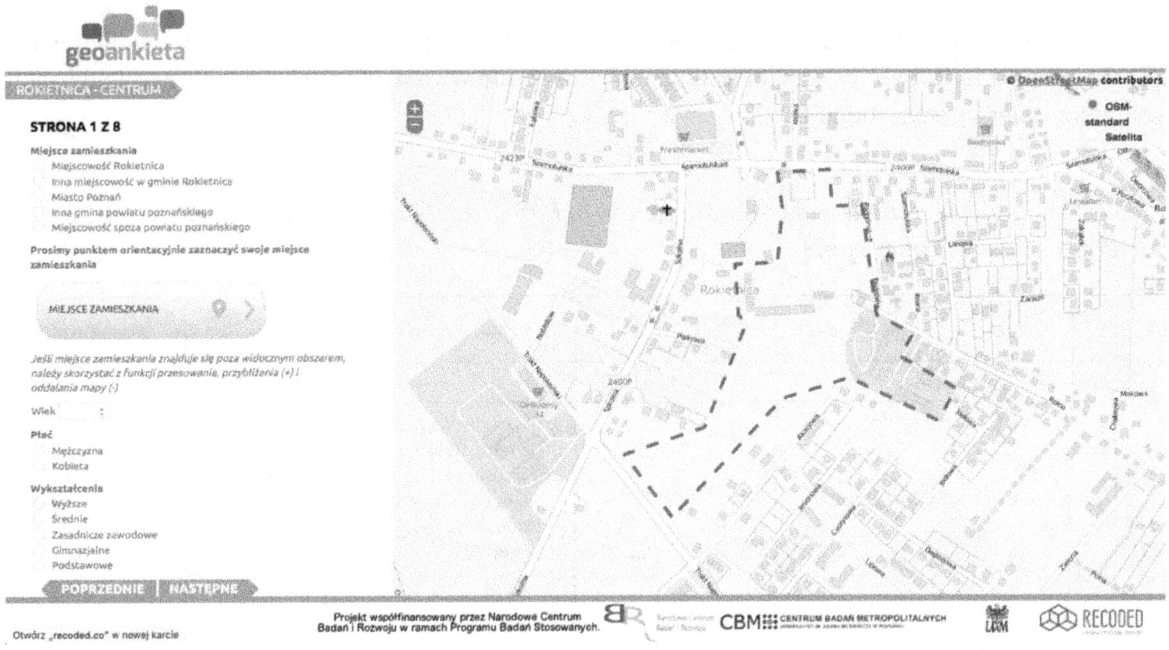

Figure 4. Maps presenting a) clusters of points, b) linear objects, c) clusters of areas indicated by geo-questionnaire respondents
Source: Kaczmarek & Bąkowska 2016a, b; Czepkiewicz & Brudka 2016b.

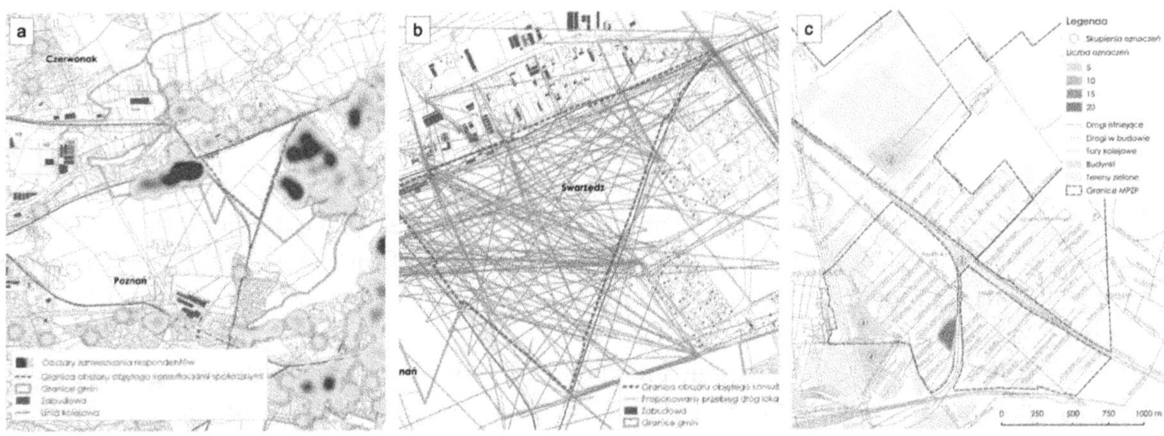

The pilot implementation of geo-questionnaire performed in Poznań and Łódź agglomerations in 2015-2017 concerned two types of case studies: legal procedure of preparing local spatial development plans and modernization of public spaces. Tab. 1 and 2 present the information describing applications of the geo-questionnaire in public consultations: space and subject of consultations, number of respondents, partners of public consultations and information on the application of results.

LOCAL SPATIAL DEVELOPMENT PLANS

The application of the geo-questionnaire in the sites of P1-Andrzejów, P2-New Center of Rokietnica and P3-Sports airport in Kobylnica was undertaken in connection with the formal procedure of local spatial development plans preparation (Tab. 1).

Table 1. Pilot implementations of geo-questionnaire in Poznań and Łódź agglomerations on the examples of local spatial development plans

ID[1]	Name	Scale	Number of respondents	Consulted issues	Partners in the process	Use of results
Local spatial development plans						
P1	Andrzejów in Łódź	district	123	– the evaluation of the quality of space in a housing estate – changes in spatial development, – preferred solutions	Municipal Authorities of Łódź, Urban Planning Office in Łódź	Project of local spatial development plan
P2	New Centre of Rokietnica in Poznań agglomeration	small urban space	435	– the density of the area, – available services – land use – transportation system	Rokietnica Commune Authorities/ external institution working out plans	Project of local spatial development plan
P3	Sports airport in Kobylnica in Poznań agglomeration	district	933	– the evaluation of current land use – postulates and new ideas related to the spatial development of the area	Municipal Authorities of Swarzędz	Pre-planning assumptions for the project of local spatial development plan

Geo-questionnaires concerning the development of drafts of local spatial development plans allows to use a map to indicate their preferences concerning localization of functions within the area covered with consultations, i.e., the transportation system, services, public areas, as well as type of buildings and construction density.

The course and results of geo-questionnaire application for public consultations regarding spatial planning are presented on the example of P2 – New Center of Rokietnica. In Rokietnica the public consultations included (1) taking decision on making local spatial development plan, (2) meeting with citizens to inform them on basic formal and spatial conditions of making such a plan, presentation of preliminary urban conception and the intention of using the geo-questionnaire, (3) realization of public consultations via the geo-questionnaire, and (4) presentation of the geo-questionnaire results at a meeting with local citizens (Fig. 5).

The public consultations on plans of a New Center of Rokietnica were addressed to the residents of Rokietnica commune, in particular Rokietnica village. The main form of spreading the information on the public consultations was to distribute leaflets to mailboxes in the whole commune area. From among the 434 respondents, 413 (95% of all respondents) were the citizens of Rokietnica commune, and 311 (72%) were the citizens of Rokietnica village. The residents of the Rokietnica commune that participated in the internet consultations made 3.3% of the commune population (persons above 10 years old). The participants from the village of Rokietnica made 6.6% of its inhabitants. Among the participants were 16 persons (3.6%) from Poznań or another commune in Poznań district, while 5 persons (1.1%) were from outside the Poznań district.

Figure 5. The process of public consultations on local spatial development plan of a new center of Rokietnica

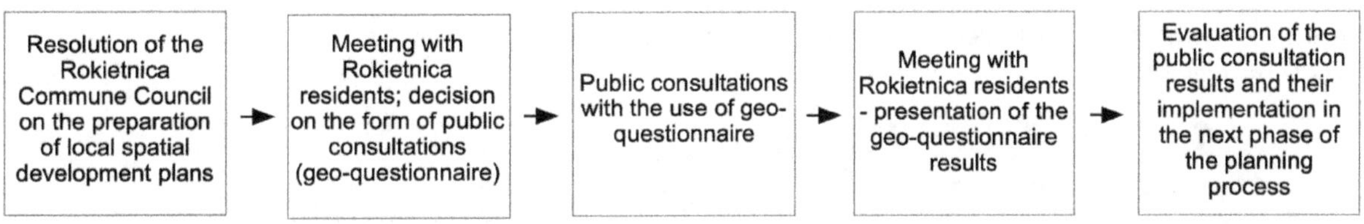

The age structure of respondents reveals the overrepresentation of persons aged 25 – 44 years and underrepresentation of the youth and the elderly (Fig. 6) with respect to the population of the commune. The probable reason of low participation of young residents is the lack of interest, while the low participation of the elderly may be related to the use of the internet. The age structure of the respondents shows that tool can be used to complement other types of civic engagement, who attract the other age groups (young and older).

Figure 6. Age structure (in %) of the participants in public consultation in comparison with the age structure of Rokietnica commune residents
Source: Own elaboration based on geo-questionnaire results and data from the Central Statistical Office

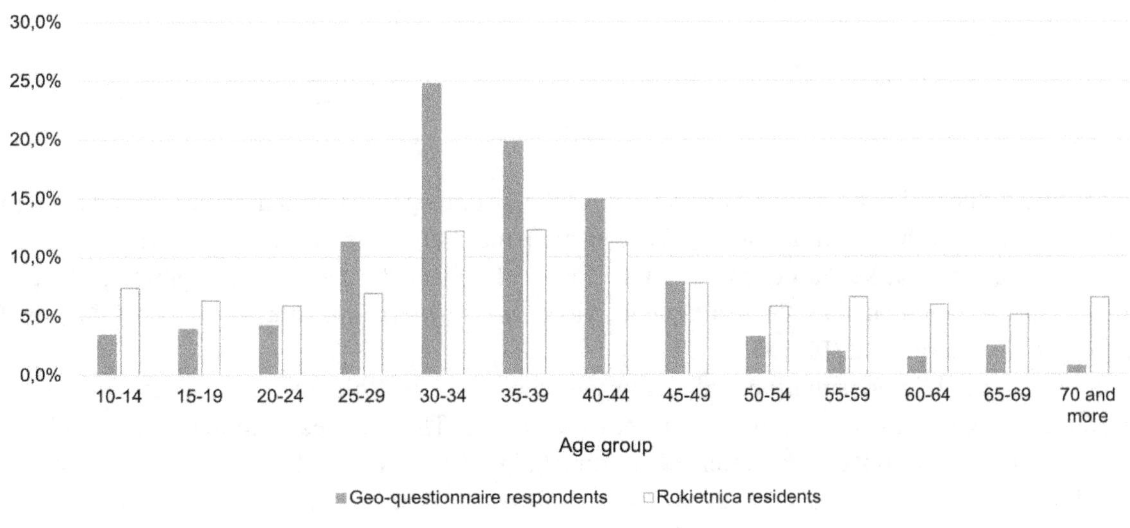

The main element of the online consultations was the possibility to express opinions, preferences and attitudes, and comments on the local spatial development plans. The commune authorities and urban planners working on the plans were aware of the preferences and opinions of local citizens. One of the main questions asked in the geo-questionnaire concerned the functions of the village center and objects that should be in the center. As Rokietnica maintained for centuries the village type of land use, there is no central square of marketplace, typical of towns. According to the respondents the new local spatial development plans should include a central square or market that would attract and integrate residents of Rokietnica (329 responses per 369 respondents, 89.2%). Figure 7 presents the preferred localization of the central square.

Figure 7. Localization of the central square preferred by respondents
Source: Own elaboration.

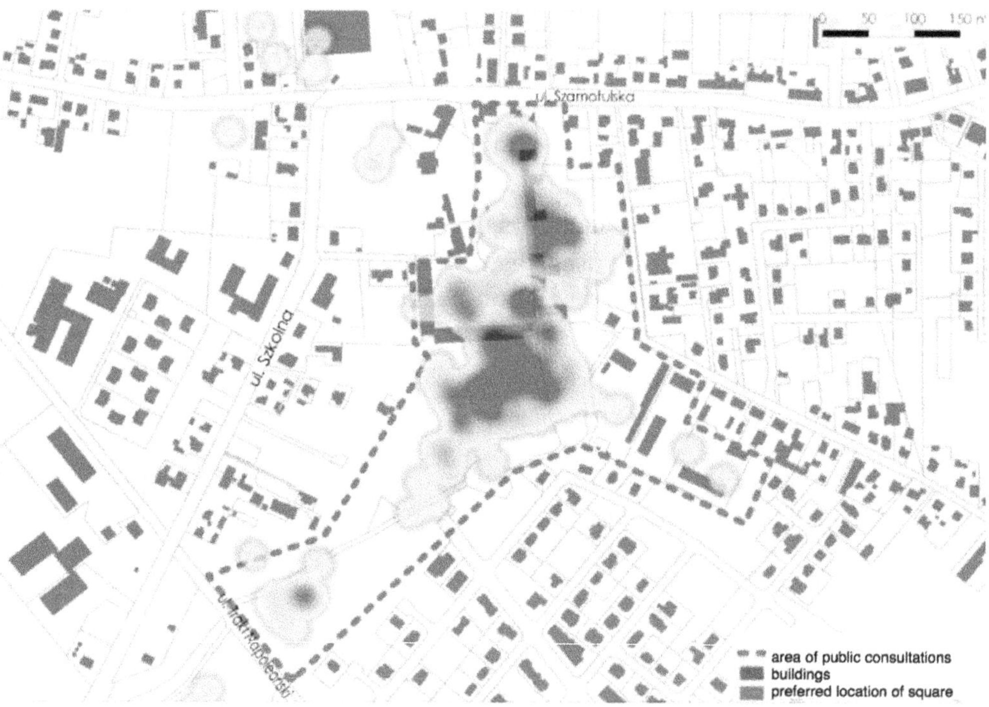

According to the majority of respondents (71.9%) the central space should include all kinds of shops and services. Against this postulate were 20.8% participants. As to the question concerning the space for local trading, e.g., market hall or market stalls for local traders or producers, the preferred localization of such a space indicated by the respondents was in the northern and southern parts of the area for planned spatial development (Fig. 8).

As a result of public consultations related to the new center of Rokietnica, local planners decided to introduce changes to the architectural concept for that area. The changes were made considering the respondents' postulates concerning preferred building height (the number of levels of buildings planned in accordance with the principles of determining the land use outline in high density areas was lowered). The draft of the local spatial development plan presented at the second stage of public consultations also considered other proposals of residents taking part in public consultations concerning the localization of selected functions, traffic calming and the course of major routes. It failed, however, to include the participants' postulate of increasing the number of parking spaces.

Figure 8. Localization of the central marketplace preferred by respondents
Source: Own elaboration.

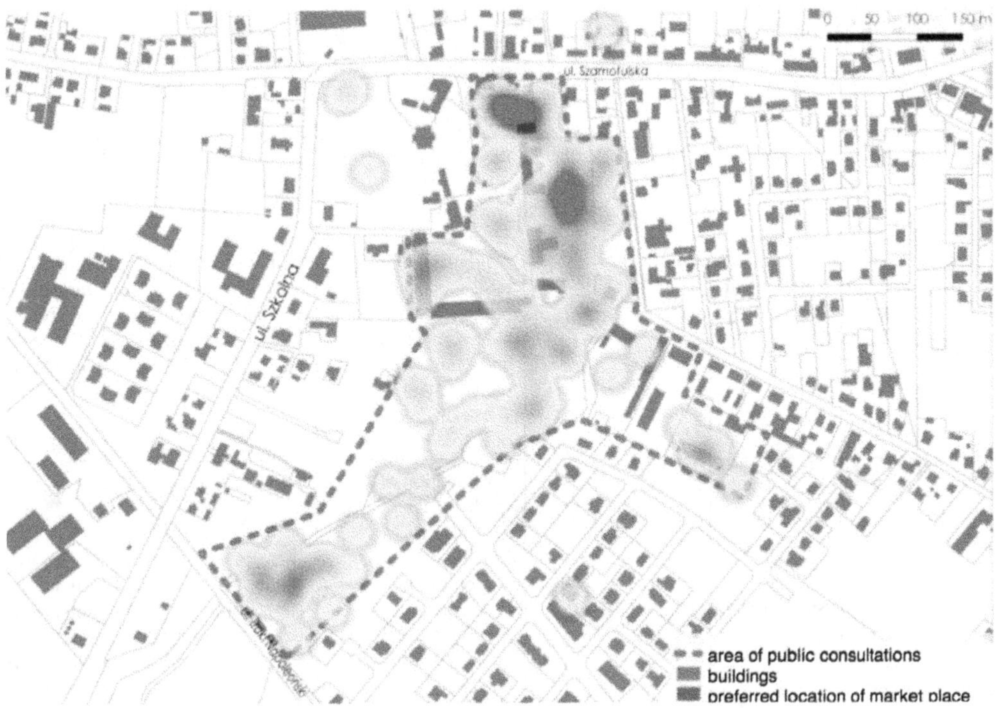

MODERNIZATIONS OF PUBLIC SPACES

Implementations M1 – Łazarski Marketplace in Poznań, M2 - Bernardyński Square, M3 - Wielkopolski Square were made to collect input from residents for spatial and functional-organizational transformations of inner-city squares and open-air marketplaces in Poznań (Tab. 2). In the case of Łazarski Marketplace in Poznań, the results were used to prepare guidelines for the architectural-urban planning competition for the concept of modernization of the area. The diagnosis prepared as part of implementations of geo-questionnaires connected with Bernardyński Square and Wielkopolski Square allowed for the determination of guidelines for further actions comprising modernization and organizational changes within the squares and the choice of one of the three proposed variants of planned changes. Geo-questionnaire implementations in these two cases were preceded with focus group of interested parties' study, which additionally widened the group of participants and the scope of information collected during the entire processes.

Table 2. Pilot implementations of geo-questionnaire in Poznań and Łódź agglomerations on the examples of public spaces modernizations

ID[2]	Name	Scale	Number of respondents	Consulted issues	Partners in the process	Use of results
M1	Łazarski Marketplace in Poznań	Small urban space	386	– hitherto forms of the market use – mobility and parking spaces – commercial services and recreation – directions of modernization	Municipal Authorities of Poznań, Łazarz Neighborhood Council, Targowiska Ltd. (administrator)	Urban planning and architectural concept of the market
M2	Bernardyński Square in Poznań	Small urban space	1102	– evaluation of market offer, – esthetics – safety – mobility of customers and parking areas – new functions, and changes in management – use of sellers' offers	Municipal Authorities of Poznań, Targowiska Ltd. (administrator)	To be used as directives for further modernization and organizational changes in the Square space
M3	Wielkopolski Square in Poznań	Small urban space	784	– evaluation of market offer, – esthetics – safety – mobility of customers and parking areas – new functions, and changes in management – use of sellers' offers	Municipal Authorities of Poznań, Targowiska Ltd. (administrator)	To be used as directives for further modernization and organizational changes in the Square space

The course and results of the geo-questionnaire application in public consultations about modernization of public spaces are presented on the example of site M3 – Wielkopolski Square. The number of respondents was 784. Fig. 9 presents the age structure of respondents, divided into 5-year groups. The age structure of respondents reflects to a high degree, but not exactly, the age structure of Poznań citizens. The respondents can be only partly considered as representative of each age group. There is distinct overrepresentation of persons of 20 to 44 years of age, which is characteristic for consultations via the internet. Persons from these age groups have both high level of digital skills and interest in local matters. The underrepresented groups were the young persons and the elderly, only 28 persons (5.4%) were above 60 years of age and only 12 (2.3%) were 15-19 years old.

The aspects of Wielkopolski Square modernization subjected to consultations were the allocation of space for recreation purposes and localization of stations of the city bicycle stations. The quality of the present space for recreation at the square was assessed as negative: according to 75% of respondents as very bad or bad, and according to the respondents' preferences, even a part of the present marketplace could be transformed into a recreation area.

Figure 9. Age structure of the respondents of the geo-questionnaire on Wielkopolski Square in relation to the age structure of Poznań citizens
Source: Own elaboration based on geo-questionnaire results and data from the Central Statistical Office.

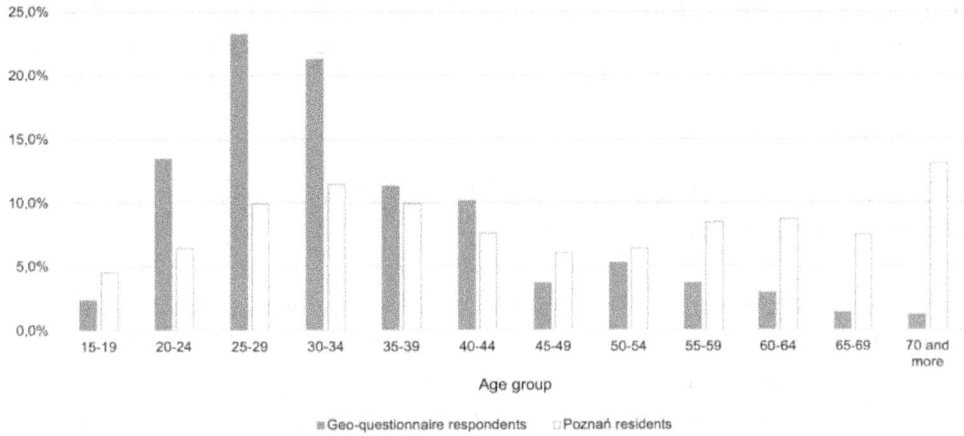

As far as the communication is concerned, a commonly accepted investment was location of city bike station, supported by 71% of respondents (Fig. 10). The preferred localization of such a station was the southern part of the Square, near the tram stop or the northern part, near the pedestrian crossing. The localizations proposed by the respondents are justified because of cumulation of traffic in these two zones.

Figure 10. Localization of city bike station preferred by the respondents
Source: Own elaboration.

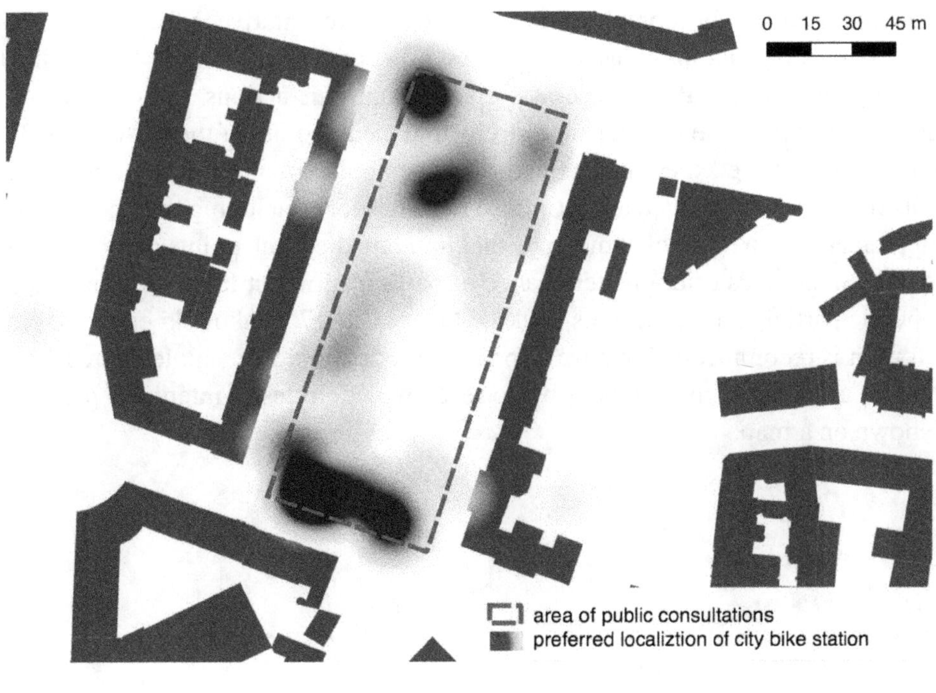

The results collected via the geo-questionnaire indicated the preferences of the users of this area regarding the modernization and localization of selected functions. The information was assessed as useful and submitted to relevant institutions including the Poznań City Urban Office. The frequency of inhabitants and positive assessment of the institution commissioning the consultations permit concluding about the usefulness of the geo-questionnaire as a tool for eliciting information on opinion and preferences on modernization of urban public space.

EVALUATION OF THE GEO-QUESTIONNAIRE AS A TOOL OF PUBLIC CONSULTATIONS

Also, the information on the preferences and opinions of local citizens was possible to collect through a geo-questionnaire (see also: Bąkowska et al. 2016; Czepkiewicz et al. 2016a, 2017; Jankowski et al. 2016, 2017a). Most respondents of the geo-questionnaire expressed positive opinions about this form of consultations (79.5% of all opinions). Respondents offering positive evaluations named such main advantages as possibility of participating in the consultations via the Internet and possibility of indicating precise information on the map. Negative opinions focused on doubts about the actual use of such consultation results in the decision-making process and reporting technical problems that needed to be improved.

According to the results of the interviews conducted with partners in the processes of public consultations (i.e., representatives of local authorities and relevant offices), one of the most important advantages of this way of consultations was inclusion of larger group of people in the consultation process, connected with the removal of the requirement of attending a public meeting at a specific place and time. They also mentioned more precise definition of needs, especially in the spatial context of area under planning consideration, as well as a possibility of making a database with quantitative and spatial information that may be used for more detail analyses. For representatives of local authorities also low cost of participation in consultations was an important aspect of the processes. The application of geo-questionnaires in public consultation processes related to various actions and plans from the field of spatial planning made it possible to determine strong and weak points of their use (Tab. 3). They were formulated on the basis of the experience gained from the application of the geo-questionnaire and the evaluation of its results. The indicated aspects make the basis for concluding that the analyzed consultation tool is supplementary to traditional methods that mitigate the weak points of the geo-questionnaire. Internet consultation methods offer greater chance of participation but fail to eliminate all the barriers appearing in public participation processes (Bąkowska et al. 2017; Jankowski et. al. 2017b; Pina et al. 2017). Therefore, it is recommended that traditional methods should be supplemented by tools such as geo-questionnaire, as the latter make it possible to reach wider circles of interested parties with precise information shown on a map.

Table 3. Advantages and disadvantages of using geo-questionnaires in public consultations

Strong points (advantages)	Weak points (disadvantages)
– it is possible to take part in public consultations from any place, at any moment during the entire consultation period, the only requirement being the Internet access – possibility of providing more precise information with the use of an interactive map – possibility of collecting opinions and preferences from a group of people that is larger and more representative than in traditional consultations – depending on the decision of the public consultations organizer, interested parties may learn about detailed summed up preferences and opinions of residents (maps, diagrams, reports) – possibility of using consultation results for other studies (diagnostic material) – possibility of combining this method with other methods (e.g., workshop methods)	– no possibility of filing motions and formal requests without engaging in a more complicated procedure based on electronic signature – longer duration of the basic planning procedure resulting from the need to include extra stages – little interest of citizens in case of large-scale studies devoted to several varied territorial units (people tend to identify less with units at higher levels of the territorial structure, such as counties) – less valuable results obtained when consultations tackle more general issues examined in larger areas (a big number of single indications) – possible doubts in the truthfulness of the geo-questionnaire results expressed by certain interest groups – possibility of occurrence of technical problems – no obligation to take the results of public consultations collected with the use of geo-questionnaires into account by the authorities

A MODEL PROCEDURE OF THE IMPLEMENTATION OF A GEO-QUESTIONNAIRE IN PUBLIC CONSULTATIONS

The experience gained from the hitherto applications of the geo-questionnaire has indicated that the effectiveness of consultations carried out with this tool needs a well-prepared multi-stage procedure. Correct preparation and realization of each stage ensures the right representativeness and frequency of participants and useful results that can be supplemented with consultations conducted with other tools (meetings, debates, workshops). On the basis of the pilot implementations of the geo-questionnaire a model procedure has been worked out. It includes nine stages as illustrated in Fig. 11.

The first stage is to analyze the issue that is to be subjected to public consultations. At this stage, the aim of public consultations, their required scope and informative content should be defined; also, specific conditions of the area, stakeholders and partners of the process (main recipients of the consultations results) are to be identified and the scope of information obtained through the questionnaire should be determined.

At the second stage, the plan for carrying out public consultations is developed. Issues to be determined at that stage are as follows: starting and ending dates for the geo-questionnaire, methods of promoting consultations among citizens, costs, additional methods of public consultations, form and date of providing a report and persons representing authorities/institutions taking part in the process.

Based on the identification of the issues to be consulted with the public, the content of the questionnaire is prepared (including the type of question - open-ended, closed-ended, map-based, etc.) (stage 3). The geo-questionnaire should also include questions enabling socio-demographic characteristic of respondents (age, sex, place of residence, education level, etc.) to determine representativeness of participants. Such characteristics are required to determine which groups need to be activated with the use of different methods.

Once the content of the geo-questionnaire is approved, it is published in the application with the use of an administrative panel of the software that allows for editing, monitoring at later stages, and viewing the results (stage 4). When the questions have been put in and the questionnaire configured, it should be tested by persons with varying Internet skills to eliminate possible errors and make corrections (stage

5). The time necessary to complete the questionnaire should also be measured. A geo-questionnaire that is too long may deter respondents, therefore it should be limited to questions that would fully address the information needed to be defined at the first stage. As indicated by Haklay & Tabón (2003), proper testing of functionality of PPGIS-based tools is essential for their proper and effective use by respondents with varied skills.

Figure 11. Procedure of the geo-questionnaire implementation in public consultations process

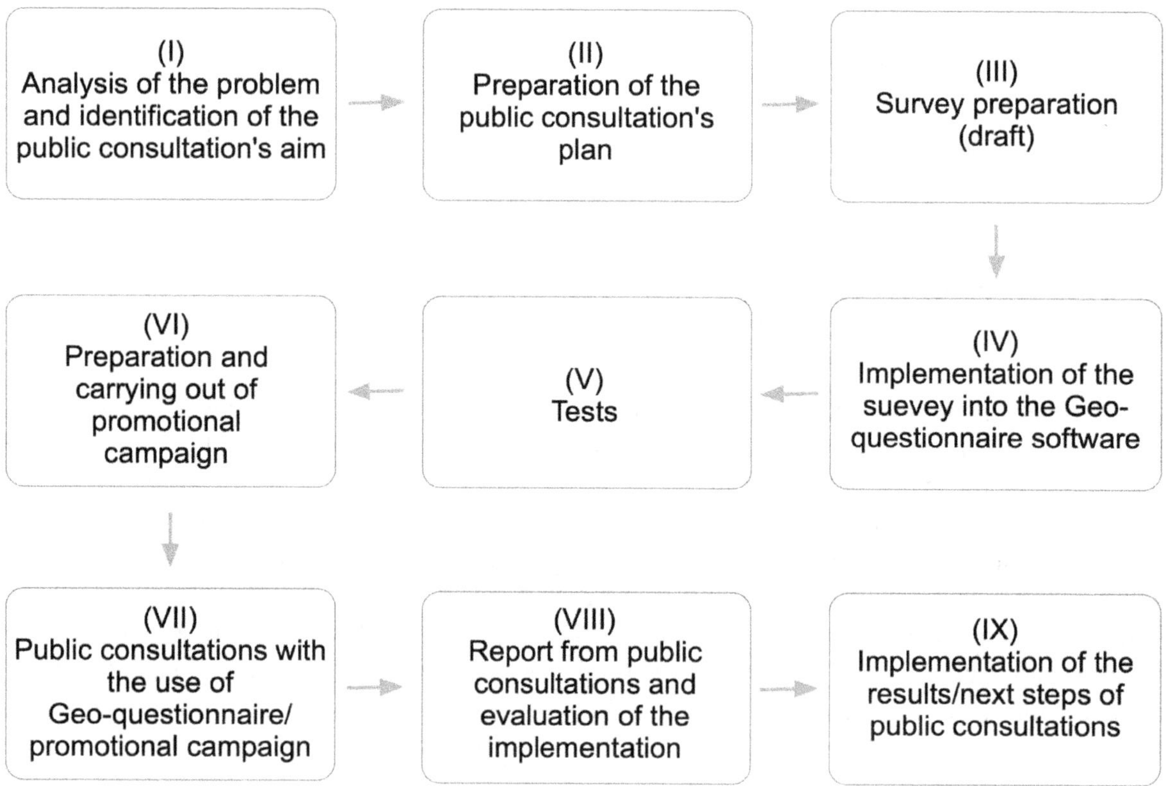

The next stage (6) is the preparation and carrying out a campaign promoting public consultations. The use of proper methods of informing residents about the planned consultations is essential for ensuring a sufficient level of participation. According to the research carried out by Jankowski et al. (2016), most efficient methods include local media news and brochures addressed to each household. On the other hand, the analysis of the survey on geo-questionnaires related to city squares in Poznań (Bąkowska et al. 2017) reveals that information posted in social media was the most effective (43.9%). The choice of methods of informing respondents about the the consultations should depend on the target group of residents, so that such information can encourage their participation. Constant monitoring of the number of a geo-questionnaire page views and the number of completed questionnaires is necessary to determine the need to repeat or widen the promotional campaign if turnout is insufficient.

Stage 7 is finally connected with the geo-questionnaire launch and making it available to residents. The time during which the questionnaire is available for filling in depends on the scale of the problem and the areas from which respondents are recruited. After questionnaire closing, a team of experts begins work on the analysis of results including preparation of maps (stage 8). Depending on the needs of the consultation organizers, the results may be prepared in the form of a report supplemented with a database in a spreadsheet format and as shapefiles, or in the form of a summary report that can be made available to residents, e.g., in a local newspaper. The consultation report should contain recapitulation of major results and recommendations for future actions, and the way in which results will be implemented.

Once the report is handed over to consultation organizers, further stages of the process depend on the specific nature of the issue and on formal requirements (for legally regulated procedures). Due to the specific features of the geo-questionnaire, which mainly enables to identify and diagnose the needs and expectations of inhabitants, it is usually the beginning of a longer consultation process leading to a draft proposal (e.g., of a local spatial development plan or urban-architectural concept, etc.). At the further stage citizens should be offered a chance to express their opinions about the proposed solutions; a particularly handy tool for that is geo-discussion, in which participants may see a project, comment on proposals, and discuss them with others[3].

RESULTS

The effects of implementation of geo-questionnaires show that the tool can be used for the processes of public consultations related to issues entirely or partly focused on spatial development. Thus, major fields in which geo-questionnaires might find their application in spatial planning in Poland works on the documents of formal nature (regulated by the law) such as local spatial development plans), and informal documents such as conceptions of spatial development or modernization of public spaces. The informal documents that may make use of the geo-questionnaire are also known as soft planning or preplanning (Mikuła 2019). This term refers to voluntary procedures regarding spatial planning, with no binding force and aimed at acquiring knowledge and consensus on spatial planning.

Insufficient participation of the general public in the process of spatial planning and urban design is one of its weak points. A limited scope of public consultations currently stipulated by the law often described as tokenism combined with little interest among residents are not conducive to developing good spatial development plans and projects. Therefore, any initiative aimed at increasing a public participation in the process of spatial planning should be perceived as even more valuable.

The response to the research question on which areas of spatial management can be addressed in the e-consultation procedure is that the Internet tool presented in this chapter, i.e., the geo-questionnaire, may be used in public consultations both for the purposes of planning documents regulated by the law, such as local spatial development plans, and for other planning studies connected with public spaces.

The above-described applications permit identification of the strong and weak points of using geo-questionnaire in public consultations, to respond to research question 2. The use of PPGIS tools in public consultations has an undoubtful advantage of engagement of wider public in the process of local spatial planning. Local citizens are a rich source of information on local matters, often unnoticed or impossible to uncover by decision-makers. Local authorities, urban planners, landscape architects and those responsible for preparation of documents of strategic significance can use PPGIS to increase the interest of public in local matters and make the decisions of higher approval of citizens. The process of decision-making also becomes more transparent. Due to the engagement of public in the process of spatial planning, it may gain a contribution of socially sensitive dimension.

Despite numerous advantages of such tools, such as availability of spatial information, wide accessibility, or possibility of interacting with other participants, weak points of PPGIS should also be taken into consideration. Among them are the requirements connected with basic skills of using Internet maps and online forms, which can considerably support participation of elderly respondents. Similarly, as in the study of Ganapati (2010) it has been shown that people with higher computer skills, younger, better educated, with higher social status are overrepresented, which is contrary to principles of inclusiveness and equal access to participation tools. As the authors are aware of the weak points of the presented tools, it is recommended to use them along with traditional public consultation methods, thus allowing the two to complement each other and mitigate their respective negative sides.

THE FUTURE OF PPGIS IN THE LIGHT OF NEW POST-PANDEMIC REGULATIONS

The COVID-19 pandemic, which is a type of phenomenon known as the black swan effect (Taleb 2010), verified at the beginning of 2020 both the individual readiness of citizens to digitize different spheres of their life and the preparation of local government administration for remote management and communication with residents. The sphere of spatial planning, especially the legal procedures, experienced a serious disturbance. Despite the available technological solutions, this situation exposed the lack of administrative and statutory preparation for the use of PPGIS in place of traditional consultation meetings. Spatial planning procedures were suspended due to the ban on assemblies in force from March 2020. It prevented the organization of public discussions in the traditional, face to face, formula, and introduced provisions temporarily suspending administrative procedures, which was regulated by one of the subsequent versions of the Polish government's "Anti-crisis shield". The first reaction of local governments to the above-mentioned restrictions was to suspend planning procedures, which was explained primarily by the inability to hold statutory public consultations. The lack of statutory empowerment to use PPGIS and e-consultation tools was generally a significant obstacle for proceeding and adopting planning documents by local governments. In the face of these challenges, numerous groups of experts and non-profit organizations had online discussions in the first weeks of the new functioning of offices on the current possibilities and the future of participation processes in the pandemic reality. They debated the new digital reality which would not only be a temporary change, but a revolution both residents and administration would have to embrace. At the same time, new technical solutions that could support public participation in spatial planning in this new virtual reality emerged. The geo-questionnaire was commonly mentioned among the listed tools which were the only ones that fully contained the spatial component. Nevertheless, many local office workers had concerns about meeting the statutory requirement to carry out public consultations through remote methods.

The question that remains open at the moment is who bears the highest cost of not being able to participate directly in the process of public consultations in spatial planning? It is reasonable to suppose that these are the elderly and those with the lowest digital skills as well as individuals unable to independently meet the requirements of administrative procedures. On the other hand, people with disabilities or those who permanently care for other people (children, the sick) received for the first time, in many cases, a set of tools to engage in consultation processes regarding their neighborhood. Both the results of the presented research and the observations of ongoing discussions confirmed the previously formulated statement that the implementation of PPGIS in decision-making processes is to the greatest extent limited by legal regulations, organizational readiness of offices and the approach of local leaders, and not by technological issues (Ganapati 2010).

Restarting the planning procedures was possible thanks to the Act of May 14, 2020 on changing certain acts with respect to protective measures regarding the spread of the SARS-CoV-2 virus (Journal of Laws of 2020, item 875). Under this Act, the administrative proceedings were reinstated. Nevertheless, a significant limitation were other regulations related to the epidemiological threat, reducing the number of interested parties in one room, which obviously made it difficult, if not impossible, to lead a public discussion on the proposed planning documents in the seat of a planning unit (Kaczmarek & Mikuła 2020). It was only the national regulations that were introduced at the end of 2020 that unequivocally made it possible to include remote solutions in the procedures of public consultations in spatial planning. These provisions enabled comments and proposals regarding planning acts and their projects to be submitted in both paper and electronic forms and public discussions to be held at a distance, also through the Internet. PPGIS have thus become an important tool that could support these processes. Changes in legal regulations and the increase of available tools and the interest in PPGIS among local governments observed since the mid-2020 allow the conclusion that despite the obvious global crisis, the COVID-19 pandemic has rapidly accelerated the possibility of popularizing the use of solutions such as geo-questionnaire. However, the new PPGIS tools offered by private entities and the level of support provided for officials raise concerns about the quality of the processes mentioned and the deterioration of standards and methodologies developed by scientists.

Development of e-consultations as an important element in making decisions on spatial planning faces some serious challenges (research question 3). Widespread adoption of such tools as the geo-questionnaire as standard elements of public consultations still requires coping with legal, organizational, technological and social barriers. The acceptance of public consultations results of such consultations depends on the good will of authorities of those territorial units in which they were held and on the adopted local law. Implementation of such tools also requires major organizational regulations connected, among other things, with improvement of employees' competencies. As far as technological and social barriers are concerned, one should bear in mind the problem of digital exclusion and varying level of competencies of residents. For this reason, it is recommended that those who may find it difficult to use these tools should be offered such access in the local office together with direct support by employees.

CONCLUSION

The effects of implementation of the geo-questionnaire in Poland show that PPGIS can be successfully used in the processes of public consultations regarding spatial context. Action Research approach was adopted as the main method in the research, oriented towards supporting local self-governments in the process of participative planning. Urban planners, authorities and local communities were involved in the processes to learn how to use online tools in the public participation process. The Action Research approach is still not common in governance in Poland (Chrostowski & Jemielniak 2011) and its use in this study indicates the significant benefits of its application for implementation of new tools in the public management process. As highlighted by Greenwood & Levin (2007) Action Research is characterized by a long-lasting and risky application process, while organizations look for quick results. This problem also concerns the public participation in spatial planning. The use of PPGIS tools in accordance with the assumptions of new spatial planning and based on the planning process on the public participation and negotiations may extend the process itself and make it more complex and complicated (Haughton et al. 2010). The success of participatory governance is therefore dependent on careful planning of the transformation process and slow implementation of changes. Changing attitudes of local authorities and administrative employees as well as planners in relation to participative planning takes time, as any change in the culture of the organization.

In this context, the use of Action Research approach fulfilled its assumptions. On the one hand, it was possible to obtain cooperation with local self-governments, trying to improve the organization of the planning process through its greater public participation with the use of PPGIS tools. The actors in this process, due to its innovative character, needed technological and organizational support as well as expert knowledge from outside sources regarding benefits and deficits of PPGIS application.

The use of online tools of public communications in the years to come should pave the way for a wider consultation process in spatial planning and in other decision-making processes at the local level. Along with the legal changes, development of new technologies, mechanisms of partnership and multi-level governance, the opportunities to participate in decision-making should increase.

ACKNOWLEDGMENT

Research described in this chapter was conducted as part of the project 'Geoportal supporting public participation in spatial planning' financed by the National Research and Development Center (contract number PBS3/A9/39/2015). The authors gratefully acknowledge the support of the Poznan City Hall, Łódź City Hall, County Office in Poznań, Rokietnica Municipal Office and Swarzędz Municipal Office.

The chapter builds upon an article published in the International Journal of E-Planning Research (IJEPR).

REFERENCES

Armitage, D. F., Berkes, F., & Doubleday, N. (2007). *Adaptive Co-management: Collaboration, Learning and Multi-level Governance*. University of British Columbia.

Atzmanstorfer, K., & Blaschke, T. (2013). The Geospatial Web: A Tool to Support the Empowerment of Citizens through E-Participation? In C. N. Silva (Ed.), *Citizen E-Participation in Urban Governance: Crowdsourcing and Collaborative Creativity* (pp. 144–171). doi:10.4018/978-1-4666-4169-3.ch009

Bąkowska, E., Kaczmarek, T., Jankowski, P., Zwoliński, Z., Mikuła, Ł., Czepkiewicz, M., & Brudka, C. (2016). Geo-questionnaire in urban planning – preliminary results of the experimental application in Poland. *Rozwój Regionalny i Polityka Regionalna, 35*, 9–26.

Bąkowska, E., Kaczmarek, T., & Mikuła, Ł. (2017). Wykorzystanie geoankiety jako narzędzia konsultacji społecznych w procesie planowania przestrzennego w aglomeracji poznańskiej. *Roczniki Geomatyki, XV*, *2*(77), 147–158.

Brown, G., & Raymond, Ch. M. (2014). Methods for identifying land use conflict potential using participatory mapping. *Landscape and Urban Planning*, *122*, 196–208. doi:10.1016/j.landurbplan.2013.11.007

Bugs, G., Granell, C., Fonts, O., Huerta, J., & Painho, M. (2010). An assessment of Public Participation GIS and Web 2.0 technologies in urban planning practice in Canela, Brazil. *Cities (London, England)*, *27*(3), 172–181. doi:10.1016/j.cities.2009.11.008

Carver, S., Evans, A., Kingston, R., & Turton, I. (2001). Public participation, GIS, and cyberdemocracy: Evaluating on-line spatial decision support systems. *Environment and Planning. B, Planning & Design*, *28*(6), 907–921. doi:10.1068/b2751t

Chrostowski, A., & Jemielniak, D. (2011). *Skuteczne doradztwo strategiczne: metoda Action Research w praktyce*. Poltext.

Czepkiewicz, M., & Brudka, C. (2016b). *Raport z badania preferencji mieszkańców na temat kierunków zmian zagospodarowania przestrzennego osiedla Andrzejów w Łodzi*. Centrum Badań Metropolitalnych UAM.

Czepkiewicz, M., Brudka, C., Jankowski, P., Kaczmarek, T., Zwoliński, Z., Mikuła, Ł., Bąkowska, E., Młodkowski, M., & Wójcicki, M. (2016a). Public Participation GIS for sustainable urban mobility planning: Methods, applications and challenges. *Rozwój Regionalny i Polityka Regionalna*, *35*, 9–35.

Czepkiewicz, M., Jankowski, P., & Młodkowski, M. (2017). Geo-questionnaires in urban planning: Recruitment methods, participant engagement, and data quality. *Cartography and Geographic Information Science*, *44*(6), 551–567. doi:10.1080/15230406.2016.1230520

De Filippi, F., Coscia, C., & Guido, R. (2017). How Technologies Can Enhance Open Policy Making and Citizen-Responsive Urban Planning: MiraMap - A Governing Tool for the Mirafiori Sud District in Turin (Italy). *International Journal of E-Planning Research*, *6*(1), 23–42. doi:10.4018/IJEPR.2017010102

Flyvbjerg, B. (2018). *Making social science matter: why social inquiry fails and how it can succeed again*. Cambridge University Press.

Ganapati, S. (2010). *Using Geographic Information Systems to Increase Citizen Engagement E-Government*. IBM Center for The Business of Government.

Geddes, M. (2006). Partnership and the Limits to Local Governance in England: Institutionalist Analysis and Neoliberalism. *International Journal of Urban and Regional Research*, *30*(1), 76–97. doi:10.1111/j.1468-2427.2006.00645.x

Glaser, B., & Strauss, A. (1967). *Discovery of Grounded Theory: Strategies for Qualitative Research*. Aldine Publishing.

Greenwood, D. J., & Levin, M. (2007). *Introduction to action research. Social research for social change*. Sage Publications, Inc. doi:10.4135/9781412984614

Gryl, I., Jekel, T., & Donert, K. (2010). Geoinformation and Spatial Citizenship. *Learning with GeoInformation*, *V*, 2–11.

Haklay, M., Singleton, A., & Parker, Ch. (2008). Web Mapping 2.0: The Neogeography of the GeoWeb. *Geography Compass*, *2*(6), 2011–2039. doi:10.1111/j.1749-8198.2008.00167.x

Haklay, M., & Tabón, C. (2003). Usability evaluation and PPGIS: Towards a user-centred design approach. *International Journal of Geographical Information Science, 17*(6), 577–592. doi:10.1080/1365881031000114107

Haughton, G., Allmendinger, P., Counsell, D., & Vigar, G. (2010). *The New Spatial Planning: Territorial Management with Soft Spaces and Fuzzy Boundaries*. Taylor & Francis.

Horelli, L., & Wallin, S. (2013). Starting points. In L. Horelli (Ed.), *New Approaches to Urban Planning. Insights from Participatory Communities* (pp. 11–22). Aalto University.

Jankowski, P., Czepkiewicz, M., Młodkowski, M., & Zwoliński, Z. (2016). Geo-questionnaire: A Method and Tool for Public Preference Elicitation in Land Use Planning. *Transactions in GIS, 20*(6), 903–924. doi:10.1111/tgis.12191

Jankowski, P., Czepkiewicz, M., Młodkowski, M., Zwoliński, Z., & Wójcicki, M. (2017a). Evaluating the scalability of public participation in urban land use planning: A comparison of Geoweb methods with face-to-face meetings. *Environment and Planning. B, Urban Analytics and City Science*. Advance online publication. doi:10.1177/2399808317719709

Jankowski, P., Kaczmarek, T., Zwoliński, Z., Mikuła, Ł., Wójcicki, M., Bąkowska, E., Czepkiewicz, M., Młodkowski, M., & Brudka, C. (2017b). *Narzędzia internetowe w konsultacjach społecznych. Idea, obszary zastosowań i wdrażanie*. Wydawnictwo Naukowe UAM.

Kaczmarek, T., & Bąkowska, E. (2016a). *Raport z konsultacji społecznych na temat kierunków zagospodarowania przestrzennego centrum miejscowości Rokietnica*. Centrum Badań Metropolitalnych UAM.

Kaczmarek, T., & Bąkowska, E. (2016b). *Raport z konsultacji społecznych poświęconych nowym kierunkom zagospodarowania przestrzennego terenu lotniska w Kobylnicy*. Centrum Badań Metropolitalnych UAM.

Kaczmarek, T., & Mikuła, Ł. (2020). Konsultacje społeczne w planowaniu przestrzennym z użyciem narzędzi internetowych PPGIS. In A. Bochentyn & J. H. Szlachetko (Eds.), *Cyfrowa czy analogowa? Funkcjonowanie administracji publicznej w stanie kryzysu* (pp. 411–435). Wydawnictwo Uniwersytetu Gdańskiego.

Kahila, M., & Kytta, M. (2009). SoftGIS as a Bridge-Builder in Collaborative Urban Planning. In S. Geertman & J. Stillwell (Eds.), *Planning Support Systems Best Practice and New Methods* (pp. 389–411). Springer. doi:10.1007/978-1-4020-8952-7_19

Kahila-Tani, M., Broberg, A., Kyttä, M., & Tyger, T. (2015). Let the Citizens Map—Public Participation GIS as a Planning Support System in the Helsinki Master Plan Process. *Planning Practice and Research, 31*(2), 195–214. doi:10.1080/02697459.2015.1104203

Kettl, D. F. (2002). *The Transformation of Governance: Public Administration for Twenty-first Century America*. John Hopkins University Press.

Kingston, R. (2011). Online Public Participation GIS for Spatial Planning. In T. L. Nyerges, H. Couclelis, & R. McMaster (Eds.), *The SAGE Handbook of GIS and Society* (pp. 361–380). SAGE. doi:10.4135/9781446201046.n19

Manzo, C. L. (2003). Beyond house and haven: Toward a revisioning of emotional relationships with places. *Journal of Environmental Psychology, 23*(1), 47–61. doi:10.1016/S0272-4944(02)00074-9

McCall, M. K., & Dunn, Ch. E. (2012). Geo-information tools for participatory spatial planning: Fulfilling the criteria for 'good' governance? *Geoforum, 43*(1), 81–94. doi:10.1016/j.geoforum.2011.07.007

Mikuła, Ł. (2019). *Zarządzanie rozwojem przestrzennym obszarów metropolitalnych w świetle koncepcji miękkich przestrzeni planowania. Studia i Prace z Geografii, 73.* Bogucki Wydawnictwo Naukowe.

Pina, V., Torres, L., & Royo, S. (2017). Comparing online with offline citizen engagement for climate change: Findings from Austria, Germany and Spain. *Government Information Quarterly, 34*(1), 26–36. doi:10.1016/j.giq.2016.08.009

Rantanen, H., & Kahila, M. (2008). The SoftGIS approach to local knowledge. *Journal of Environmental Management, 90*(6), 1981–1990. doi:10.1016/j.jenvman.2007.08.025 PMID:18556113

Rhodes, R. A. W. (2000). *Debating Governance.* Oxford University Press.

Schroeder, P. (1996). Criteria for the design of a GIS/2. *Specialists' meeting for NCGIA Initiative 19: GIS and society.*

Sieber, R. (2006). Public Participation Geographic Information Systems: A Literature Review and Framework. *Annals of the Association of American Geographers, 96*(3), 491–507. doi:10.1111/j.1467-8306.2006.00702.x

Siemiński W. (2014). Partycypacja spontaniczna a kształtowanie przestrzeni. *Partycypacja społeczna w planowaniu przestrzennym, Konferencja Towarzystwa Urbanistów Polskich Oddziału w Warszawie oraz Biura Architektury i Planowania Przestrzennego Urzędu m.st. Warszawy.*

Sullivan, H., & Skelcher, C. K. (2002). *Managing across Boundaries.* Palgrave Macmillan. doi:10.1007/978-1-4039-4010-0

Taleb, N. N. (2010). The Black Swan: Second Edition: The Impact of the Highly Improbable. Random House.

Wiench J. (1991). *Metody i techniki konsultacji społecznych w planowaniu przestrzennym.* IGPiK.

Yin, R. K. (2003). *Case study research. Design and Methods Applied Social Research Methods Series.* International Educational and Professional Publisher.

ENDNOTES

1. P – local spatial development plans.
2. M – modernization of public spaces.
3. Geo-discussion, unlike geo-questionnaire, enables interaction between its participants - respondents, decision-makers, planners, etc. – and therefore makes it possible to obtain feedback on the proposed solutions and clarify doubts related to e.g. legal regulations as early as at the consultation stage.

Chapter 3
Evaluating Urban Sensing Within Engaged Planning and Design

Andrew Mondschein
University of Virginia, USA

Mona El Khafif
University of Virginia, USA

ABSTRACT

Civic technology models propose that urban environmental sensing and analysis can inform community-based decision making, but technology-driven planning and design is still subject to systemic challenges including a lack of community trust and local government capacities. This chapter examines the problem of integrating urban sensing into engaged planning and design. The authors assess the outcomes of two planning and design efforts in Virginia, USA utilizing the community-centered urban sensing (CCUS) and engagement platform. Developed by planners, urban designers, landscape architects, and data scientists at the University of Virginia, CCUS addresses the need for actionable information on the urban environment through community-engaged urban data collection and analysis. The cases provide evidence for the importance of engagement around sensing itself. Based on these findings, the authors propose a conceptualization of urban sensing applications that accounts for a range of local contexts from high capacity and trust to dysfunction and mistrust.

INTRODUCTION

The emergence of the "smart cities" paradigm of urbanization has amplified the longstanding tension between democracy and technocracy in planning. The promise of new infrastructures, data, and analytics to address urban problems is tempered by the possibility that these systems might not equitably benefit all members of the community and, in fact, may perpetuate ongoing inequalities in cities (Green, 2019; Vanolo, 2014). Even in cities with significant public and private resources directed towards urban technology, significant opposition has emerged to investment in new urban technologies (Mann et al., 2020; Valverde & Flynn, 2020). Most cities around the globe have yet to grapple with the purposes and implica-

DOI: 10.4018/978-1-7998-9090-4.ch003

tions of data, networking, and analytics within planning processes that aspire to community engagement and localized democracy. In fact, smart city initiatives frequently come in the context of longstanding, ongoing stresses on social cohesion and civic decision-making. Given this challenge, we examine the problem of integrating urban sensing, a major element of smart cities, into an engaged planning process. Smart cities research has been paralleled by an increased focus on engagement within planning, valuing local input and community-based approaches to neighborhood design (Slotterback & Lauria, 2019; B. B. Wilson, 2018). In this context, we investigate a smart, engaged planning and design process called Community-Centered Urban Sensing (CCUS). As a sensing protocol and online engagement tool, CCUS can flexibly support the formulation of concerns, the design of the urban sensing technology, the collection of quantitative and qualitative data, and ultimately decision-making and local action.

Like any planning tool, CCUS, must function within distinctive social and governmental contexts, and the deployment of CCUS in two cases examines how a technology-driven process might operate under conditions of social and civic dysfunction. We describe and assess the outcomes of planning and design efforts in two cities in Virginia in the United States: the City of Charlottesville and the City of Waynesboro. Charlottesville, a small city hosting a large public university, seeks to both adopt new, technologically-driven planning methods while also addressing longstanding inequities between city and university, black and white, and wealthy and low-income residents. Waynesboro, a largely working-class city in the Shenandoah Valley seeks to address longstanding infrastructure needs in some of its neighborhoods. We examine the effects of Community-Centered Urban Sensing in these two cities. CCUS is an urban sensing toolkit and process developed by a team of urban planners and designers, landscape architects, and information technologists at the University of Virginia. CCUS seeks to address the need for actionable, spatialized quantitative and qualitative information on the urban. The outcomes of CCUS deployments in Charlottesville and Waynesboro serve as the basis for our findings.

This chapter begins with a review of key concepts and precedents from the literature, including smart cities and urban governance, as well as modes of participation in urban sensing. We then describe the CCUS program and present the engagement process and community response to CCUS case studies. Our findings address challenges in how technological urbanism moves – often nonlinearly – from data to action, as well as the potential for increased marginalization of some populations. Building on outcomes in Charlottesville and Waynesboro, we discuss a conceptualization of smart and engaged planning and design that accounts for urban dysfunction and recommend modes of engagement in technological urbanism that vary from oppositional and insurgent to systematic and administrative. If new urban technologies are to become more than top-down interventions, then the smart cities paradigm must expand to encompass modes and methods that function even when local urban systems lack consensus and cohesion on the purposes of data, analytics, and planning and design more broadly.

RESEARCH CONTEXT: SMART CITIES AND CITIZEN PARTICIPATION

Researchers, as well as public- and private-sector actors, utilize the smart cities concept to encompass the array of technological innovations being applied to urban systems, environments, and society in the early 21st century (Batty, 2013a; Caragliu et al., 2011). The literature on smart cities is expanding quickly, and applications of the term range from practical to critical. Much of the research on smart cities focuses on defining the systems and technologies that comprise urban innovation (Yigitcanlar et al., 2018). Smart cities, according to Albino (2015), are generally thought of as having key shared characteristics, including networked infrastructure, social inclusion, and environmental sustainability as objectives, typically with a business-led approach to urban development. Critical perspectives often frame smart cities projects and initiatives as part of longstanding urban trends such as the privatization of urban systems and spaces or the loss of social vibrancy and civic participation through optimization (Fard, 2018; Krivý, 2018; Yates, 2017). In addition, smart cities and their boosters have been characterized as hubristic, assuming that prior lessons from planning research and practice are obviated by new technology (Green, 2019). Within that context, the way that technology-driven systems and practices interface with planning practice and varied socio-political contexts remains under examined.

Smart Cities and Governance

Addressing the role of the community in the design, deployment, and utilization of urban technology is imperative if smart cities are to be understood as part of democratic governance. Many smart cities projects have so far emphasized the transformation of a city's systems, rather than focusing on the role of communities and community-members in the development of new systems (AlAwadhi & Scholl, 2013). When connected to resident outcomes, smart city projects often emphasize changing citizen behavior, rather than engaging citizens in city governance (Granier & Kudo, 2016). Even when smart city projects include engagement, it may still be viewed as "top-down," intended primarily to notify residents and allow superficial input, rather than facilitate community-centered decision making about plans and technologies (Goodman et al., 2020). In a review of smart cities literature, Yigitcanlar et al. (2018) find that governance issues within smart cities research remain relatively neglected by researchers and professionals, who instead focus on smart cities technologies and systems.

Local conditions and context are a key factor in smart cities governance, including "less predictable" factors such as demographics, social pressures, community activism, and administrative cultures (Ruhlandt, 2018, p. 9). Meijer (2016) emphasizes the importance of contextual factors in smart cities implementation, including the willingness of stakeholders to take part in "collective problem-solving." For example, residents may be resistant to engagement, resulting in the diminishment of civic and political aspects of smart cities (Cowley et al., 2018). Nilssen (2018) proposes that smart cities not only introduce technological innovations but also organizational, collaborative, and experimental innovation, including citizen-directed action with cities functioning as "urban living labs" as proposed by Bulkeley et al. (2016). Batty argues that the use of big data to address urban issues may shift the emphasis of urban governance from long-term planning to short-term management of urban issues (2013b). However, the same transition from planning to management can be portrayed as insidious, where political action is edited-out of the city in favor of environmental and behavior control (Krivý, 2018). Therefore, planning must take an active role in the development of smart cities, if they are to be sustainable and equitable (Batty, 2014; Murgante & Borruso, 2015).

Engaging Communities in Data-Driven Planning and Action

Engagement practices in planning frequently make use of technology, such as geographic information systems (GIS) and web-based information and communications systems (Elwood, 2002; Goodchild, 2007; Goodman et al., 2020). There are numerous approaches to collecting information about urban environments, varying in terms of technologies used, the active or passive involvement of citizens, and the purpose and use of the collected information (See et al., 2016). Within the framework of "crowdsourcing," Certomà et al. (2015) classify technology-driven planning projects across two dimensions, where the first dimension defines data sources from sensors to "people engagement techniques" and the second dimension defines data availability from closed to open. They argue that participation in data collection must be fundamentally linked to participation in governance, stating that:

[Crowdsourcing] can be adapted to different contexts according to specific environmental, social, political, and economic needs, as long as they are aimed at advancing participatory governance and the related goals of recognition, participation, and redistribution (Certomà et al., 2015, p. 103).

They advocate for flexibility in data-driven planning processes given the diversity of circumstances under which they occur. Grey et al. (2017) portray collaborative urbanism as the outcome of top-down and bottom-up urban practices, both informed by "big and small data," where small data is the qualitative, contingent information that characterizes local conditions. Beyond data-driven planning research, however, planners acknowledge that not all participation is part of a sanctioned process (Lydon & Garcia, 2015). Instead, in instances where social or governmental processes are failing to meet community needs, tactical or do-it-yourself approaches that operate outside of traditional bounds may be necessary (Iveson, 2013).

Overall, linking spatial-data and crowdsourcing methodologies to longstanding planning challenges remains a central concern for smart city research (Roche, Nabian, Kloeckl, & Ratti, 2012). The concept of the "democratization of data" has existed in urban planning since at least the 1990s, acknowledging the challenges communities have faced when seeking to make data collection and the use of data accessible to diverse community groups rather just governments and dominant private actors (Sawicki & Craig, 1996). In some cases, the implementation of civic technologies, such as the sensor-based Array of Things in Chicago, can effectively link planners, technologies, and community members around shared visions of the use of technology in the city (B. Wilson & Chakraborty, 2019). However, significant challenges still persist in facilitating effective, equitable deployment and utilization of urban data systems, including barriers to data access, limited openness, and limited capacities within governments and communities for using these new data (Yoon & Copeland, 2020). Technologies and practices within planning and design that generate new information while being flexible enough to address a wide range of socio-political contexts remain an important area for research.

TOOL: COMMUNITY-CENTERED URBAN SENSING

Community-Centered Urban Sensing (CCUS) is a toolkit for collecting and disseminating quantitative and qualitative spatial information about urban environments (El Khafif et al., 2018; Zhang et al., 2018). As a multidisciplinary team including urban planners and designers, landscape architects, and data scientists based at the University of Virginia, we designed CCUS to facilitate participatory urban sensing, allowing residents and community-based organizations to contribute to urban planning and design in their neighborhoods. Participatory urban sensing can be defined as a social practice, where urban inhabitants *purposively* use sensing technologies to collect and share information about their surroundings (Lane et al., 2008). As a customizable toolkit, CCUS offers multiple avenues of engagement, including quantitative data collection, qualitative photo and text inputs, and integrative analytics. All inputs and analyses are spatialized through geolocation and online mapping. Because each community will have varying levels of interest in and capability to potentially code, build and use sensing devices, and analyze spatial data, CCUS to allows different elements of the system to be carried out by different actors, including individual residents, community organizations, city staff, and intermediaries such as non-governmental organizations (NGOs) or universities.

The toolkit is composed of two primary components, a portable sensor array and an interactive platform for contributing information and visualizing and analyzing spatial patterns and relationships. The toolkit is built on prior precedents in urban sensing and crowdsourced, volunteered geographic information (VGI). Table 1 identifies precedents for CCUS. The precedents in Table 1 are those that specifically include functionality that directly informs CCUS and are only a small subset of the many sensing and VGI projects developed by public and private actors during the 2010s.

Table 1. CCUS Direct Precedents

Project	*Functionality*	*Components*
AirCasting (2018) (http://aircasting.org/)	Measures air quality and spatial location (2.5 micron particulate matter) and posts data and mapping to open web.	• Arduino-based air quality sensor • Mobile app for recording and uploading data • Website for sharing data
Darkness Map (Hoffman, Indiana, & Madsen, 2010) (https://www.instructables.com/id/Darkness-Map-Data-Collection-Device/)	Collects light levels and spatial location to measure nighttime street lighting.	• Arduino-based light sensor
OnTheLine (El Khafif & Przybylski, 2014) (http://datalab.uwaterloo.ca/project/ontheline)	Synthesizes transit information and crowdsourced destination information to create an activity guide for transit users.	• Web-based photo contributions • Integrated transit data
Safecast (2018) (https://blog.safecast.org/)	Measure radiation level, air quality and spatial location to mitigate inefficiency of public data.	• Arduino based radiation meter, GPS, • Online mapping, • Vibrant volunteer community.

CCUS seeks to facilitate local planning and design that is accessible to a diverse populace, customizable to local concerns, and incorporates spatially precise quantitative data and analysis with qualitative contributions. Therefore, we prioritized low-cost technology, as well as functionality that allows a wide range of inputs, including sensor data, photos, and text, all of which can be presented on a customizable web interface. Figure 1 presents the components of CCUS as a system. The sensor-generated data listed in the figure are those specified for addressing the issue of nighttime street lighting in Charlottesville, Virginia, USA, the case we describe in greater detail in the next section. However, the sensors are customizable depending on community concerns.

The sensing device is built on the low-cost, small-footprint, low-energy Arduino computing platform. Figure 2 illustrates its components, including GPS, luminosity sensor, light color sensor, microphone, CO2 sensor, and data logger with microSD memory card. The entire device can be operated from a standard 9V battery or off-the-shelf external smartphone battery power bank. Environmental data and location are logged roughly once per 200 milliseconds, providing dense and spatially precise information whether being carried on foot, by bike, or attached to a car. The data footprint is small, allowing for days of data collection without the need to offload data to the CCUS webserver. The other component of

CCUS is its website and backend server. The website is programed in HTML 5, JavaScript and Google Map API (application programming interface) and all of the content and analytic tools are open-source or free-to-use. CCUS is designed to operate at a neighborhood scale, though zooming in to street level and out to larger geographies is possible. Functions on the website include sensor data visualization, overlaying GIS layers including infrastructure and socio-demographic data, and an interface for uploading geolocated photo and text comments, as well as loading photos and text into the map.

Sensor data can be collected by individuals, moving on foot, bike, or car. Depending on the interests and objectives of the community, collection can be undertaken by community members, city staff, or intermediaries such as an NGO or university. We established protocols for using the device, particularly setting sensor height between 1m and 2m off the ground during collection to allow for consistent readings across individuals and transport modes. Graphical instructions for using the sensor device are included when the device is shared with individuals (see Figure 3). Residents contribute photo and text comments via the CCUS website.[1] The website is designed to be used either on mobile devices or computers, so that a smartphone camera can be directly activated from the website or previously taken photos can be selected for upload later on.

Figure 1. Community-Centered Urban Sensing System

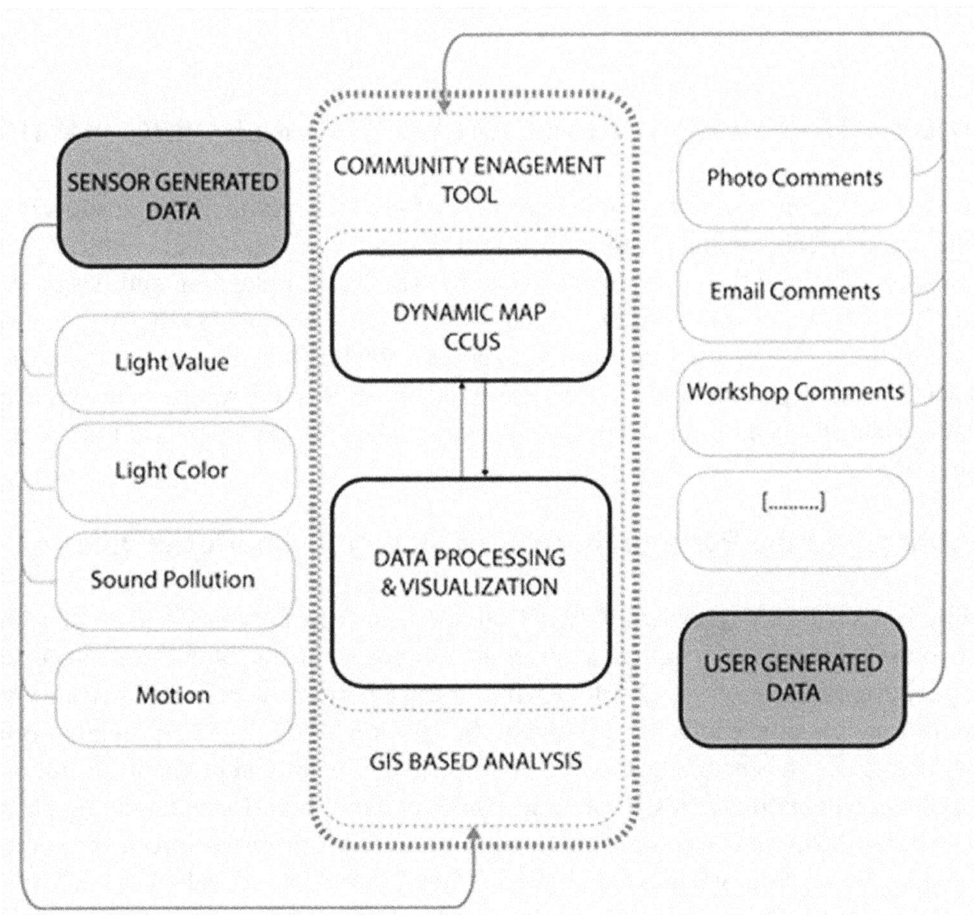

Figure 2. CCUS Arduino-based sensing device

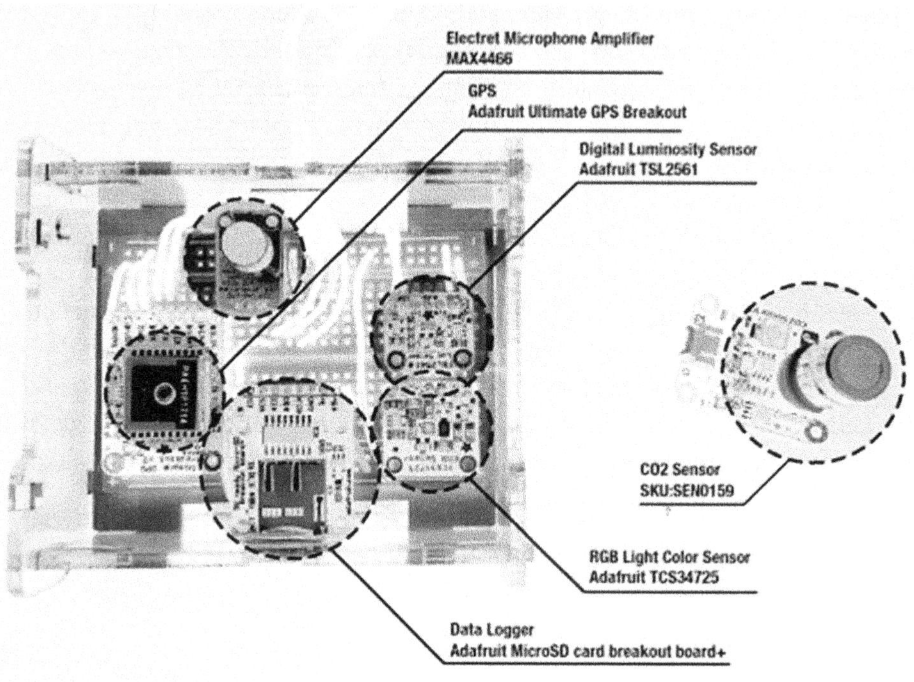

CASE STUDY: STREET LIGHTING IN CHARLOTTESVILLE, VIRGINIA, USA

Case studies in urban planning and design can be powerful tools for explaining how abstracted processes, such as participatory urban sensing, manifest in complex socio-spatio-political contexts such as cities and neighborhoods (Birch, 2012). Particularly when the results of a planning and design process are dependent on interactions between multiple causal factors, case studies can help generate new insights about why that process may or may not work as expected. We begin by presenting the socio-political context of CCUS implementation in Charlottesville and then describe the results of the yearlong project to collect data, establish mapping and analysis tools, and engage the community on the issue of night-time street lighting.

Street Lighting and the Perpetuation of Inequality in Charlottesville

Charlottesville is a small city (population 49,000 in 2019) in the southern US state of Virginia. The city is home to the University of Virginia, a university of approximately 24,000 students in addition to a major hospital and medical school. Charlottesville's built environment reflects its history as a segregated southern American city, with separate neighborhoods for white high-income, white working class, and African Americans. In general, the city did not invest significantly in public infrastructure such as streets and lighting, with narrow roads, limited sidewalks, and low quality street lighting. Particularly in traditionally African-American neighborhoods, investment in public infrastructure has been more limited and poorly maintained than in white areas (Yager, 2021). Municipal redevelopment efforts have also actively worked against African American neighborhood stability, and in the 1960s the City demolished one significant African American neighborhood, replacing it with a major road and shopping center (Saunders & Shackelford, 2005).

Figure 3. Participant instructions for carrying the sensor device for data collection

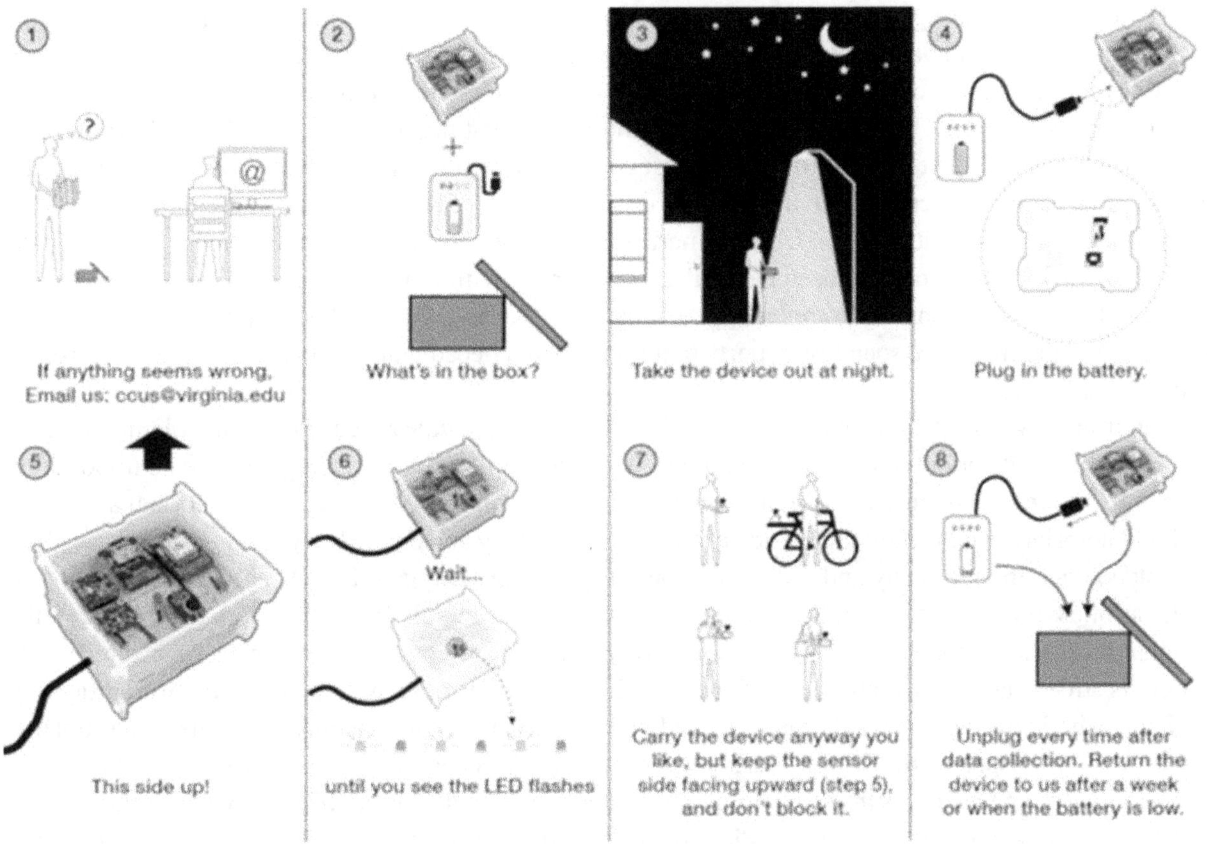

In more recent history, street lighting has remained limited and uneven throughout the City, and studies have demonstrated continuing deficiencies even on major streets (RK&K Consultants, 2016). Motivations for outdoor lighting regulation, design, and investment in Charlottesville continue to vary, with some advocating for darker streets to prevent light spillover into the sky and into people's homes, while others consider increased lighting necessary for safer walking and bicycling, as well as to contribute enhanced perceptions of security. Even light color, spanning from redder to bluer light, has become an issue since the American Medical Association found that the bluer light from modern LED lighting can also disrupt sleep patterns and long term health (Stevens et al., 2013). Charlottesville has uniquely become well known in American media over the past decade for nighttime safety deficiencies (Associated Press, 2017; Washington Post, 2014). At the same time, however, a perception of bias towards student and higher-income, white residents' concerns, reinforced by data showing police bias against Charlottesville African-Americans by the City, has exacerbated the sense that nighttime streets are unsafe or off-limits for most residents (Suarez, 2017).

Charlottesville and its neighborhoods have made efforts to improve lighting and help address nighttime safety and livability issues. The City passed its first lighting ordinance in the 1990s, primarily regulating light spillover between properties and preserving dark skies (Charlottesville Outdoor Lighting Ordinance, n.d.). Since that time, the City has made only limited efforts to address lighting in its planning and investment. For example, its most recent Comprehensive Plan, the document which drives Charlottesville's planning and investment choices, makes only one mention of lighting, and solely in reference to park maintenance, as opposed to on-street lighting (City of Charlottesville, 2013). However, in 2016, the City hired a consultant to complete an assessment of street lighting in the City, in response to a growing sense that City streets were unsafe and insecure at night, particularly for pedestrians and bicyclists (RK&K Consultants, 2016). The assessment, however, only covered a limited number of intersections in the City center, leaving lighting issues for most of the City and its neighborhoods unaddressed. Fifeville, a traditionally African-American, working-class neighborhood of the City adjacent to the University's hospital, has been particularly concerned with lighting, and in fact a neighborhood-scaled lighting assessment was completed by a consultant (Kurasz, 2006). However, the president of the Fifeville Neighborhood Association reports that more than a decade later, none of the identified issues had been addressed.

Importantly, Charlottesville has recently been confronted by a wide range of civic and planning issues that have drowned out urban investment and design questions like street lighting, despite its importance to many of its residents. In 2017, Charlottesville was again confronted by a crisis of national and international notoriety, when white supremacists marched in the City and on the University's campus, leading to violence that injured many and killed one counter-protestor (Duggan, 2018). This crisis, brought on by Charlottesville's efforts to remove statues of Confederate generals from two of its parks, represents for many residents the City's inability to come to consensus and control its own built environment: its parks, its streets, and its neighborhoods (Spencer, 2018). Furthermore, as gentrification of traditionally African-American neighborhoods has increased, the sense that the City cannot address its own problems has become widespread. In addition, residents complain that recent planning efforts in Charlottesville do not truly engage residents in the planning of its future (Hays, 2018). In 2021, the City remains mired in efforts to update its Comprehensive Plan to address equity, housing affordability, and gentrification with consensus remaining elusive (O'Hare, 2021).

Working with City and Community to Introduce Community-Centered Urban Sensing

Within this context, the CCUS effort in Charlottesville provides a tool by which community members and municipal officials can address nighttime street lighting, a significant issue for local planning and design. At the outset of our engagement with this issue, we hypothesized that better quantitative data and qualitative inputs, aggregated through low-cost, readily available technologies, could provide a substantive basis for responsive urban planning and design. We envisioned CCUS as a way to bridge the technocratic need for better data on street lighting with the democratic requirement of an engaged planning process that can facilitate community input, participation, and even control during each step of a planning and design process.

In early 2017, we began by meeting with City staff to describe CCUS and discuss whether and how the toolkit might facilitate addressing lighting issues in the City. Staff members were generally supportive of deploying CCUS, and a staff engineer recognized CCUS as an instance of smart cities approaches to planning, something that the City was seeking to embrace. Staff recommended that we begin by focusing on two neighborhoods within Charlottesville, Fifeville and North Downtown. Fifeville, as described above, is a traditionally African-American community that is gentrifying and has had longstanding concerns with lighting issues (Kurasz, 2006). North Downtown is a wealthy, predominantly white neighborhood, which staff indicated had made frequent complaints about poor lighting, particularly between its residential area and the pedestrian-oriented commercial district in the center of the City. For this deployment of CCUS, we planned to collect the quantitative data ourselves, using a team of University students and collect qualitative inputs – geolocated photos and text comments – from community members.

Before collecting lighting data, we presented the CCUS concept to the neighborhood associations in both North Downtown and Fifeville. The presidents of both neighborhood associations readily invited us to present at their monthly board meetings. We first presented CCUS to the North Downtown Residents Association. We described what CCUS would do, collecting nighttime light levels, light color, noise, and carbon dioxide levels on streets in the neighborhood, mapping those data on the CCUS website, and inviting geotagged photos and comments from the public as well. We also invited residents to walk around the neighborhood and collect data using the Arduino sensing device. Overall, meeting attendees were supportive of the project and our objective of collecting fine-grained environmental data in the neighborhood. However, some attendees did raise the question of personal privacy. The privacy concern was, in general, based on a lack of familiarity with environmental sensing (an issue in both neighborhoods). We explained that the sensing device would only capture very specific data and not record audio or video, attendees were generally satisfied with our response. One concern raised by several residents stood out: Will the City act on these data and improve lighting in our neighborhood? Will they actually act on them? We relayed our prior conversations with City staff but could also not promise that the data would automatically lead to new planning or investment.

In Fifeville, we presented the same information about the project. Many in the room were supportive of the plan. The neighborhood, already in the midst of a neighborhood planning process, observed that not only were there areas that were poorly lit, but newer developments were also over lit, spilling too much light onto the street and homes. Some distinctive opposition to the project stood out. One attendee described the Arduino sensing device as a "spycam," and was not convinced by our explanations of what the device would really do. Others expressed the concern that collecting noise data as well as light data could also be a way to further police African-American social activities that might occur on the street at night. In general, there was a more heated discussion among community members about the benefits of collecting environmental data, and the sense that the City would not act on the data anyway was raised again, as in North Downtown. Furthermore, our position as faculty and students at the University of Virginia was a significant problem for several meeting attendees. Two attendees raised the fact that University students were part of the noise and livability problem themselves, partying loudly and leaving garbage in the neighborhood. Although quantitative data collection was being undertaken by University students, one resident was also interested in participating in the quantitative part of the process, and we facilitated her involvement by providing the device and instructions (see Figure 3). Despite the more significant concerns among Fifeville residents at the meeting, ultimately the leadership of Fifeville neighborhood association supported us proceeding with the project.

Working with the single Fifeville volunteer and students from the University, we deployed the Arduino device throughout Fifeville and North Downtown, collecting detailed data on nighttime light levels, light color, noise, and carbon dioxide. The environmental data were collected to ensure consistently valid data (following protocols for consistent horizontal lux measurement) at a minimum density of one data point per linear meter of sidewalk or street (in the absence of sidewalks) in the study areas. Figure 4 shows the results of the data collection in Fifeville, focusing on light levels. As an interactive map, the CCUS interface can display light (and other) data in multiple representations depending on the viewers' interest. Figure 4 displays a sample of all data, showing original lux readings for one out of ten points selected sequentially along streets. We also provide a "heatmap" visualization, which averages data from all readings within a given radius. The data highlight the overall low levels of lighting in Fifeville, which we compared to North Downtown as well as lighting around the University of Virginia. The data show locations where no street lighting is available at all for nighttime activities.

Figure 4. Light levels (lux) in Fifeville neighborhood, city of Charlottesville

Following collection of the quantitative environmental data, we returned to Fifeville to present the quantitative maps and conduct a workshop with community members to train and encourage them to take geotagged photos and text comments using the CCUS participation tool. Six community members joined the workshop, a smaller group than our first presentation to the community. Those at the workshop were, as before, thoroughly convinced of the importance of street lighting as an issue and agreed that the data and comments collected by CCUS could be used to make better planning, design, and investment decisions by the City. Even after the workshop, we found participants were slow to participate in submitting photos and comments, though CCUS in Charlottesville is an open process that continues to allow input over time as residents learn more about the initiative and observe issues in their neighborhood. Overall, CCUS engaged with residents in the Charlottesville neighborhoods through three modes of participation, (1) shaping the toolkit and process, (2) collecting quantitative data, and (3) collecting qualitative data. Notably, participation was strongest in the first phase, with approximately eight North Downtown and fifteen Fifeville residents contributing feedback on how CCUS should be configured and deployed of CCUS. We did not emphasize resident participation in quantitative environmental data collection, though a single Fifeville resident volunteered to collect data. Finally, six to eight residents in each neighborhood participated in focused workshops around collecting qualitative data, and geolocated photos and comments continue to be submitted to the CCUS website. The website itself is public and continues to be used by residents and city staff. The Fifeville data was used to inform a general street lighting policy for the new neighborhood plan. The data were not used, however, to identify specific sites for investment. Instead, the plan calls for yet another lighting study in the neighborhood. As of 2021, that plan is under consideration for adoption by the City of Charlottesville.

CASE STUDY: STREET LIGHTING IN WAYNESBORO, VIRGINIA, USA

In 2019, following our work in Charlottesville, we collaborated with the Planning Department in the City of Waynesboro, Virginia to implement CCUS. Waynesboro is a small city (population 22,000 in 2019) in the Shenandoah Valley of Virginia. Waynesboro's economy has traditionally centered around industrial production, but as factories have closed or downsized, the City has worked to reorient its economic more towards tourism, taking advantage of its close proximity to Shenandoah National Park and the Appalachian Trail. Waynesboro also has a history of differential investment the built environment across neighborhoods. One neighborhood in particular, called Basic City, has historically been a low-income community, with employment highly dependent on the diminished industrial economy and other families descended from smallholders evicted from land when Shenandoah National Park was established in the 1930s (L. Juday, personal communication, June 9, 2021). Basic City has significant underinvestment in urban infrastructure, including sidewalks and street lighting.

The City of Waynesboro has sought state funding to improve the built environment across the City, including in Basic City. In order to identify locations with deficient street lighting, the Planning Department worked with the CCUS team to collect lighting data and make it available for information and comment. As part of an ongoing planning effort in Waynesboro, the CCUS team did not lead engagement but supplied data to the Planning Department who led engagement. In this case, the planning process was far more traditional, in the sense that the effort to remediate the problem was led by primarily by the planners rather than the community itself. No informational meetings were held with community members regarding the data collection effort or the technologies being used to collect environmental data. This top-down effort stands in contrast to the approach in Charlottesville.

The quantitative lighting data was collected by University students using the same sensor kit used in Charlottesville (see Figure 5). On first night of data collection, the students were confronted by local residents as they collected data, including some verbal threats. The residents who confronted the students did not accept that the students were coming into the neighborhood for positive purposes and did not trust that the data collection would lead to improvements in local street lighting. Because of the confrontation, data collection was completed the subsequent week with the assistance of an escort by the police and the Director of the Planning Department. While the data were uploaded to the CCUS online interface to facilitate community engagement in addition to quantitative data collection, the lack of engagement meant that no text or photo comments were submitted. The Planning Department themselves have used the data collected to identify sites for street lighting investment, which as of 2021 in ongoing.

Figure 5. Light levels (lux) in basic city neighborhood, city of Waynesboro

The Waynesboro case stands in contrast to the Charlottesville experience. While the data had a clear and ready purpose, the community was not engaged in any significant way regarding the overall planning process or the means of data collection. The engagement methods we used in Charlottesville may not have guaranteed that the confrontation during data collection would not have occurred. However, if the residents had had more information, as well as opportunity to shape or even engage in the process, a different outcome might have occurred. Despite this, Waynesboro is actually further along in making necessary street lighting investments, with sites already determined by the Planning Department. In contrast, in Charlottesville the engagement around street lighting has led, primarily, to a call for another street lighting plan to be completed subsequent to adoption of the broader Small Area Plan in Fifeville.

FINDINGS: MOVING FROM DATA TO ACTION, EQUITABLY

The two case studies in Charlottesville and Waynesboro each demonstrated different approaches to deployment of the CCUS tool, demonstrating its ability to inform planning processes in varying socio-spatial contexts. In Charlottesville, we find that the promise of participatory urban sensing to build consensus and serve as the basis for planning and design is highly contingent on local conditions. We observe that community context had a significant impact on CCUS during its initial deployment. Despite agreement among City staff and most, if not all, community members that better data and input are needed for addressing street lighting issues in Charlottesville, the participatory aspects of CCUS were difficult to implement.

Two key issues, repeated consistently by community members, arose during the engagement process. First, and most broadly, residents asked whether the data and input CCUS provides would actually lead to the City taking action to improve lighting in their neighborhoods. The logic of using better information to make planning and design choices was clear to everyone, but many were not confident that the City-driven processes for planning, design, and investment are capable of responding to data. The historical lack of action on lighting and the current dysfunction in the City apparent from daily news reports make the residents' general assumptions difficult to dispel. Therefore, the common response from community members is understandable: interest in and support for CCUS but a lack of willingness to invest their own time into the project. This, certainly, is not unique to technology-driven engagement practices but planning engagement generally. The data collected with the technology, however, was not used to accelerate local action. Two years after the planning process in Fifeville, the local plan is still under consideration by the City, and the new plan only *calls for* another street lighting plan that would eventually lead to investment in street lighting, despite the availability of the CCUS data to make decisions now. In Waynesboro, by contrast, the lack of engagement was clearly problematic, but the City is proceeding only a year later in making lighting investments, eschewing further community-based planning for investment based on available data.

The second issue we observed is more distinct to Fifeville than Waynesboro, but it raises a critique not just of the process but also the objectives of CCUS. Specifically, localized collection, analysis, and sharing of data may further the marginalization and oppression of some communities, rather than empower them. CCUS's approach was seen as threatening by some Fifeville residents, collecting detailed spatial information about a neighborhood and sharing those data openly with those inside and outside the community, including City staff and University members. The particular example of mappable noise data stands out as a possible means of identifying and repressing African American social life. However, in a changing neighborhood, where not all residents agree on what its future should look like, even lighting data can be used as a marginalizing tool, where a more powerful constituency can use the data to foster its vision of the future over the vision of a less powerful group. In Fifeville, while most residents agreed that street lighting is a major neighborhood issue, some were more concerned with a lack of security on dark streets while others saw spillover light from new developments as the key issue. The lighting data could be used to advocate for either position, depending on who controls neighborhood decision making. The critique, however, could not even arise without providing community members with legitimate venues for learning about urban sensing, analysis, and their relationship to neighborhood wellbeing. In that way, CCUS, as a practice and toolkit makes dialogue, even if conflictual, possible.

The experience in Waynesboro represents an alternative circumstance, where City officials chose not to utilize the engagement components of CCUS, or a separate engagement process, to conduct the study. Any assumptions about the attitudes towards street lighting and other infrastructure investment in Basic City remain unheard. Furthermore, the social and technical practices of collecting data to understand local needs remain opaque and – in the case of at least some residents – seen as a threat. In Waynesboro, this challenge was addressed by using police presence to ensure the safety of the data collectors. While effective, it is reasonable to assume that those witnessing the data collection with police presence would further associate urban sensing and planning analysis with top-down action.

Community concerns about municipal inaction or ineffectiveness, potential misuse of sensing data, or fundamental alienation from urban sensing, require framing CCUS and smart, engaged planning within a broader context than just the application of information technologies to community planning and design processes. Instead, contextual issues observed in Charlottesville and Waynesboro require that we consider urban dysfunction as an inherent challenge in any engaged planning that cannot necessarily be fixed with smart city strategies. In order to be smart, technological approaches to city planning and design must directly account for the level and types of dysfunction in a city's social fabric and governmental processes. Enhanced data and analytics – including urban sensing – can make a difference, but effective "smartness" in dysfunctional contexts should be more than simple process improvement. Instead, it requires knowledge of where interventions in existing processes are likely to be effective and when interventions should potentially occur outside normal, sanctioned planning processes.

DISCUSSION: SMART CITIES UNDER THE CONDITION OF DYSFUNCTION

Building on our findings in Charlottesville and Waynesboro, we seek a range of potential breakdowns that "smart" planning and design practices due to local dysfunction in planning, design, and governance regimes. Critically, we observe that dysfunction is not a singular characteristic but take on different features depending on location. Therefore, smart and engaged planning and design must be able to flexibly address a wide range of socio-technical challenges, each with specific responses. We conceptualize those responses along a continuum from insurgent and oppositional to administrative and systematic. Where a technology-driven planning initiative fits along the continuum should be determined by the type of dysfunction in existing social and governmental systems in the locale. Figure 6 illustrates how this continuum can deploy data-driven technologies to address a variety of approaches to engagement. The continuum builds on the "tactical spectrum," from unsanctioned to sanctioned, proposed by Lydon and Garcia in *Tactical Urbanism* (2015). Rather than emphasize physical interventions, we focus on how urban data and analysis can support varied approaches to resident participation, contingent on local conditions. At its most insurgent, data can be used for opposing existing social and governmental structures, such as using environmental data to pursue legal action. At its most administrative, cities can collect urban environmental data with passive approval or no awareness at all by citizens, such as in projects like Chicago's Array of Things (Mone, 2015). Between are variations of participatory urban sensing and analytics that oppose, support, or operate outside of existing systems.

CCUS, in concept, is designed to mediate between "top-down" and "bottom-up." In this case, "top-down" emphasizes the actions of municipalities, their consultants, and private companies engaged in urban infrastructure, while "bottom-up" encompasses community member actions towards addressing local conditions. In the case of Charlottesville, where the planning process has continued for years and little action has been taken, a more insurgent approach to CCUS may have been appropriate to local conditions. The data and analysis collected by CCUS could be used to make an argument for change in the absence of a functional planning process. Notably, more insurgent approaches must also be more specific in the changes being proposed, so that change can occur with as little friction as possible. Drawing on tactical urbanism, demonstrations and do-it-yourself interventions should be directly linked to data and analytics. In Waynesboro, where city officials proceeded with data collection without substantive community engagement, the top-down nature of the process gave no outlet for community concerns about technical practices or planning outcomes, and has potentially engendered additional distance and alienation among community members. In this case oppositional approaches to city actions may be more likely in the future.

Figure 6. Participatory dimensions of smart planning interventions

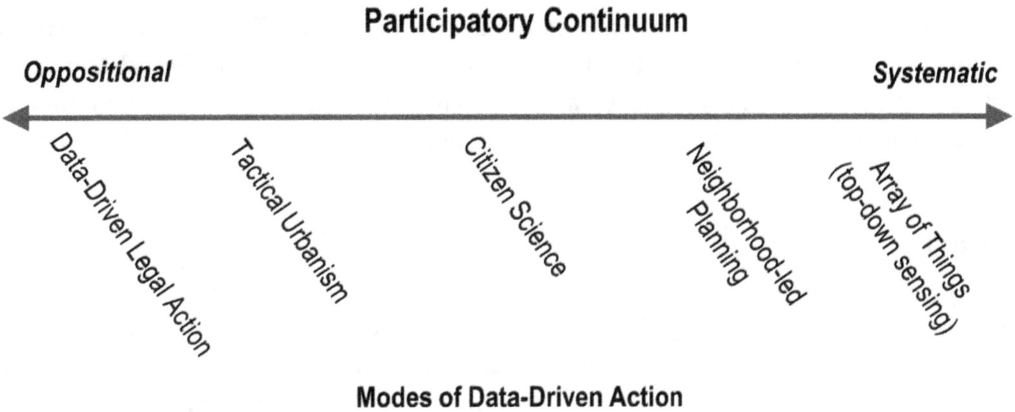

Insurgent uses of urban sensing are likely to come from community-members, not governments or public-private partnerships. However, the emphasis should focus primarily on whether technology is being used as part of an existing urban process or in order to fundamentally change that process. If we seek to establish smart planning and design that effectively fosters urban change, CCUS highlights several important considerations: (a) technical complexity, (b) openness and flexibility, (c) integration with municipal systems, and (d) the role of intermediaries. The technical complexity of an urban sensing toolkit such as CCUS may not be excessive compared to large-scale engineered systems, but it does involve coding and physical computing (HTML 5, JavaScript, web services, and Arduino programming and assembly) that few community members may have prior experience with. Unless the initiative contains an explicit "citizen science" objective, technical complexity should be minimized. However, minimizing complexity may entail tradeoffs with other considerations, such as system openness and flexibility and the role of intermediaries. Making the system open, that is non-reliant on proprietary software or hardware and making data freely available to community members, may increase technical complexity. For example, CCUS is based on open-source software, despite the fact that proprietary software, such as that provided by Environmental Systems Research Institute (ESRI), the makers of industry-standard ArcGIS, can reduce the coding burden on community members. Despite increasing technical complexity, using open-source software and data repositories also keeps the financial burden low.

Smart cities tool builders should consider whether a system is designed to function within or outside existing municipal data and analytics platforms. CCUS was built separately from the municipal GIS platforms, which allows flexibility in the types of data collected and the analytics provided but also means that it exists outside of municipally sanctioned datasets. Operating outside of the City's servers may hinder its use in sanctioned planning processes. Building a tool within a city's existing systems may provide administrative and political validation but may also represent co-optation of the process to community members opposed to city actions. Finally, the role of intermediaries in a smart planning process should also be carefully considered. As a part of the University of Virginia, we are frequently seen as powerful actors or even outsiders on issues of neighborhood livability. On the one hand, those building smart and engaged tools and systems must have legitimacy in the eyes of community members. When technological tools can only be developed by outside experts, whether academics, consultants, or non-governmental advocates, community members need to have some level of trust in system developers. On the other hand, outside experts must account for the social dimension of technological innovations, fully considering ethical implications of smart tools and technologies. This requires smart cities tool builders to form teams of diverse expertise in multiple dimensions, some technical and some social. Still, despite significant community outreach and transparency in our actions, some community members remain resistant to a University role in local planning and design, even if our objectives are shared.

CONCLUSION

Smartness is contingent on the socio-political context of the city where planning and design processes are being implemented. Participatory urban sensing, and smart cities planning generally, has been framed primarily in terms its ability to enhance existing planning processes. However, in many situations, engaged technological practices may be more effective when deployed in opposition to current practices. The experience of deploying Community-Centered Urban Sensing in Charlottesville and Waynesboro to address nighttime street lighting issues demonstrates that urban technologies integrated into existing planning systems may not be able to overcome dysfunction inherent to those systems. However, they may be able to change those processes from the outside, if the participatory elements of the system are flexible and take into account considerations such as system complexity, openness, integration with municipal systems, and the role of intermediaries. In principle, collaborative approaches to planning and design that allow individuals to contribute their piece, whether via data collection or simply contributing a geotagged comment, should allow the burden of engagement to be reduced and distributed more evenly across the community, whether within an existing planning process or as a tactical process outside of sanctioned engagement.

Future research should examine whether alternative approaches to local planning and design address the shortcomings we identified in this research. Our findings highlighted that the lack of knowledge and trust in technical systems and processes was a critical stumbling block to consensus and acceptance of urban sensing-driven planning in both Charlottesville and Waynesboro. Our two case studies also demonstrated that traditional engagement practices based on workshops and community meetings are inherently time- and cost-prohibitive, and even in the best case are unlikely to reach the broadest cross-section of any community. Therefore, we are integrating data and analytics with physical interventions in public space that make sensors, data, and implications for community members evident in the built environment (El Khafif et al., 2021). This approach, which couples traditional engagement practices with intentionally non-threatening, informative experiences of urban sensing technologies may help address the burden of explaining new technologies in a community engagement process.

Future research should also investigate more flexible approaches to urban planning and design that allow for more continuous changeability in the built environment. Flexible places, such as streets with variable, programmable street lighting, are made possible by new technologies but require new approaches to community engagement so that environmental changes are directed from within the community, rather than by external actors. While Charlottesville and Waynesboro may be a unique cases, many cities around the world face dysfunctional planning regimes due to administrative deficiencies, social conflict, or a lack of resources. Enhanced data and analysis capabilities provided by smart technologies may not lead to constructive action. If these hurdles cannot be overcome, we risk building ever increasing inequality within and among cities, between places with the cohesion and resources to act upon smart technologies and those places that cannot.

ACKNOWLEDGMENT

We appreciate the participation and input of the residents and city staff of the City of Charlottesville and City of Waynesboro, who seek a brighter future for their community. We also thank Gabriel Andrade, Matthew Hunt, and Taylor Thompson for their work on the project. This research was funded by the University of Virginia School of Architecture and the National Science Foundation through a seed fund provided by MainStreet 21.

This chapter builds upon a article published in the International Journal of E-Planning Research (IJEPR).

REFERENCES

AlAwadhi, S., & Scholl, H. J. (2013). Aspirations and realizations: The smart city of Seattle. *2013 46th Hawaii International Conference on System Sciences*, 1695–1703. 10.1109/HICSS.2013.102

Albino, V., Berardi, U., & Dangelico, R. M. (2015). Smart Cities: Definitions, Dimensions, Performance, and Initiatives. *Journal of Urban Technology*, *22*(1), 3–21. doi:10.1080/10630732.2014.942092

Associated Press. (2017). *Rolling Stone slams defamation claim in UVa rape story*. The Daily Progress. https://www.dailyprogress.com/news/local/rolling-stone-slams-defamation-claim-in-uva-rape-story/article_e6bc4080-cbff-11e7-bed8-0b3c378ba1bc.html

Batty, M. (2013a). *The new science of cities*. MIT Press. doi:10.7551/mitpress/9399.001.0001

Batty, M. (2013b). Big data, smart cities and city planning. *Dialogues in Human Geography*, *3*(3), 274–279. doi:10.1177/2043820613513390 PMID:29472982

Batty, M. (2014). Can it happen again? Planning support, Lee's Requiem and the rise of the smart cities movement. *Environment and Planning. B, Planning & Design*, *41*(3), 388–391. doi:10.1068/b4103c2

Birch, E. L. (2012). Cities, People, and Processes as Planning Case Studies. In R. Crane, R. Weber, & E. L. Birch (Eds.), *The Oxford Handbook of Urban Planning*. Oxford University Press. https://www.oxfordhandbooks.com/view/10.1093/oxfordhb/9780195374995.001.0001/oxfordhb-9780195374995-e-14

Bulkeley, H., Coenen, L., Frantzeskaki, N., Hartmann, C., Kronsell, A., Mai, L., Marvin, S., McCormick, K., van Steenbergen, F., & Voytenko Palgan, Y. (2016). Urban living labs: Governing urban sustainability transitions. *Current Opinion in Environmental Sustainability*, *22*, 13–17. doi:10.1016/j.cosust.2017.02.003

Caragliu, A., Del Bo, C., & Nijkamp, P. (2011). Smart Cities in Europe. *Journal of Urban Technology*, *18*(2), 65–82. doi:10.1080/10630732.2011.601117

Certomà, C., Corsini, F., & Rizzi, F. (2015). Crowdsourcing urban sustainability. Data, people and technologies in participatory governance. *Futures*, *74*, 93–106. doi:10.1016/j.futures.2014.11.006

Charlottesville Outdoor Lighting Ordinance. Chapter 34, Article IX, Division 3, Section 34 Chapter 34, Article IX, Division 3, Section 34. https://library.municode.com/va/charlottesville/codes/code_of_ordinances?nodeId=CO_CH34ZO_ARTIXGEAPRE_DIV3OULI_S34-1000PUIN

City of Charlottesville. (2013). *2013 Charlottesville Comprehensive Plan*. City of Charlottesville, Virginia. http://www.charlottesville.org/home/showdocument?id=35059

Cowley, R., Joss, S., & Dayot, Y. (2018). The smart city and its publics: Insights from across six UK cities. *Urban Research & Practice*, *11*(1), 53–77. doi:10.1080/17535069.2017.1293150

Duggan, P. (2018, October 2). Four alleged members of hate group charged in 2017 'Unite the Right' rally in Charlottesville. *Washington Post*. https://www.washingtonpost.com/local/public-safety/federal-officials-to-announce-additional-charges-in-2017-unite-the-right-rally-in-charlottesville/2018/10/02/60881262-c651-11e8-9b1c-a90f1daae309_story.html

El Khafif, M., Mondschein, A., & Rosado Murillo, L. F. (2021). *Ostenda illuminata: A socio-ecological 1:1 prototype for networked public space*. Architectural Research Centers Consortium.

El Khafif, M., Zhang, Z., & Mondschein, A. (2018). Mapping the Nocturnal City: Community-Centered Urban Sensing in Charlottesville, Virginia. *Proceedings of the CityStreet3 Conference*.

Elwood, S. A. (2002). GIS Use in Community Planning: A Multidimensional Analysis of Empowerment. *Environment and Planning A. Economy and Space*, *34*(5), 905–922. doi:10.1068/a34117

Fard, A. (2018). Not Sharing: Urban Techno-Colonialism in the Age of Big Data. *PLAT*. https://www.platjournal.com/articles/not-sharing-urban-techno-colonialism-in-the-age-of-big-data

Goodchild, M. (2007). Citizens as sensors: The world of volunteered geography. *GeoJournal*, *69*(4), 211–221. doi:10.100710708-007-9111-y

Goodman, N., Zwick, A., Spicer, Z., & Carlsen, N. (2020). Public engagement in smart city development: Lessons from communities in Canada's Smart City Challenge: Smart city public engagement. *The Canadian Geographer. Geographe Canadien*, *64*(3), 416–432. Advance online publication. doi:10.1111/cag.12607

Granier, B., & Kudo, H. (2016). How are citizens involved in smart cities? Analysing citizen participation in Japanese "Smart Communities". *Information Polity*, *21*(1), 61–76. doi:10.3233/IP-150367

Green, B. (2019). *The Smart Enough City: Putting Technology in Its Place to Reclaim Our Urban Future*. MIT Press. doi:10.7551/mitpress/11555.001.0001

Grey, T., Dyer, M., & Gleeson, D. (2017). Using Big and Small Urban Data for Collaborative Urbanism. In C. Certomà, M. Dyer, L. Pocatilu, & F. Rizzi (Eds.), *Citizen Empowerment and Innovation in the Data-Rich City* (pp. 31–54). Springer International Publishing. doi:10.1007/978-3-319-47904-0_3

Hays, E. (2018, August 7). City commission to bring community leaders into Comprehensive Plan process · Charlottesville Tomorrow. *Charlottesville Tomorrow*. https://www.cvilletomorrow.org/news/article/31273-city-to-bring-community-leaders-into-comp-plan/

Iveson, K. (2013). Cities within the City: Do-It-Yourself Urbanism and the Right to the City. *International Journal of Urban and Regional Research, 37*(3), 941–956. doi:10.1111/1468-2427.12053

Krivý, M. (2018). Towards a critique of cybernetic urbanism: The smart city and the society of control. *Planning Theory, 17*(1), 8–30. doi:10.1177/1473095216645631

Kurasz, G. (2006). *Fifeville Neighborhood Lighting Assessment*. Kurasz & Associates.

Lane, N. D., Eisenman, S. B., Musolesi, M., Miluzzo, E., & Campbell, A. T. (2008). Urban Sensing Systems: Opportunistic or Participatory? *Proceedings of the 9th Workshop on Mobile Computing Systems and Applications*, 11–16. 10.1145/1411759.1411763

Lydon, M., & Garcia, A. (2015). *Tactical Urbanism: Short-term Action for Long-term Change*. Island Press. doi:10.5822/978-1-61091-567-0

Mann, M., Mitchell, P., Foth, M., & Anastasiu, I. (2020). #BlockSidewalk to Barcelona: Technological sovereignty and the social license to operate smart cities. *Journal of the Association for Information Science and Technology, 71*(9), 1103–1115. doi:10.1002/asi.24387

Meijer, A. (2016). Smart City Governance: A Local Emergent Perspective. In J. R. Gil-Garcia, T. A. Pardo, & T. Nam (Eds.), *Smarter as the New Urban Agenda: A Comprehensive View of the 21st Century City* (pp. 73–85). Springer International Publishing., doi:10.1007/978-3-319-17620-8_4

Mone, G. (2015). The New Smart Cities. *Communications of the ACM, 58*(7), 20–21. doi:10.1145/2771297

Murgante, B., & Borruso, G. (2015). Smart Cities in a Smart World. In S. Th. Rassia & P. M. Pardalos (Eds.), *Future City Architecture for Optimal Living* (pp. 13–35). Springer International Publishing., doi:10.1007/978-3-319-15030-7_2

Nilssen, M. (2018). To the smart city and beyond? Developing a typology of smart urban innovation. *Technological Forecasting and Social Change*. Advance online publication. doi:10.1016/j.techfore.2018.07.060

O'Hare, E. (2021, June 10). Future Land Use Map conversations cause confusion, controversy. *Charlottesville Tomorrow*. https://www.cvilletomorrow.org/articles/future-land-use-map-conversations-cause-confusion-controversy/

RK&K Consultants. (2016). *City of Charlottesville Pedestrian Lighting Study Improvement Plan Report*. http://www.charlottesville.org/home/showdocument?id=48739

Ruhlandt, R. W. S. (2018). The governance of smart cities: A systematic literature review. *Cities (London, England), 81*, 1–23. https://doi.org/10.1016/j.cities.2018.02.014

Saunders, J. R., & Shackelford, R. N. (2005). Urban Renewal and the End of Black Culture in Charlottesville, Virginia: An Oral History of Vinegar Hill. McFarland.

Sawicki, D. S., & Craig, W. J. (1996). The Democratization of Data: Bridging the Gap for Community Groups. *Journal of the American Planning Association, 62*(4), 512–523. https://doi.org/10.1080/01944369608975715

See, L., Mooney, P., Foody, G., Bastin, L., Comber, A., Estima, J., Fritz, S., Kerle, N., Jiang, B., Laakso, M., Liu, H.-Y., Milčinski, G., Nikšič, M., Painho, M., Pődör, A., Olteanu-Raimond, A.-M., Rutzinger, M., See, L., Mooney, P., … Rutzinger, M. (2016). Crowdsourcing, Citizen Science or Volunteered Geographic Information? The Current State of Crowdsourced Geographic Information. *ISPRS International Journal of Geo-Information, 5*(5), 55. doi:10.3390/ijgi5050055

Slotterback, C. S., & Lauria, M. (2019). Building a Foundation for Public Engagement in Planning. *Journal of the American Planning Association, 85*(3), 183–187. https://doi.org/10.1080/01944363.2019.1616985

Spencer, H. (2018). *Summer of Hate*. University of Virginia Press. https://www.upress.virginia.edu/title/5323

Stevens, R. G., Brainard, G. C., Blask, D. E., Lockley, S. W., & Motta, M. E. (2013). Adverse Health Effects of Nighttime Lighting: Comments on American Medical Association Policy Statement. *American Journal of Preventive Medicine, 45*(3), 343–346. https://doi.org/10.1016/j.amepre.2013.04.011

Suarez, C. (2017, August 23). Stop-and-frisk data show African-Americans disproportionately detained. *The Daily Progress*. https://www.dailyprogress.com/news/local/stop-and-frisk-data-show-african-americans-disproportionately-detained/article_1b606292-886e-11e7-960b-57d65b2b63d4.html

Valverde, M., & Flynn, A. (2020). *Smart Cities in Canada: Digital Dreams, Corporate Designs: Independent experts analyze often-controversial schemes from Nunavut to Montreal to Toronto's failed Sidewalk Labs waterfront scheme*. James Lorimer & Company.

Vanolo, A. (2014). Smartmentality: The smart city as disciplinary strategy. *Urban Studies (Edinburgh, Scotland), 51*(5), 883–898.

Washington Post. (2014). Timeline: The search for Hannah Graham. *Washington Post*. https://www.washingtonpost.com/apps/g/page/local/timeline-the-search-for-hannah-graham/1332/

Wilson, B., & Chakraborty, A. (2019). Planning Smart(er) Cities: The Promise of Civic Technology. *Journal of Urban Technology, 0*(0), 1–23. https://doi.org/10.1080/10630732.2019.1631097

Wilson, B. B. (2018). *Resilience for All: Striving for Equity Through Community-Driven Design*. Island Press.

Yager, J. (2021). *Mapping Cville*. https://mappingcville.com/author/jordyyager/

Yates, J. (2017). Saving the Soul of the Smart City. *The Hedgehog Review: Critical Reflections on Contemporary Culture, 19*(2), 18–35.

Yigitcanlar, T., Kamruzzaman, Md., Buys, L., Ioppolo, G., Sabatini-Marques, J., da Costa, E. M., & Yun, J. J. (2018). Understanding 'smart cities': Intertwining development drivers with desired outcomes in a multidimensional framework. *Cities (London, England), 81*, 145–160. https://doi.org/10.1016/j.cities.2018.04.003

Yoon, A., & Copeland, A. (2020). Toward community-inclusive data ecosystems: Challenges and opportunities of open data for community-based organizations. *Journal of the Association for Information Science and Technology*, *71*(12), 1439–1454. https://doi.org/10.1002/asi.24346

Zhang, Z., Mondschein, A., & El Khafif, M. (2018). Human-Tool Assemblage: Designers in the Big Data World. *Journal of Digital Landscape Architecture*, *3*, 397–405.

ENDNOTE

1 https://web.arch.virginia.edu/urbansensing/ccus/

Chapter 4
Citizen Commitment in City-Organized Air Quality Monitoring

Gwendoline l'Her
UMR Ambiances Archtectures Urbanités CNRS-ENSAN-ECN, France

Myriam Servières
UMR Ambiances Archtectures Urbanités CNRS-ENSAN-ECN, France

Daniel Siret
UMR Ambiances Archtectures Urbanités CNRS-ENSAN-ECN, France

ABSTRACT

Based on a case study from Rennes, the chapter presents how a group of urban public actors re-uses methods and technology from citizen sciences to raise the urban air quality issue in the public debate. The project offers to a group of inhabitants to follow air quality training and proceed PM2.5µm measurements. The authors question the impact of the ongoing hybridisation between citizen science and urban public action on participants' commitment. They present how the use of PM2.5-sensors during 11 weeks led to a disengagement phenomenon, even if we observe a strong participation to workshops. Finally, the authors present a spatial analysis of the measurements produced and analyse how the inhabitants use the sensors in their daily lives. These results come from an interdisciplinary methodology using observations, interviews, and data analyses.

INTRODUCTION

The fast and widespread use of smartphones emerged when urban public actors, such as local authorities and urban administrations, realised the possibilities of changing citizen participation. Research in urban sociology analyses the impact of information and communication technologies (ICTs) on participative democracy and urban e-governance: some authors note in particular a transcription for digital technologies

DOI: 10.4018/978-1-7998-9090-4.ch004

of some classical civic tools, like participatory budgeting, collaborative maps, and public debate platforms (Douay 2016; Ottaviano 2015). This digital adaptation benefits the change of scales – geographic scale, time scale, number of participants, etc. –, characteristic of ICTs. It enhances the possibility to produce and to share urban data on the field. An enterprise sector structures itself – the Civic Tech – in order to provide public institutions with technical solutions. Private actors represent values stemming from digital culture (Turner, 2012), among which some are now in tension with political expectations regarding the digital tools of participatory democracy (Mabi, 2021).

Other uses of digital technologies, especially mobile technologies, are observed and involve citizen participation: crowdsourcing, voluntary geographic information, citizen sciences (See et al. 2016). Some convergence points appear between citizen participation organized by cities on environmental issues and citizen sciences approaches. New digital technologies also lead to many experiments in participatory science projects, and they result in reconfigurations of historical amateur practices (Heaton & al. 2018).

In this chapter, we study the effect of the ongoing hybridisation between a citizen science project and urban public action on the commitment of participants in a citizen-sensor-network project based on a case study in Rennes (France). By hybridization we mean the exchange of modality of participation, like the task or purpose that a resident has to do to be considered as a member of the participative process. In the case of citizen-sensor-network studied, the main modality of participation was sensing the air. How did local authority become an organizer of citizen metrology and how was the participant's experiences affected? How were the sensors used in a metrology project, whose main objective is not the creation of knowledge but education?

Firstly, we review literature on citizen participation and metrology and we propose an analytical framework of commitment based on Thévenot's theory (Thévenot 2006). Next, we describe the case study on urban air pollution and the political issue that induced a municipality to make a citizen metrology project, and we present the city's expectations of the participants' contributions. Then, we present the plurality of the participants' commitment during the project and the difficulties to keep participating due to technical failure and misunderstanding of the purpose of the project. Finally, we show how the data set produced by the participants can be used to study the practices of citizen metrology.

LITERATURE REVIEW

Rick Bonney (Bonney 1995) defined citizen sciences as scientific methods that involve non-scientists. Another definition of citizen sciences was given by Alan Irwin (Irwin 1995) and refers to the democratisation of science and citizens' relations with science governance. In both cases, the public participates in a collective action including field observation, inventory, and metrology that describe the environment and spur discussions. With the rise of low-cost air pollution sensors (Kumar et al. 2015), and the *citizen as a sensor utopia* – in other words, the possibility to create a network of a six billion humans using ICTs to measure and to share situated knowledge with the aim to create a common good (Goodchild 2007) – citizen metrology becomes a suitable way of action for cities.

Environmental metrology aims to quantify chemical components, species, among other attributes of the environment to describe and study phenomena. Metrologies are structured by methods and protocols recognized by scientific communities. Metrology is sometimes used by citizens to produce street science in local communities when environmental controversies emerge (Corbun 2005). Those citizen metrology projects can be managed by scientist-activists (Topçu 2008). Historically, metrologies of air quality have covered several purposes: 1) to produce a map to monitor the state of the environment, 2) to affect the responsibilities of pollutions, 3) to share a discomfortable, health issues, 4) to be a place for negotiations and for the implementation of a new governance regarding the environment (Charvolin & al. 2015). The emergence of electronic microsensors and biosensors – lichens, tree bark – represents new possibilities for the production of knowledge on the environment (Gramaglia & al. 2017).

The number of alternative metrologies on air pollution that are outside of an institutional framework and without certified experts for monitoring air quality is increasing in Europe – e.g. Montre verte, Mobicitair, Ambassad'air in France; Smartcitizen in Spain and in the Netherland; Luftdaten in Germany; Bruxsel'air in Belgium (ADEME et al. 2017). There is a large scientific literature on sensors and data quality, such as sensitivity of the sensors to varying environmental conditions (Broday & Citi-Sense_Project_Collaborators 2017), and measurement bias, such as the impact of siting on data (Morawska et al. 2018; Miskell et al. 2017). We noted that the bias due to the use by non-expert has not been studied, although the large widespread use of air quality sensors has increased the number of citizen-sensor-network projects. The ethnomethodology, initiated by H. Garfinkel, shows how non-experts define specific strategies, techniques and methods, called ethnomethods, when they have to proceed with some tasks (Garfinkel 1967). The research related to this chapter aims to show the existence of ethnomethods in a citizen-sensor-network community, that we name alternative metrology. Other studies focus on the use of low-cost sensors to enhance air quality models to provide insights into patterns of pollutant emissions (Popoola et al. 2018), to enhance measurement errors in epidemiological studies (Jerett et al. 2017), and to produce information for heath applications (Huck et al. 2017).

In this chapter, we present a case in which low-cost air quality sensors are used in urban public action by local authorities with the participation of citizens. This institutionalization of alternative metrology takes part in an ongoing phenomenon of sharing practices between citizen science and citizen participation led by local authorities. V.M. Eitzel (Eitzel et al. 2017), for example, mentions the case of the city of Boulder (US), which organized citizen science projects to enlist residents in the development of an urban resilience strategy. The use of sensors as part of the participatory science project affected the participants' individual knowledge and attitudes. Nowadays, the question of its expansion to a regional level is being raised (Schaeffer & al. 2020). Our case study, the Ambassad'air project of the City of Rennes (France), provided a group of residents in two urban districts with air quality training and gave them 2.5μm particulate matter sensors during the wintery pollution periods. The main objective of the project was to alert the general public to environmental health issues through peer education (Chobeaux, 2012). Recent studies on education by public health peers showed that such approaches encourage the development of self-esteem and independence in individuals (Gillaizeau, 2019). This approach is close to awareness and information campaigns in the context of health democracy, with the intent of increasing individual responsibility on health issues (Hache 2007). In our case study, the creation of new knowledge on air quality was secondary for local authorities.

ANALYTICAL FRAMEWORK

A number of authors study citizen participation in science projects looking at the integration of the citizen to knowledge creation during the research project (Haklay 2015; Houllier 2016). This framework is similar to Arnstein's ladder (Arnstein 1969), which comes from studies in urban planning and citizen participation. The ladder is composed of eight levels of citizen participation defined by organizer's intention: *manipulation, therapy, informing, consultation, placation, partnership, delegated power* and *citizen control.* The main difference is that the scale's integration is based on the decision-making process in a context of participatory democracy. The hybridisation of the project between citizen science and citizen participation in public action makes the usual analytical frameworks on participation hard to use, because the purpose of the participation and the objectives of the project were unclear and not shared by all the members. As an alternative to measuring citizen participation with Arnstein's ladder or a similar theoretical framework, we focused on the commitment of the volunteers during the project, and not on the influence they had in regard to the purpose of the project.

The notion of commitment covers a wide range of definitions, including: the implication in a common cause, the obligation bound to a promise, and the way people are bound to a project or a situation. Using Ervin Goffman's microsociology, Howard Becker formalized an analytical framework for this notion in the 1960s (Becker 2006). He defined commitment as a side-bet made in another situation before the action, which must be respected by a person in order to keep the coherence of his behaviour in the ongoing action.

For Laurent Thévenot (Thévenot 2006), people are made of several commitments, which are combined and mixed in time and space. With the notion of regime of commitment (*régime d'engagement*), the sociologist considers the variability of behaviour during the action to adapt to the material environment and to the behaviour or actions of other people. He draws up a typology of commitment to make the distinction between the regime of familiarity (*régime familier*), the regime of regular action (*régime du plan*) and the regime of public justification (*régime de justification publique*). The regime of public justification describes the commitment, in which an actor justifies his or her action by their participation in a common good and by following principles and values. The regime of familiarity describes the commitments that come under routines, habits, and personal actions. In the regime of regular action, the actor is considered as a rational, autonomous from and unattached to their environment, whose commitment to a common action aims to realise their own agenda. With this framework, the sociologist shows that a commitment depends on a person's capacity to coordinate with other actors and with their environment during an action.

CASE: AIR POLLUTION ISSUE AND THE CITY'S ACTION

The Ambassad'air project in Rennes was supported by a large range of stakeholders, including: the municipality, several local associations dedicated to environment protection and digital education, the air quality monitoring authority (AASQA), called Airbreizh for the Brittany Region, and the French Environment and Energy Management Agency (ADEME). In this section describing political science concepts and environmental issues, we present the context of air quality issues in France and the response of the City of Rennes with citizen metrology.

A Growing Preoccupation for Urban Air Pollution and the Obligation for the City to Act

The medical community agrees that air pollution has substantial effects on health. Since 2013, the World Health Organization (WHO/Europe 2013) classifies outdoor air pollution as a leading environmental cause of cancer deaths. The consequences can also be daily discomforts and illnesses, severe diseases of the respiratory system, the cardiovascular system and related organs. Urban air pollution, and in particular PM_{10} and $PM_{2.5}$, became a public health problem, and awareness campaigns are regularly carried out. During the Ambassad'air project, a short training was given to the participants by a local expert on air quality monitoring. He presented on the physical and chemical aspects of air pollution, monitoring protocols, and health risks. To describe the risks, the expert used striking numbers:

Just some figures to visualize them better. In France, it is said that 48,000 people die prematurely. I am not saying that they died of air pollution. A premature death due to air pollution. The European Commission refers to 700,000 people in Europe… It is also said that, depending on the cities, pollution within cities can reduce the life expectancy from three to eight months.

This number of 48,000 deaths is frequently reported by the media. It comes from a study published in 2016, led by the French National Public Health Agency – Santé Publique France (Bourdillon 2016). It is usually associated with the cost of air pollution, estimated at 101,3 billion euros per year in France by a parliamentary inquiry committee (Commission 2015). Those impressive human and financial costs put the urban air pollution issue at the top of environmental priorities in the public debate.

According to the health and environment service of the City of Rennes, this international and nationwide coverage has had an impact on the risk perception among citizens. Information requests made by citizens have increased and the issue of air quality has appeared in urban planning processes and in consultation meetings. An actor summarized, "There is no complaint, but preoccupations definitely increase".

This increasing public debate on the health environmental issue takes part in a large policy transformation on outdoor air quality. The evolution of public instruments has changed with the enlargement of targeted pollutants sources, enabled by the development of measurement networks associating metrologists and epidemiologists (Lascoumes 2007). The main sources of pollution have been emissions from industrial activities. For those pollutants, the strategy used by public policy is to reduce emissions with technological improvement, and to isolate the sources of emissions. Nowadays, they are still monitored, but ordinary emissions from transport and domestic heating systems, are also considered important pollutant sources. Controversies on diesel use for car engines show the limits of technological improvement strategies. Other modes of action have been experimented with by citizens. It can be restrictive regulations, like urban congestion charges and low emission zones, or mobilisations and information, as in our case study. This evolution of policy is also a result of the transposition of the European directive from 1990 into the French law, the "Loi Barnier" (1995). It established the participation principle, which stipulates that every citizen has the right to access environmental information.

In this context of public health concerns, cities remain in a leading position to act. Since 2016, the City of Rennes and the organization for environment and consumption in Rennes (MCE) have been acting on urban air quality with citizen metrology. During the first years, the purpose was to measure the 2.5µm particle matter levels on two districts, with a fleet of mobile sensors.[1] The project, named Ambassad'air, takes part in a voluntarist policy led by a group of municipal councillors. They chose to fight against urban air pollution by mobilising inhabitants, instead of directly running a restrictive policy. Originally, the aim was for citizen monitoring to "change behaviour" and push stakeholders to act on this health environment issue. This method embodies the hypothesis expressed by the political scientist F. Boutaric (Boutaric 2014), on the evolution of French public policies on urban air pollution:

[...] Public action wants to question citizens in order for them to adapt their behaviour to imperatives regarding air quality improvements – due to the multiplicity of stakeholders, to the difficulties of acting at different scales, to the weights added by economic actors, who are in position to get their points of view and interests across. (Boutaric, 2014, p.15)

Soon after the beginning of the project, the expectation of the city changed: the main goal was to give participants the role of air quality ambassadors who are in charge of sensitising their relatives, their colleagues, and their neighbours.

Citizen Metrology or City Metrology: A Fragile Balance Between Public Action and Citizen Sciences

Using citizen metrology is not common for local authorities. The measurement process is a tool to describe the environment used by scientists and experts. It is also a tool for environmental policy; guideline values, warning levels, critical thresholds are fixed by regulation to evaluate and define objectives. The French regulation follows the European Union directive (2008), which defines WHO recommendations as an objective. To perform this kind of policy, common protocols and metrology conventions have to be established (Desrosières 2008). An expert has to be authorized to proceed with the certified measurements. In France, the AASQA is devoted to this mission at the regional level. This accredited metrology gives decision makers the possibility – and the obligation due to national law and EU-directives – of acting on a short-term scale, such as instituting alternate traffic procedures, and through long-term actions, like urban planning guidance. With the mandatory disclosure of the environmental data, citizens have access to the measurement values produced by the AASQA. However, measurements are difficult to interpret, to verify or to contest for those who do not know the techniques and methods. The ATMO scale, created in 2004, transcribes the measurements of four pollutants (O_3, SO_2, NO_2, PM_{10}) into one value between 1 and 10, from "very good" to "very bad". The daily worst pollutant-level defines the ATMO. Unlike the PM_{10} level, the $PM_{2.5}$ level is, however, just informative and not yet restrictive.[2] The ATMO aims to inform, and takes part in the emergency air quality alert for pollution events. Nowadays, this strategy is controversial because some epidemiological studies show that long-term exposure to a middle-high background pollution level is more harmful than punctual and short peaks. In this situation of uncertainty, discussions on public action strategies and on technical democracy arise (Callon et al. 2001).

Technological evolution has enabled the development of low-cost sensor networks. One consequence is an important increase in alternative metrology. This practice comes from activist circles, who use it to put institutional metrology to the test when sociotechnical controversies appear. It is currently being adopted by other publics (Chateauraynaud & Debaz 2013). Those citizen networks belong to a field of citizen science defined by Alan Irwin (Irwin 1995). Linda Carton and Peter Ache (Carton & Ache 2017) analyse two citizen-sensor-networks in the Netherlands that use sound metrology and seismic metrology to protest against public action. They argue, in particular, that it has a strong impact on local environmental governance by creating a boundary-object (Star & Griesemer 1989) that brings together all the stakeholders. A boundary-object is an artefact around which actors form heterogeneous social worlds with different points of view – in those projects: urban managers, scientists, activists, neighbours, among others– manage to coordinate themselves.

In Rennes, a group of councillors decided to replicate a citizen-sensor-networks approach from Barcelona, Spain, which appeared in Carton and Ache's inventory. According to an urban actor in Rennes, there is a risk that the project, led by the local authority, could be turned against itself by some participants, driven by activist intentions. One consequence of this political risk is that the stakeholders present the measurement as a pedagogical tool and insist on the poor quality of the sensors, compared to expert devices: "It is not a citizen observatory, it is not citizen science" reminds a member of the project. This leitmotiv came from the confusion for the public on the purpose of this hybrid project and volunteer experience: talking about air quality and doing measurement.

The Effect of Sociotechnical Representations on the Choice of Measurement Tool

The quality of the sensors was not the only decision criteria, and the choice of the measurement tools was affected by several sociotechnical representations. We recount below the process of decision about tools selection.

During the project preparations in 2015, the ambition of the local authorities and partners was to build the complete infrastructure on their own: a mobile $N0_2$ sensor. It was originally not a $PM_{2.5}$ one, as finally decided, designed and made in the FabLab subsidised by the City and connected to a low rate frequency network, which is currently experimentally deployed in some areas of Rennes. The demands also called for a real-time visualization of the measurements and the data sent to an open platform. Several arguments were raised to justify the "sensors from Rennes". Firstly, a locally designed infrastructure, with an open licence, could guarantee independence from the digital space. Private actors, for example Internet service providers, digital companies – GAFA or start-ups, are joining a growing urban data market, in particular on air quality and environment monitoring. Secondly, building a tool with the participants was considered as a way to trigger curiosity about metrology. Finally, in June 2016, the Do It Yourself approach, with a FabLab collaboration, was abandoned by the stakeholders, because it was "technically infeasible". They discovered that a reliable and cheap NO_2 sensor, which fits a small and mobile box, does not exist. Furthermore, the councillors and the ADEME – the nation's expert agency and financial partner – had set the project's agenda to begin during autumn 2016, and it would have been too short in time to build a prototype. A participant, member of an open-technology activist association, who wanted to design a sensor, explained the sudden turnaround of the initial project:

No one had measured the entailed technical complexity. Emmanuel deconstructed each step of the project, and we ended up with a dozen steps. Each one would mobilize someone, between the building of the case, information dissemination, data formatting, storage on the IT servers, and visualization by the users... It is phenomenal!

After a technological monitoring and a set of trials on several optical sensors that measured $PM_{2.5}$ levels, the stakeholders decided to buy a batch of Airbeam sensors, and as many smartphones. This sensor is made by a non-profit environmental health justice organization called HabitatMap, based in New York. A Bluetooth connection links both devices and the data transfer to AirCasting servers, supported by the Internet. A website offers visualization of all the data shared by the community of Airbeam users, and it gives the possibility to download the measurement files. All the technology is open-source and the user interface of the application, even if it is in English, is considered easy to use by the participants. In the scientific literature, a study compares some low-cost sensors, including the Airbeam (Karagulian & al. 2019; Jiao et al. 2016). However, an actor noted that there is a severe lack of information and specification on the operating conditions, uncertainly interval, and so on, on low-cost sensors.

Figure 1. Screenshots of the AirCasting application

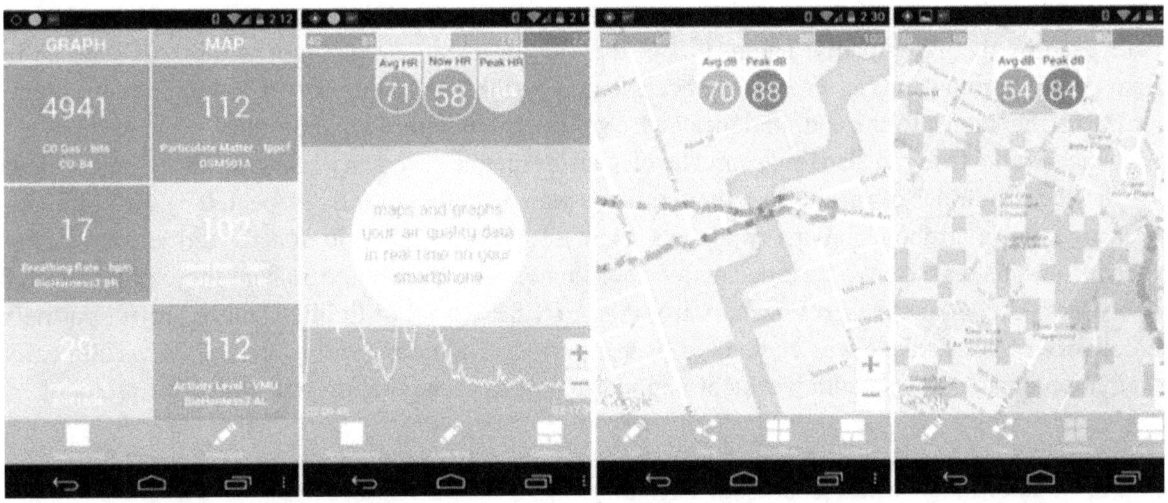

By organizing citizen metrology, the urban public actors wanted to mobilise and sensitise the citizens, offering residents the opportunity to take measurements that give them the legitimacy to become an air quality ambassador. What we observed is that the stakeholders used the data from the sensor only to verify the concordance of the citizen measurement and regional expert's monitoring. The use of the data to produce local knowledge was not considered because of "the subjectivity and the non-scientific aspect of the data" according to an organizer.

Methods

This study was based on an ethnography started at the beginning of the project in November 2016. In this chapter, we present the results of the first campaign, in 2016-2017. To analyze the participation process, we observed face-to-face workshops organized during the project. The participants were committed for a long period, for over eight months, and had to follow a two hour-long training sessions on air quality and on metrology with low-cost sensors. Other optional workshops, like mapping, viewing of films[3] with debate, and data-analysis workshops were organized. The measurement period covered eleven weeks in January, February and March, because it is in the winter that particle matter peaks are observed.

To study the participants' commitment and their perception of the project, we conducted nine semi-structured interviews with residents engaged in the first campaign of the Ambassad'air project. About twenty participants took part in the project for the 2016-2017 campaign – twenty were registered at the first meeting in October, but only sixteen continued with sensor readings in January. We contacted all of the participants for interviews with an e-mail and presented information during the workshop sessions. Five thematic categories of questions were posted to gauge their commitment to participating in the project, their perspective on environment issues, with their familiarity with the technology, their orientation toward citizen participation in urban public action, and their experience with $PM_{2.5}$ measurements with the sensors. We used the multi-thematic coding method to analyze the interviews (Ayache & Dumez 2011). The method consists in the use of a large number of heterogeneous themes that can be partially crossed. It aims to look at research materials according to several systems of similarity and difference to limit the risk of circular reasoning.

To complete the study of the participation, we have started to analyze the data set provided by the sensors used by the participants. The results of the measurement campaign of winter 2017 are an important dataset: around 700 .csv files have been shared on the AirCasting server. Every measurement, a record from the launch to the end, includes $PM_{2.5}$ values, temperature and humidity, with an interval of almost one second. One can also find sound level measurements, based on the smartphones' microphone. However, the conditions of the measurement process were usually not fully respected by the volunteers: they carried the smartphone in bags or pockets. Figure 1. presents the ratio of activated sensors for each day, based on this dataset. We included sixteen sensors carried by volunteers, and three others used by stakeholders (City employees, researchers from the EHESP school of Public Health, project partners). The delivery of the devices was conducted over a period of ten days from January 23rd. The peak of participation during the eleventh week corresponds to the implementation of a collective protocol.

Figure 2. Number of activated sensors for each day

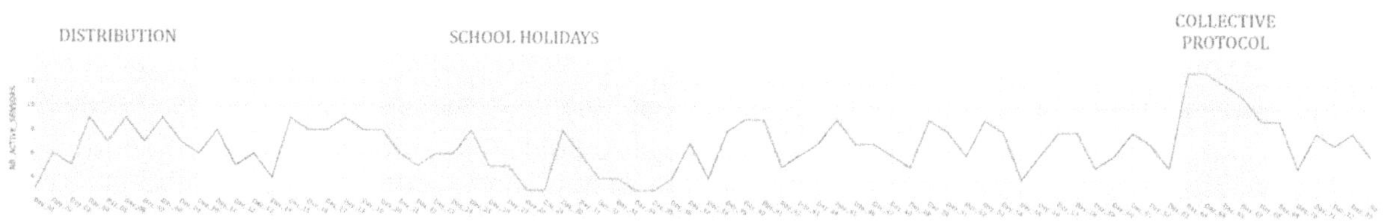

Results on the Participants' Commitment

The group of residents that carried sensors were named "*les volontaires*" (the volunteers) by the facilitator. This word puts free will and their own choice to participate as preliminaries to any action. At every step of the project (e.g. workshops, design of measurement protocols, etc.), the facilitator reminded everyone that "it has to stay a voluntary action, and not too constraining". However, we observed deep commitment. The first step was a small mediatisation of the call for volunteers: it was published in the municipal newspaper, and the information was sent to the mailing list of the *maison de quartier* of one of the targeted districts. Then, a public meeting was organized to present the main objectives of the project. Anyone from the two districts could volunteer by signing a commitment charter. With this small mediatisation, the public was also restrained to people who were already interested in local public action. Twenty inhabitants from the two districts responded. In this section of our chapter, we review different moments during the process (e.g. becoming a volunteer, staying a volunteer) and different situations (e.g. participation with or without measurement) to show the plurality of commitment in time and space.

Becoming a Volunteer: The First Step of Commitment

When we asked during the interviews about reasons to participate in the project at the beginning, four major points appeared, which we can analyze with Thévenot's framework on commitment: regime of familiarity, regime of regular planned action, and regime of public justification.

Firstly, several participants mention a civic action (regime of public justification). They considered participation in the project as an action that contributes to improving the environment. A participant explained that he was trying to follow and educate his children about "the environmental spirit", by which he meant:

[Interviewee]: Everything we could do ourselves, on our level, whether it is transportation, living in the city to own only one car for the whole family. We are not purists, but we're trying to do everyday things.

[Interviewer]: Assimilating eco-gestures?

[Interviewee]: Exactly. Embarking on this project interested me in the first instance. And this offer to measure pollution, I thought: that's really good.

Secondly, a few participants mentioned a contribution to the community (regime of public justification). One reason to commit was the particularities being a volunteer could bring to the group. For example, a resident who lived near the overhead railway, in service since 2002, and the ring road, justified her commitment by sharing an interesting observation specific to her location:

I hesitated before applying. I did it because I thought I was privileged, seeing what traffic can do to the air. Sometimes, there's a lot of traffic, which causes pollution. What about the underground railway? We try to clean the balconies, but they keep getting darker. There is black dust, which was not here before.

Thirdly, some participants mentioned personal use of the sensor (regime of regular planned action). With reflection about his or her own environment, volunteers attempted to verify and control personal exposure. The use of technologies is similar to quantified self-movement. The followers of this trend collect measurement about their daily exercise routine or daily nutrition to evaluate their performance and keep healthy. However, most volunteers, driven by this reason, also systematically mentioned relatives or friends who had respiratory diseases, or that belong to a vulnerable population defined by the medical community (regime of familiarity). The metrology process gave them more information about daily risks.

Fourthly, most of the participants mentioned a personal interest in participating (regime of regular planned action). Curiosity about air pollution phenomena and socialisation within the group were common reasons described by volunteers.

Finally, we noted that not just volunteers used the sensors. Some of their relatives, colleagues and the stakeholders themselves took part indirectly in the metrology stage. A member of the urban service explained this phenomenon by attraction: "When you see them, you want to try". Nineteen sensors were in activity between January 23rd and April 6th. Sixteen were with volunteers.

Staying Volunteers: The Side-Bet and Technological Failures

Between October and April, seven workshops, meetings or trainings were organized, some of them on weekday evenings, others on Saturday afternoons. They took place either at the MCE in the city centre or in both districts. Volunteers appreciated the diversity in the schedule, which allowed them to incorporate the project among other obligations, professional and extra professional activities, family time, etc. In fact, volunteers said that the air-monitoring project was a low priority and had to be compatible with other activities.

In contrast with strong participation in the face-to-face workshops, we observed an important disengagement process during the metrology stage, due to technical difficulties. A participant relates his dismay:

In fact, smartphones are great, apps are great, utility items are great, but when they crash, it's not easy. I would have liked to measure more, but sometimes it crashed. I don't know why. [...] I tried to reset, it lasted for twenty minutes, then twenty minutes again... After a while, it is annoying! Just that: when it works, it's great.

Memory failures, untimely updates, connection problems – in reaction, some participants put the sensors aside for a week, or more. Sometimes, they definitively abandoned the idea of taking measurements. Another volunteer, who had knowledge in computer science, decided to completely reprogram the sensor. He corrected software malfunctions and translated the application into French. In the end, his goal was to improve the sensor by adding a NO_2-sensor. We noticed that the disengagement process due to bugs and inabilities to manage them provoked a discrepancy with the feeling of obligation due to their commitment to produce some measurements and to contribute to the collective action. A participant said:

I have some insecurities because I didn't manage to commit, it's sad. I hope the others have contributed to the project.

However, bugs are not the only issue with technology that caused disengagement. For several volunteers, bringing contributions to the group collectively was not enough to inspire follow through with a daily commitment for eleven weeks. The sensors generated data files for each measurement, and the mobile application allowed the user to visualize only their own measurements one-by-one; they could not easily compare their own data with the one of other participants. A volunteer, for example, created a datasheet that compared the average value of her measurements with the value of the AASQA. She explained:

It allowed me to have a more concrete view of what I was measuring. Without it, I would take measurements, look at the results at night, but after two or three weeks, I would have given up. With the datasheet, it was a little more... I don't know... concrete.

Another volunteer recorded every measurement and observations about the context of that measurement in a notebook. Those additional inscription techniques are essential to the production of meaning, and are basic scientific tools (Latour 1985). Other volunteers also had a small science project (e.g. observation of the traffic impact on air quality, of the floor of the apartment on exposure) that they did not achieve, in particular because they did not have a way to do it. The lack of technology to access the data and transform it into information (Joliveau 2004), and to give sense to the action of measuring, had consequences on the long-term commitment process.

Being Involved in the Project: Different Participations, but Only One is Perceived by the Volunteers

The cross-analysis of interviews and observations of the workshops revealed that the meaning of participation was a focus for volunteers. However, the stakeholders did not have high expectations of the quality of the metrology and the specific knowledge or methods that could change in air quality monitoring. A participant, who stopped the measurements because the Bluetooth connection between smartphone and sensors was too instable, said:

[Interviewee]: It's too bad, because I really wanted to do something, but it ended up really annoying me. It saddens me, that's too bad. At least I learned a lot.

[Interviewer]: On the technique?

[Interviewee]: Yes. I took part in all the meetings, because I commit, I really do it.

Participation was also defined by other aspects of action. As we detailed above, the project had a pedagogical purpose. Air quality, environmental regulation, metrology, and technology were subjects that came up in daily discussions during the measuring period. All the interviewees agreed that "we are amongst convinced people", convinced by the necessity to act against air pollution, and they came here to learn more about the subject. A question was raised regularly: How to reach other audiences? Children, relatives, colleagues, and neighbours were identified by the volunteers as potential public. This position of relay, experienced by the participants, even if they did not see the purpose of their contribution to the project, is reflected in the project named "Ambassad'air", contraction of the French words "ambassadeur" (ambassador) and "air". Here is the response of one participant, when we asked him about this role:

[Interviewer]: Do you think you played the part of an ambassador for your relatives or colleagues?

[Interviewee]: I had experiences like that. It's not at work, because I work from home, but rather when I met people while taking measurements. It sparked the discussion. The hardest thing was to have enough time to explain all the aspects. [...] On a large scale, I think it could be positive, because it would involve working people, and beyond.

[Interviewer]: Members of associations?

[Interviewee]: Yes, that's why I'm referring to working, active people, because we need people to spread the information to the general public.

On some occasions, the measurements illustrated a specific phenomenon and residents brought their observations to the expert on air quality. For example, on March 16th, for about ten hours, an intense increase was detected by the citizen-sensor-network. The AASQA did not notify the public about this peak, because the air pollution period must last at least 24 hours to be reported, and it is based on a day-before broadcast. The model, crossing metrology and simulation, did not foresee the pollution event. However, two volunteers observed it:

We had two measurements with Patrick, which matched with March 16th. It was interesting. When I saw the data on my phone, because it was working at that point, I thought: That's weird, is it working well? That's when knowing the source and checking with the Airbreizh data is interesting. [...] We haven't had feedback on that yet. It's interesting to witness something like that, not to wait for the sign: ATTENTION, POLLUTION PEAK.

With the sensors, they witnessed an event they could not perceive otherwise, and it gave them the possibility to ask the air quality monitoring expert for clarifications about alert procedures and air quality broadcasts. This relation between citizens and experts initiated a hybrid forum (Callon et al. 2001). However, the stakeholders of the project did not take the opportunities to propose a role of observers of pollution or local phenomena to the participants, due to the poor credibility of the sensor and the unverifiable context of measurement. We noticed that being involved in the project reflected different types of participation and commitment that were not always perceived by the participants: crowdsourcing, relay and citizen expertise.

Taking Measurements: A Search for Air Pollution Based on Perceptions And Representations

The sensors made a physical description of the visible environment: a more or less dense presence of particular matter larger than 2.5 µm in a small volume of air. The metrology can be considered as a strategy for use of an instrument and the translation of commitment into action. In the case of the ASQAA, the mission of air quality monitoring leads to developing a network of stations processing continual measurements. The strategies elaborated by volunteers are different because the device is mobile and the participant can turn it on or off if he or she wishes. To analyze the practices and the translation of commitment, we propose first to focus on the relation between what the participant perceives and what the sensor shows the participant. At some point, we observed that the resident questions his or her certainty of presence of air pollution. Verifying these perceptions and representations drives the metrology process. We present four ordinary situations that we observed:

Situation 1: *The participant sensed pollution through various visual or olfactory cues; he or she took a measurement, and recorded a high $PM_{2.5}$ level.* Air pollution is perceived via a number of sensitive cues: smelling chimney fire or muck spreading, seeing someone holding a lit-up cigarette, noticing bad visibility at the horizon, having some irritations or difficulties to breathe for the most sensitive people, etc. Those cues provoke the intention to measure.

[Interviewee]: At the moment, every time we're in the fields, it's horrible, with the spreading. I didn't measure anything, just noticed the phenomenon. It was this morning. Horrible! I need to measure it.

[Interviewer]: How did you notice it?

[Interviewee]: From the smell. Yes, the smell. Just by smelling, I didn't do anything special. I don't know how it affects me. At the same time, in the morning, it's so wet that I don't know if my measurements are reliable. I should check. Whereas in town, it's less important.

Metrology can confirm the participants' sensations. However, sometimes, the certainty of the presence of air pollution is refuted by the measurements.

Situation 2: *The participant perceived pollution; he or she took a measure, and recorded a low $PM_{2.5}$ level.* This situation is often observed with transportation: diesel cars, charged intersections, ring way proximity, traffic lights, parks and rides, etc. Transport is an important source of air pollution in urban environments, but as the ASQAA repeats: "Fine particles are not a marker of traffic, unlike NO_2". In other words, transport contributes to air pollution, but the difference between direct proximity and hundreds of meters is not significant when one measures $PM_{2.5}$. The discrepancy between measurement and representation troubled some volunteers. We observed the same phenomenon, with the comparison between outdoor $PM_{2.5}$ level and indoor $PM_{2.5}$ level. Most of the participants noticed that the level was higher outside than inside their home. Some of them concluded that their observation was in contradiction with the recommendations of opening the windows every day. Faced with this situation, which is certainly confusing, the stakeholders had to remind them of basic metrology principles: a sensor is made for one component, and air quality depends on multicomponent levels. Throughout the project, the idea of the necessity to find a multicomponent sensor increased in participants' discourses.

Situation 3: *The participant recorded a high $PM_{2.5}$ level, and he or she observed the environment to find a cause.* In this case, knowing that the level was high was not enough. The cause of pollution had to be identified. At times, the explanation can be found: canyon effect, temperature inversion, visible sources, cigarettes, etc. Other times, the representation of sources and conducive weather conditions with high $PM_{2.5}$ level did not match with the environment of measurements. A participant said, for example, that she was surprised that the sensor showed a higher $PM_{2.5}$ level when she was running along the river than in another place on a path. In this case, a hypothesis came out after a collective discussion, that the optical sensor overvalued $PM_{2.5}$ levels when the humidity was high.

Situation 4: *The participant did not perceive pollution, so he or she did not take a measurement.* This situation resulted from the idea, shared by several participants, that the metrology entailed a search for air pollution that could assist in identifying sources. An example of this situation is narrated by a volunteer:

I took a few roaming measurements, given that I have a dog, I need to take her out often. Every time we went out, I measured. I did not do it when it was raining, because it was useless, the fine particles rates were very low.

We can see with these four situations of interactions with the sensor that there is a difference in metrology strategies between the air quality monitoring, as the expertise proceeded, and the citizen metrology observed, based in particular on perceptions and representations. However, we could not characterize citizen metrology in this program based on isolated situations alone (e.g. application of the regime of familiarity in Thévenot's typologies), because protocols were proposed, and sometimes followed too (regime of regular planned action).

Taking Measurements: Routines and Protocols Followed by the Participants

During a meeting presenting the sensor device, advice on metrology was given, which was necessary for later analysis of the aggregated data: "take long and repeated measurements", "take fixed measurements rather than mobile ones," and "take measurements in teams". Those pieces of advice corresponded to the basic conventions used in air quality monitoring. During an interview carried out at the beginning of the first campaign, the project leader explained why no specific protocol other than the advice had been given: first, too strict a protocol could restrain participation. This would have been perceived as a top-down and not a bottom-up project, and the protocol should be decided with participants. The last argument was that participants get a lot of ideas, and the diversion (De Certeau 1990) of the sensors from outdoor air quality measurement (e.g. indoor measurement, cigarette or barbecue effects) could be interesting to include in the project.

We observed that along the set of meetings and workshops, the advice to take long, fixed and regular measurements – was repeated often to frame (Goffman 1974a) the metrology practice. Categorizations were also determined between participants and stakeholders to facilitate the re-use of data: fixed or mobile, indoors or outdoors, on foot, on bicycle, on car, the districts, the time of the day. Those descriptors had to be added to the file name when the measurement was saved, because the app did not have the option to characterize the file. We noticed that every participant encoded the metadata differently, and it made the design of an automatized process for classification complicated.

An analysis of the measurement duration with the sensor distribution showed that the participants also had different practices, as we can see in Figure 3.

Throughout the study, we noticed that most of the volunteers declared having the intention to observe a specific situation they considered a potential issue, or they believed to have an influence on air quality. These intentions included the desires to evaluate the impact of local traffic on air quality; to control the incinerator emission; to prove the potential of $PM_{2.5}$ captures of lobed-leaved trees based on a lecture of a scientific article (Janhäll 2015); to evaluate the exposure of urban cyclists; and to verify the hypothesis that the $PM_{2.5}$-level varies depending on the altitude, such as between two floors in a tower, and so on.

To run their experiments, some participants created individual protocols, and "programs" that they attempted to follow (regime of regular planned action). Here is one example:

If it had worked, I would have taken measurements for a week every morning and night on my balcony. But I had trouble making myself available to turn it on every day at the same time, because there's always something going on. […] I had planned to do a week at such hours to compare the different days of the week, to compare Wednesdays to other days, Fridays and Mondays, which are not always the same when it comes to traffic.

Another participant had the intention to follow the third piece of advice: "take measurements in teams", she said:

With my neighbour, we're going to try to measure the effect of direct proximity with the incineration plant and the boulevard. We live in an L-shaped building. She is taking care of the boulevard, and the plant. I have the rue d'Alsace. We decided to take the measurements together.

Figure 3. Distribution of measurement durations (in seconds) for the fleet of sensors

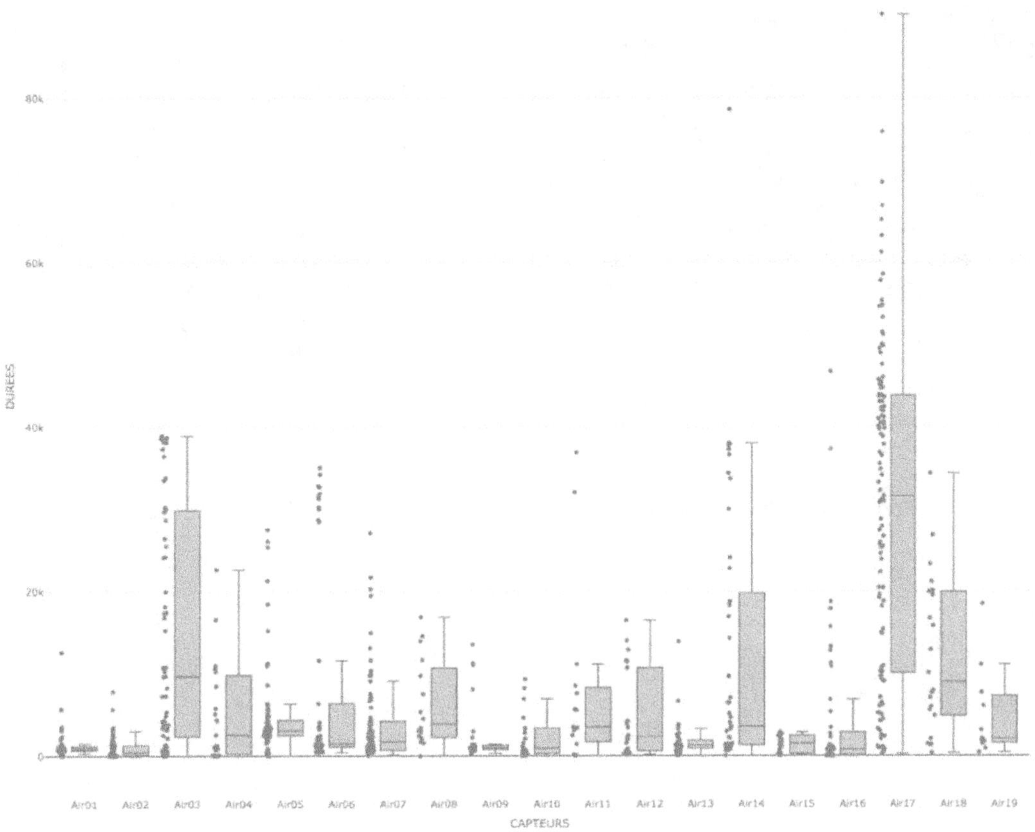

At the end of the measurement stage, during a workshop about data analysis and sharing experiences – in majority, isolated measurements – the group decided to go on an additional week to follow a collective protocol (regime of public justification). The instruction was to turn on the sensor between 7pm and 10pm at least and to measure the outdoor air quality. This collective action appeared by a bottom-up approach and came at the end of the metrology stage. The mobilization increased, as we can see in Figure 2.

In addition to those intentions of measurements, and the individual or collective trials, carried out or not, we noticed that turning on the sensor became a routine for the participants (regime of familiarity). This is transposed in the interviews with sentences like: "every time I go out, I take the sensor". The metrology practice is inserted into the ordinary day, mainly on weekdays, during one or two regular activities: home-to-work commutes, sport sessions, daily walks, etc. Fixed measurements were usually made during night-time and go on until the next morning. In Figure 3, one can identify them when the duration is between 30k and 40k seconds – approximatively 8 and 11 hours.

The routine measurement produces repeated measurements and patterns, like the map in Figure 4.

Ultimately, the density of measurements carried out in motion shows that the volunteers used the microsensors in their residential neighborhoods, as well as on ordinary trips, such as work-home journeys or towards the city center (see figure 5). We also note one-off walks in parcs and urban gardens, and on paths known for walks along waterways.

Figure 4. Examples of routines – traces produced by daily walks (left), by daily cycling paths (right)

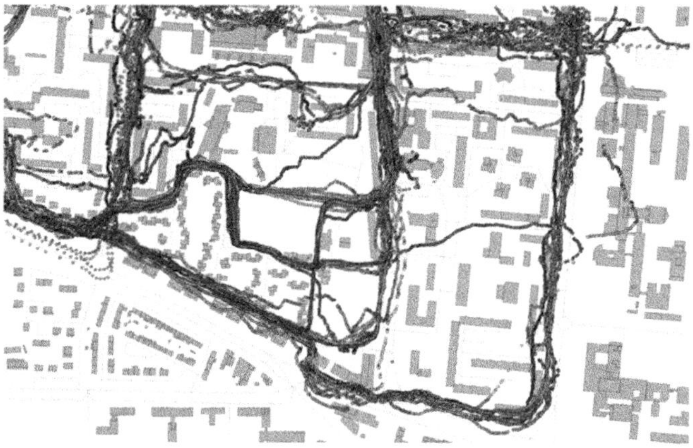

Figure 5. Map of measurements in motion

DISCUSSION AND CONCLUSION

The hybridisation between science and politics has been a question raised by sociologists since the early 20th century with the text *Politics as Vocation* by Weber (1917). Nowadays, we observe a new kind of hybridisation with the emergence of initiatives promoted and organized by municipalities, which take the forms and tools of citizen science without any scientific or research approach to act on public issues. The Ambassad'air citizen metrology project in Rennes (France) succeeded in mobilizing stakeholders and inhabitants on air quality issues using AirCasting devices. It can be considered as an urban public action instrument (Lascoumes & Le Galès 2005), whose principal aim is to inform and to raise awareness about air pollution within local publics. Political risks of counter-expertise controversies and low-cost sensors with uncertain reliability led the project to a pedagogical purpose and to minimize the importance of further scientific uses of the data. Since the beginning of the Ambassad'air project, several alternative metrology initiatives have been launched in France, usually accompanied by regional experts. They belong to a movement of renewal of risk governance (Boutaric 2021). However, alternative metrologies reduced to one pollutant raise the question of cocktail effects and of the exposure to low doses described in works on ecotoxicity (Boudia & Jas 2014). The city of Rennes increased the project of CO_2 sensors and is considering adding radon sensors.

The results of the case study show that the volunteers have plural commitments, and they were often mismatched with the city's expectations of the participants' contributions – the air quality ambassador position. Beyond their personal commitments categorised as the regime of familiarity and regime of regular planned action (Thévenot 2011), some of the interviewed participants wondered at the end of the project what the purpose of taking measurements was, the data was not perceived as having produced local, collective knowledge. Beyond the dysfunction of the devices, we also noticed that the impossibility of processing data easily triggered disengagement in several participants: the technology did not give them a way to follow the individual intentions and protocols. With the AirCasting tools, the participants could not proceed with the logical operation of data analysis to study air pollution phenomena. Discussions about hypotheses, methods or results during the workshops rapidly turned to limitations of the sensor devices. However, some participants had high scientific education (e.g. a researcher in ecology science, an engineer) and during the interviews, we noted that the interviewees were aware of the basic principles of environment science (e.g. to control the sensing condition). The design of programs with low-cost sensors should be done with careful reflection on the collaborative analysis process and not only on data visualisation.

However, citizen metrology as citizen participation (Blondiaux & Sintomer 2002) was a success according to participants and stakeholders, because it generated a public arena: discussions on health environment, on urban strategy are appearing even if the project did not integrate a decision-making process related to urban planning. In the long term, the question is how this hybrid project and the useless data – from the collective point of view – can have an influence or effects on the public debate and how the process is perceived by the participants, by the stakeholders and by outsiders of the project. The scientific literature proposes several models, such as the boundary-object (Star & Griesemer 1989) and the micro-world (*micro-monde*) (Chateauraynaud & Debaz 2017). A micro-world is characterised by a large autonomy from other communities' habits: 1) with its hybrid shape the project Ambassad'air is different from other typical citizen science projects, because it did not have the ambition to be recognised by the scientific community; 2) with its objective the project Ambassad'air is different from other typical citizen participation in urban public action, because it did not act directly on the decision-making process of urban planning.

Lastly, the practice of measurement made by citizens with a fleet of sensors, and the question of the uses of this data, raise the need to think about methods and models of metrology based on a fleet of uncertain and intermittently functioning sensors. In fact, with this study on citizen metrology and commitments, we observed that the use of sensors is based on strategies that were influenced by perceptions of the urban environment, representations of air pollution, and ordinary routines. To be able to study the practices of citizens as sensors (Goodchild 2007), we suggest broadening the focus of Human Computer Interactions, with the Human Environment Interactions, and Social Interactions (Goffman 1974b).

ACKNOWLEDGMENT

The authors are grateful to all the participants and stakeholders that are committed with the Ambassad' air project. The authors are also grateful to Sophie Provost for her editing and translation assistance.

The chapter builds upon an article published in the International Journal of E-Planning Research (IJEPR).

REFERENCES

ADEME. (2017). *Etudes des liens entre données individuelles de la qualité de l'air, changements de comportement et mises en oeuvre de pratiques favorables à l'air*. Academic Press.

Arnstein, S. R. (1969). A ladder of Citizen Participation. *Journal of the American Planning Association, 35*(4), 216–224.

Ayache, M. & Dumez, H. (2011). Le codage dans la recherche qualitative : une nouvelle perspective ? *Le libellio d'Aegis, 7*(2), 33–46.

Becker, H. (2006). *Sur le concept d'engagement. Sociologies Découverte.*

Blondiaux, L., & Sintomer, Y. (2002). L'impératif délibératif. *Plitix Revue des Sciences Sociales du Politique, 57*, 17–35.

Bonney, R. (1995). Citizen Science: A Lab Tradition. *Living Bird, 15*(4), 7–15.

Boudia, S., & Jas, N. (2014). *Powerless Science? Science and Politics in a Toxic World*. Berghahn Books.

Bourdillon, F. (2016). *Santé Publique France - Une année au service de la santé des populations*. Academic Press.

Boutaric, F. (2014). *Pollution atmosphérique et action publique*. Editions Rue d'Ulm.

Boutaric, F. (2021). *L'art de gouverner la qualité de l'air: l'action publique en question*. Presses Universitaires de Rennes.

Broday, D.M. & Citi-Sense Project Collaborators. (2017). Wireless Distributed Environmental Sensor Networks for Air Pollution Measurement - The Promise and the Current Reality. *Sensors (Basel)*, 17.

Callon, M., Lascoumes, P., & Barthe, Y. (2001). *Agir dans un monde incertain*. Editions le Seuil.

Carton, L., & Ache, P. (2017). Citizen-sensor-networks to confront government decision-makers: Two lessons from the Netherlands. *Journal of Environmental Management, 196*, 234–251.

Charvolin, F. (2015). *Un air familier? Sociohistoire des pollutions atmosphériques*. Presses des Mines.

Chateauraynaud, F., & Debaz, J. (2013). *De la métrologie en démocratie. La nouvelle vage des capteurs citoyens*. https://socioargu.hypotheses.org/4505

Chateauraynaud, F., & Debaz, J. (2017). *Aux bords de l'irréversible : Sociologie pragmatique des transformations*. Editions P.

Chobeaux, F. (2012). L'éducation par les pairs: Attentes et limites. *La Santé de l'Homme*, (421), 15–16.

Commission, P.I. (2015). *Rapport fait au nom de la commission d'enquête sur le coût économique et financier de la pollution de l'air - remis le 8 juillet 2015*. Academic Press.

Corbun, J. (2005). *Street Science: Community Knowledge and Environmental Health Justice*. The MIT Press.

De Certeau, M. (1990). *L'invention du quotidien* (Vols. 1-2). Editions Gallimard.

Desrosières, A. (2008). *Pour une sociologie historique de la quantification*. Presses des Mines.

Douay, N. (2016). La numérisation des dispositifs de participation de la Mairie de Paris - Le cas du budget participatif et de la plateforme "Madame la Maire, j'ai une idée!" *NETCOM, 30*(3/4), 249–280.

Eitzel, M. V. (2017). Citizen Science Terminology Matters: Exploring Key Terms. *Citizen Science: Theory and Practice, 2*(1), 1–20.

Garfinkel, H. (1967). *Studies in Ethomethodology*. Academic Press.

Gillaizeau, I. (2019). *Evaluation d'un programme de prévention par les pairs des consommations de substances psychoactives*. Santé Publique France.

Goffman, E. (1974a). *Les cadres de l'expériences*. Editions de Minuit.

Goffman, E. (1974b). *Les rites d'interaction*. Editions de Minuit.

Goodchild, M. F. (2007). Citizen as Voluntary Sensors: Spatial Data Infrastructure in the World of Web 2.0. *International Journal of Spatial Data Infrastructures Research, 2*, 24–32.

Gramaglia, C., & Dauphin, C.E. (2017). Toucher la pollution industrielle du doigt grâce au lichen. *Technique et Culture, 68*.

Hache, E. (2007). Is responsibility a tool of neo-liberal governmentality? *Raisons politiques, 28*, 49–65.

Haklay, M. (2015). *Citizen Science and Policy : A European Perspective*. Woodrow Wilson International Center for Scholars.

Heaton, L., Millerand, F., Dias Da Silva, P., & Proulx, S. (2018). *La reconfiguration du travail scientifique en biodiversité*. Les Presses de l'Université de Montréal.

Houllier, F. (2016). *Sciences participative en France - État des lieux, bonnes pratiques et recommandation*. Mission Sciences Paricipatives.

Huck, J. J. (2017). Combining physiological, environmental and locational sensors for citizen-oriented health applications. *Environmental Monitoring and Assessment, 189*(114).

Irwin, A. (1995). *Citiern Science: A Study of People, Expertise ans Suistainable Development*. Routledge.

Janhäll, S. (2015). Review on urban vegetation and particle air pollution - Deposition and dispersion. *Atmospheric Environment, 105*, 130–137.

Jerett, M. (2017). Validating novel air pollution sensors to improve exposure estimates for epidemiological analyses and citizen science. *Environmental Research, 158*, 286–294.

Jiao, W. (2016). Community Air Sensor Network (CAIRSENSE) project: Evaluation of low-cost sensor performance in a suburban environment in the southearstern United States. *Atmospheric Measurement Techniques, 9*, 5281–5292.

Joliveau, T. (2004). *Géomatique et gestion environnementale du territoire. Recherche sur un usage géographique des SIG. HDR*. Université de Rouen.

Karagulian, F., Barbiere, M., Kotsev, A., Spinelle, L., Gerboles, M., Lagler, F., Redon, N., Crunaire, S., & Borowiak, A. (2019). Revew of the Performance of Low-Cost Sensors for Air Quality Monitoring. *Atmosphere, 10*, 506.

Kumar, P. (2015). The rise of low-cost sensing for managing air pollution in cities. *Environment International, 75*, 199–205.

Lascoumes, P. (2007). Les instruments d'action publique, traceurs de changement : L'exemple des transformations de la politique française de lutte contre la pollution artmosphérique (1961-2006). *Politique et Sociétés, 26*(2–3), 73–89.

Lascoumes, P., & Le Galès, P. (2005). *Gouverner par les instruments*. Presses de Sciences Po.

Latour, B. (1985). Les "vues" de l'esprit - Une introduction à l'anthropologie des sciences et des techniques. *Culture Technique, 14*, 4–30.

Mabi, C. (2021). La "civic tech" et "la démocratie numérique" pour "ouvrir" la démocratie? *Réseaux. La Découverte, 1*(225), 215–248.

Miskell, G., Salmond, J., & Williams, D. E. (2017). Low-cost sensors and crowd-sourced data: Observations of siting impacts on a network of air-quality instruments. *The Science of the Total Environment, 575*, 1119–1129.

Morawska, L. (2018). Applications of low-cost sensing technologies for air quality monitoring and exposure assessment: How far have they gone? *Environment International, 116*, 286–299.

Ottaviano, N. (2015). Mécaniques et postures d'un dispositive numérique de projet urbain. *Revue Lieux Communs*, 17.

Popoola, O. (2018). Use of networks of low cost air quality sensors to quantify air quality in urban settings. *Atmospheric Environment*, *194*, 58–70.

See, L. (2016). Crowdsourcing, Citizen Science or Volunteered Geographic Information? The Current State of Crowdsourced Geographic Information. *International Journal of Géo-Information*, *5*(55), 1–23.

Star, S. L., & Griesemer, J. R. (1989). Institutional Ecology, "Translations" and Boundary Objects: Amateurs and Professionals in Berkeley's Museum of Vertebrate Zoology. *Social Studies of Science*, *19*(3), 387–420.

Thévenot, L. (2006). *L'action au pluriel : sociologie des régimes d'engagement*. Edititions la Découverte.

Thévenot, L. (2011). Grand résumé de L'Action au pluriel. Sociologie des régimes d'engagement, Paris, Editions La Découverte, 2006. Sociologies, 1–13.

Topçu, S. (2008). Le nucléaire à l'épreuve de l'expertise citoyenne. Quels savoirs? Quels enjeux. In S. Topçu, C. Cuny, & K. Serrano-Verlade (Eds.), *Savoir en débat. Perspectives franco-allemandes* (pp. 146–169). Editions l'Harmattan.

Turner, F. (2012). *From counterculture to cyberculture. Stewart Brand, the Whole Earth network, and the rise of digital utopianism*. University of Chicago Press.

Weber, M. (1917). *Le Savant et le Politique*. 10x18.

WHO/Europe. (2013). *Review of evidence on health aspect of air pollution - REVIHAAP Project*. Author.

ENDNOTES

1. Since 2016, the Ambassad'air project has gradually been deployed across the whole city. This change of scale led to a significant reconfiguration of the system of participation. The workshops meant for volunteers gave way to open sessions for large audiences, and the loan periods for the sensors have been reduced to four weeks in order to share the material between the inhabitants.
2. Since 1 January 2021, the French ATMO index has aligned with the thresholds of the European Environment Agency and now integrates $PM_{2,5}$ fine particles.
3. The films were produced by the French Ministry of Ecology and the air quality monitoring authority for the North-Pas-de-Calais Region.

Chapter 5
Creating Collaborative Environments for the Development of Slum Upgrading and Illegal Settlement Regularization Plans in Belo Horizonte, Brazil

Rogério Palhares Zschaber de Araújo
Federal University of Minas Gerais, Brazil

Ana Clara Mourão Moura
Federal University of Minas Gerais, Brazil

Thaisa Daniele Apóstolo Nogueira
https://orcid.org/0000-0001-5020-486X
Federal University of Minas Gerais, Brazil

ABSTRACT

Slum upgrading comprehensive plans and urban regularization plans are two planning tools used to promote integrated interventions in Brazilian slums and illegal settlements. Aiming at urban improvements as well as land regularization and community development, these plans have been, however, criticized for being too technical, time consuming, expensive, and top-down oriented, lacking sufficient participation to achieve community consensus on priorities, under severe budget restrictions in complex and fast changing realities. This chapter discusses the results of recent experiences in Belo Horizonte, Brazil using Geodesign framework and geovisualization strategies for collaborative planning in two illegal settlements: Maria Tereza (2016) and the Dandara (2018). A methodology for regularization plans was developed, and improvements to the Geodesign interface were tested with the use of open source Web Maps. Both experiences brought evidence on how the use of the proposed framework may enhance citizens' participation and improve planning methods based on social and community values.

DOI: 10.4018/978-1-7998-9090-4.ch005

INTRODUCTION

According to the 2010 Brazilian latest demographic census, around 11.4 million people live in the 6.329 slums identified in 323 of the 5.567 existing municipalities in Brazil[1]. This is how over 6% of the country's urban population has historically found a place in the city in face of the insufficient social housing provision for those unable to access the formal real estate market and also not eligible for government housing programs (IBGE, 2010). Being almost totally oriented to housing ownership through subsidized credit, a massive housing deficit that is concentrated below poverty levels has never been seriously tackled by national public policies, which have benefited mostly middle class segments. This way, favelas and illegal settlements cannot be considered just a planning problem but a matter of shelter provision, an expression of the right to the city, a housing solution for many. Not to mention the fact that many of them are very old and consolidated settlements, being very well located within the urban structure, with good access to services and job opportunities.

However, the recognition of a slum (favela) as a solution to be upgraded and not as an unsolvable problem to be removed and replaced by social housing projects in the far away outskirts of the city is somehow recent. The first efforts to protect these communities as part of the urban fabric and to keep them where they are date from the late eighties (mostly from the nighties) when a specific zoning category was created to somehow reduce real estate pressures over them and to assign upgrading and regularization governmental programs and projects to those settlements. These initiatives began through local planning policies and experimental housing programs in municipalities under progressive administrations such as Recife, Porto Alegre and Belo Horizonte, but only in 2001 became a national directive under The City Statute (Federal Law 10.257, 2001). Among other obligations to municipal master plans, it established the definition of Special Social Interest Zones, and the development of urban improvements and land regularization policies for slums (favelas) and illegal subdivisions corresponding to these zones.

Consequently, specific planning tools have been designed for each of these two kinds of informal settlements, which can be defined as the following. A favela is characterized by an illegal occupation of someone else's vacant land (public or private), lacking, at the beginning, public essential services such as sewers, waste management, and public facilities. They usually show an organic morphology and dense built environment with little open space and green areas, being settled on improper sites, steep slopes, flood prone areas or even environmental protected areas where formal real estate markets cannot legally be present.

Illegal settlements may have very similar urban infrastructure and housing conditions compared to slums, only they occupy land, which has been previously divided into parcels (lots resulting from a given design), which have been sold in the informal market. This means families who occupy that land paid for it and feel they are owners, even though they don't have legal ownership documents. Being implemented without going through regular planning permit procedures, they do not follow official design criteria (minimum lot area, minimum street width, maximum slope and so on), do not obey environmental constraints (steep slopes, forested areas, springs and water bodies, flood hazard areas, etc.), nor have the basic urban infrastructure that, according to Brazilian urban legislation, is due to real estate developers to implement.

Differences between these two typologies of illegal settlements are summarized by the following chart (***Table 1***) and can be clearly perceived by the compared images in Figure 1, which shows the urban morphology, the density patterns and the insertion of each of these settlements in the city context.

Table 1. Main characteristics of precarious informal settlements by type: favelas and illegal settlements

	Favelas	Illegal Settlements
Origin	Occupation of public or private urban land, followed by informal market purchase and selling transactions.	Purchase transactions from informal real estate agents, without title deeds warranty.
Parcels (lots)	Irregular shapes and dimensions, resulting from the informal subdivision of the occupied area.	Regular dimensions arising from a previous design noncomplying to legal criteria and subdivision regulations.
Location	Central city, surrounding neighborhoods or peripheral areas.	Peripheral areas where land prices are low.
Design and infrastructure	Organic layout, predominance of pedestrian alleys and staircases. Absent or insufficient urban utilities, community facilities, services and open spaces.	Regular urban layout (lots and roads). Absent or insufficient urban utilities, community facilities, services and open spaces.
Housing conditions	Self and collaborative construction with poor materials (plastic canvas, tin, plywood), later evolving to unfinished masonry, unsafe locations (geologic and flood hazard areas), structural problems, unhealthy conditions.	Self and collaborative construction on unfinished masonry and low-quality materials. May also have structural problems but tend to have better healthy conditions.

Source: Adapted and translated from Ministério das Cidades (2010, p. 20)

Figure 1. Comparing satellite images of illegal settlements in Belo Horizonte
Source: Retrieved from © 2017 Google.

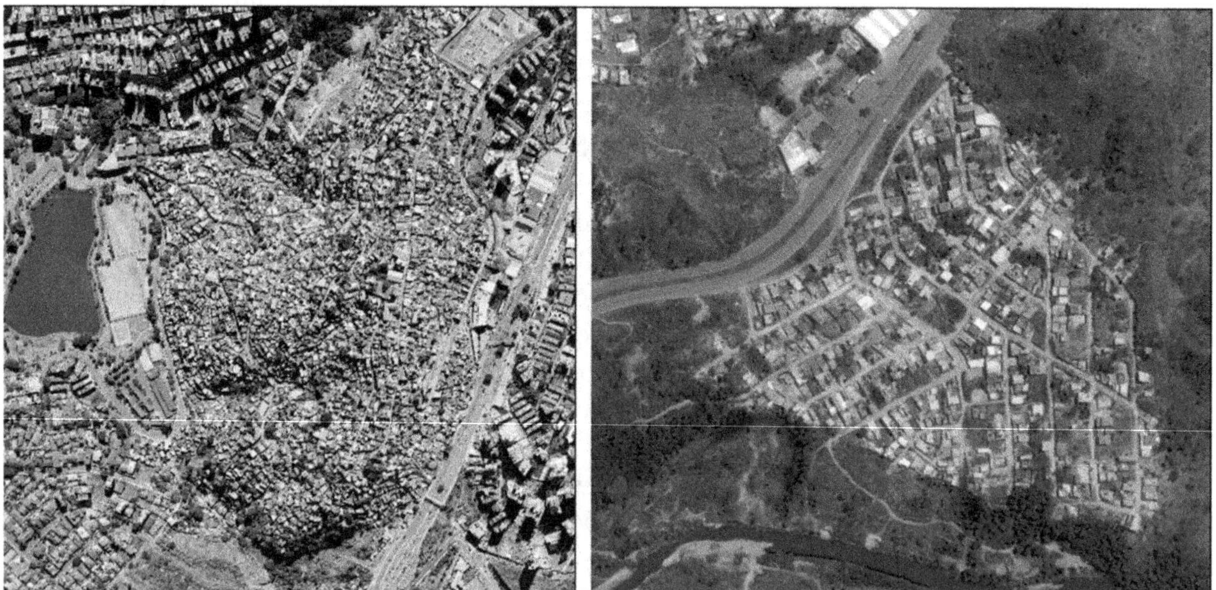

Favela: Morro do Papagaio community, central city Belo Horizonte	Irregular settlement: Maria Tereza neighbourhood, peripheral Belo Horizonte.

In 2017, another national directive was enacted (Federal Law 13.465, 2017), enabling municipalities to establish their own criteria and carry out land regularization processes. In 2020, a national program has been launched aiming at land regularization processes for more than 100,000 properties in vulnerable communities by the end of 2021, of which around 20,000 will also receive housing improvements.

Critics to this program highlight the absence of planning directives and funding that would allow local governments to carry out land regularization processes within a larger scope that involves urban and environmental improvements besides including low-income families in the real estate credit market (Observatório das Metrópoles, 2020). Within this recent context, the development of planning tools for integrated structural interventions in illegal settlements became even more relevant, so that local governments may carry out more effective land regularization initiatives.

Thus, the creation of collaborative environments for the development of planned and participatory urban upgrading and land regularization actions both in Brazil as well as other developing countries is an urgent issue. This chapter has the following objectives to be fulfilled:

- **Objective One:** Present and discuss Belo Horizonte's experience with the development and implementation of slum upgrading comprehensive plans (PGE) and urban regularization plans (PRU), with regards to its main results and improvement needs.
- **Objective Two:** Present and discuss two cases of the use of the Geodesign method as a planning support tool for the development of PRUs based on two experiments carried out in Belo Horizonte: the Maria Tereza and the Dandara communities.
- **Objective Three:** Explore the potentialities and limits of Geodesign as a methodological approach to enhance participatory decision-making processes in informal settlements and slum urban upgrading and land regularization plans.

BACKGROUND

The Local Context

Belo Horizonte, with an estimated population of 1,500,000 inhabitants (IBGE, 2020), has 186 favelas classified as ZEIS – Special Social Interest Zones and 123 illegal settlements classified by AEIS-2 – Special Social Interest Areas, according to the local zoning ordinances (Belo Horizonte, Municipal Master Plan Act 11.181/2019, 2019, p. 29,). This means they are supposed to stay where they are and be subjected to rehabilitation programs, following two main planning instruments: Specific Comprehensive Plans or PGE, as a short for Planos Globais Específicos (in Portuguese), and Urban Regularization Plans or PRU, as a short for Planos de Regularização Urbanística. PGE and PRU are the current tools to plan integrated interventions and decide on priorities for these areas.

Figure 2 shows the location of these settlements within the municipal territory of Belo Horizonte. A clear concentration of irregular settlements (AEIS) in the far north and far south portions of the city can be noticed, which supports the characteristic of peripheral location of this type of occupation. On the other hand, favelas tend to have a more scattered distribution throughout the whole municipality, including a few larger ones in very central locations.

Slum Upgrading and Urban Regularization Plans

Aiming at urban and environmental improvements such as sanitation infrastructure upgrading, geologic risks and flood hazard control, as well as housing tenure regularization and socioeconomic community development such as increasing jobs opportunities and income generation, PGE have been developed for more than 30 years now by means of traditional thematic mapping method for both data collection/analysis and proposal strategies. It consists of three main steps as the following:

- **Step 1:** Base map updating
- **Step 2:** Data collection, inventory mapping and analysis
- **Step 2:** Proposed interventions

Figure 2. Distribution of ZEIS (favelas) and AEIS (illegal settlements) in Belo Horizonte
Source: Adapted from Belo Horizonte Municipal Master Plan, 2019

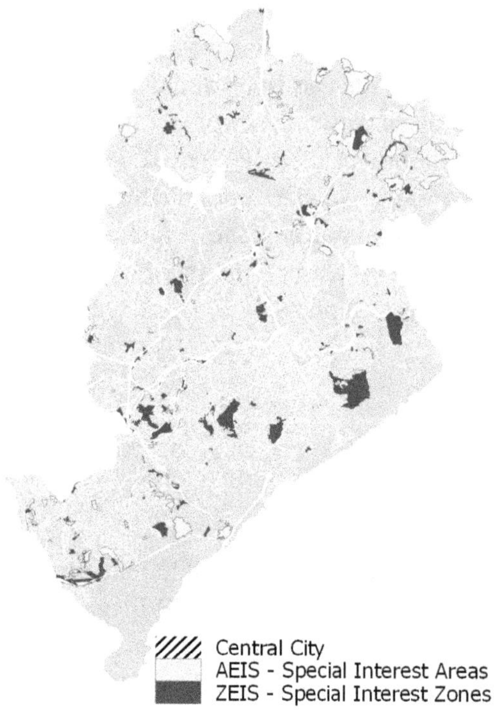

A multidisciplinary team is involved in the process, including architects, urban planners, geographers, economists, engineers, geologists, lawyers, sociologists, social service and communication specialists, among others.

The first planning step is focused on the development of topographic and built structures and urban infrastructure base map. Usually developed from aerial photographs and topographic field surveys, it becomes the base map on which all collected data is registered. Community residents are usually involved in building up this base map, both as a way of accessing less evident areas and built structures as well as a capacity building strategy, aiming at future reading and understanding of the plans.

The second step refers to the development of the detailed inventory and analysis of the settlement. For this, field surveys and interviews with residents are carried out, as well as data collection from all available secondary sources. The diagnosis of the study area is built by relating sectorial data according to three integrating themes:

- **Iteration 1:** Urban-environmental issues such as lack of basic sanitation, public spaces and community facilities, basic sanitation
- **Iteration 2:** Socioeconomic and organizational community profile
- **Iteration 3:** Legal land ownership status of the settlement and of its dwellers

An important product of this second phase is the evaluation of different settlement areas, according to the complexity for regularization purposes due to both urban environmental conditions as well as legal status. This results a map that shows areas, which are considered "able to regularization", areas that depend on interventions to be regularized, classified as "able to regularization under conditions" and those where there is severe hazard or legal impediments to occupation, which are considered "not able to regularization".

In the third and last step, interventions meant to achieve the settlement's urban and legal regularization are proposed and detailed. This process is based on the analysis of different future planning scenarios designed according to the level of physical change proposed for the settlements and their related impacts (Drumond, 2017). At this point, the community is called to participate and express their opinions on the presented scenarios. Discussions among consultants responsible for the plans, community representatives and public officials from the municipality lead to a chosen scenario. This is then developed as a final plan that comprises the detailed selected interventions, the priorities for implementation, time frames and budgets. This plan is then used to inform and guide public works in the community, being an important instrument in searching for financing, such as the participatory budget process as well as other funding sources.

Throughout this traditional planning process, citizen's participation occurs mainly through the so-called Reference Group (GR). This group of community residents is formed at the beginning of the planning process, and is composed of formal and informal community leaders, members of local social entities (citizen's associations, sports, cultural and religious groups) and also residents in general who are interested in participating. The GR is supposed to act as an intermediary body between planners and the community as a whole. They are responsible for sharing information with the other residents, making sure the main local demands are being covered by the plan and, mainly, for approving each of the planning phases (Drumond, 2017).

Figure 3 illustrates the sequence of the mentioned steps according to this traditional comprehensive planning methodology.

Up to 2020, 70 PGEs have been accomplished, encompassing 90 favelas and benefiting 259.761 inhabitants. Other two PGEs are in the process and three more are planned to begin (Belo Horizonte, 2020). PRU, which are the plans designed for illegal settlement, started to be carried out only in 2016 as a result of a more recent policy following the same methodology applied for PGEs. Since then, 9 PRUs have been accomplished and 2 are currently being developed, benefiting 32.658 inhabitants (PBH, 2021).

Although this method has been successfully applied for many years, it has been strongly criticized for being too technical, too time consuming, expensive and top-down oriented (Araújo, 2018; Kaap et al., 2012; Salomao, 2016, Monteiro et al, 2018). Conti (2004) and Bedê (2015) also argue that one of the main shortcomings of the PGE methodology is its lack of flexibility. Besides, the three stages of the plan are completely independent, and there is no feedback among them, which makes it difficult to have it updated during and after the planning process.

Figure 3. PGE development steps according to the traditional comprehensive planning methodology
Source: Material from the authors, 2016.

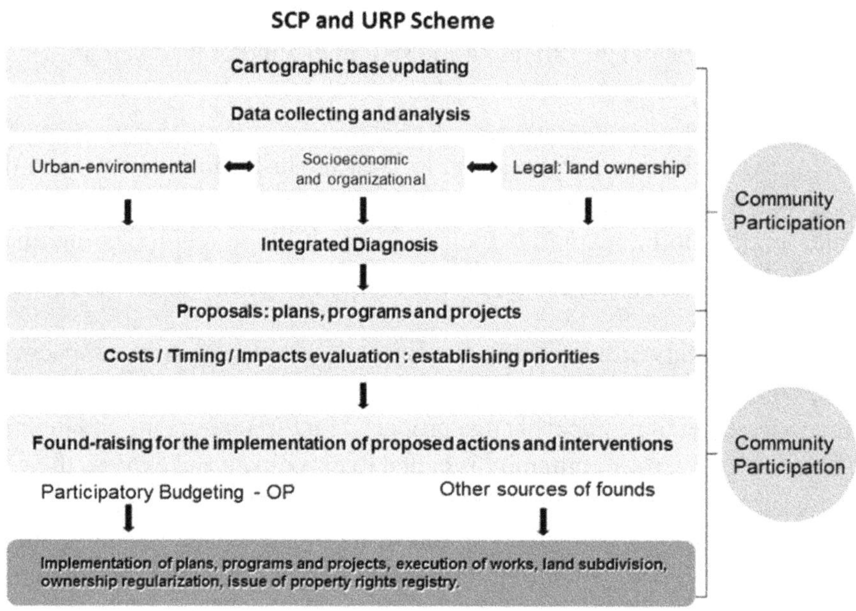

It is worth mentioning that when the municipality launches a PGE or a PRU, a large expectation is generated in the community regarding infrastructure improvements, but also and most importantly, aiming at obtaining land tenure documents. This ends up attracting more people to live in those settlements, encouraging the densification and the expansion of built areas through the occupation of other vacant spaces in surrounding areas. Since the planning process lasts about three to five years, informal real estate speculation within the community often materializes before the interventions take place. According to Conti (2004), the lack of efficient strategies to stop densification and growth, as well as methods to update plans as the dynamics of the territory changes, is what makes PGE become obsolete so quickly. Bedê (2015) highlights the lack of federal funding as an obstacle for the fast implementation of those plans, which also demotivates community participation and leads to discontinuity of the planning process.

Other authors (Araújo, 2018; Kaap et al., 2012; Salomao, 2016, Silva, 2013 Mello, 2009) also point out that the first two steps of the planning process take too much time and money, since they require very broad surveys and an extremely comprehensive set of data, some of which are not essential to support the proposals. This way, they argue that PGE has become a collection of extensive analytical reports, followed by intervention projects, which are very superficial and poorly developed. Inventory and diagnosis phases drain most resources and time, leading to a shortage of time and money for the proposals, which is frustrating for both planners and the community.

Besides, although there is a great effort in the analysis of the territory, the cartographic bases are produced individually in CAD platform, which makes it difficult to merge all the information on a single geo-referenced base that would allow for synthesis maps and a comprehensive view of the area's problems and potentialities (Salomao, 2016, Mello 2009).

With regards to citizen's participation, it is widely perceived that there is a great difficulty in engaging the Reference Group on a continuous basis as a voluntary activity in such a long process (Silva,

2013). Since participatory events are not permanent and the interval between meetings is usually long, it is common for the group of residents to lose interest in the plans and stop participating. There is no guarantee the GR communicates well and broadly with the whole community, which also generates misunderstandings and false expectations towards the content of the plans.

In addition, the very complex subjects and the technical language that characterizes the content of these plans require a decoding ability that community members do not have (Freitas, 2020, Salomao, 2016, Nascimento, 2011). Especially when taken into account the low level of education of slums and informal settlements residents as well as the social vulnerability situation in which they live (Araújo, 2018).

Citizen's Participation, Geovisualization, and Geodesign Strategies

Planning processes in areas of social vulnerability carried out in Brazil have been basically guided by the Rational-Comprehensive Planning tradition with some Advocacy or even Negotiation Planning actions according to classification proposed by Khakee (1998). In the Rational-Comprehensive processes, a broad stage of inventory and analysis is carried out, involving an exhaustive production of technical documents, followed by plans drawn up by planners with some level of citizen's participation. Efforts to include citizens in collective decisions have bureaucratic and political barriers to be overcome (Villaça, 2006; Bueno and Cymbalista, 2007; Cymbalista and Santoro, 2009; Santos and Montandon, 2011; Freitas and Bueno, 2018), but a lot must be also improved as far as information and communication strategies.

Within the context of information and communication technology development, new instruments for spatial analysis, evaluation and representation have become essential tools for territorial planning and management. Among examples of new technologies applied to cartography, GIS mapping and geovisualization stand out for their availability, usability and accessibility, resulting in an important breakthrough for information sharing in planning processes (Masala & Pensa, 2016, Magalhães e and Moura, 2020).

Representing a given territory is the first step to get to know it (Moura, Marino, Ballal, Ribeiro & Motta, 2016). With regards to urban and regional planning, geovisualization can also be understood as a tool to broaden the possibilities of community participation in decision-making processes. As a means to enhance spatial understanding, it allows for higher alignment and leverage of information among different stakeholders within a participatory planning process. It is, then, essential to provide voice to those so far excluded from planning decisions due to the absence of technical knowledge or for not feeling able to contribute.

Van Den Brink and Dane (2007) proposed a synthesis comparing different typologies of participation based on the works of Arnstein (1969), Edelenbos & Monnikhof (1998) and a few governmental guidelines for participatory planning, including the European Commission's (2002). They range from manipulative passive processes to more interactive and collaborative planning experiences, depending on the degree of power distribution and stakeholders' involvement in the decision-making process.

Different levels of participation consequently relate to different requirements for the use of digital media as a planning support tool. Streich (2004) proposed three increasing degrees, from information supply and consultation, somehow applied in Brazil in participatory local Master Plans and public hearings (Santos, 2011), to active participation, involving co-producing and co-deciding planning strategies which also relates to a growing number of actors eager to influence the contents of plans.

The prevailing methods in participatory planning in Brazil follow two basic steps. The first consists of technical inventories and analysis made by specialists resulting in comprehensive reports that often contain thematic maps. The second relates to the discussion of the technical understanding of the study area's main issues with community members through participatory workshops and public hearings, with the help of the cartographic material, which has been produced (Ministério das Cidades, 2004). The discussion process, however, is often deficient and does not achieve expected goals as far as effectively involving and communicating with the general public.

Peet & Leach (2000) have already highlighted the growing reticence among citizens and NGOs to use institutional participatory instruments also due to bottlenecks arising from the use of technical language and sectoral approaches to describe a growing complexity of planning issues. Masala, Pensa & Lami (2013) emphasize the use of geovisualization in order to stimulate communication and discussion among actors involved in the decision-making process and as a means for allowing the planning process to be more effective towards the promotion of more sustainable urban development policies.

Aiming at more inclusive and interactive participatory processes, geo-visualization approaches seem to increase opportunities for citizen participation in spatial planning as an essential tool to create collaborative environments for co-producing planning strategies and decision-making processes, which are also the main goals of Geodesign.

According to Steinitz (2012) Geodesign is a methodology that provides a design framework and supporting technology to leverage geographic information, resulting in designs that more closely follow natural systems. It is, in fact, an invented word to describe a planning method based on communication strategies aimed at sharing knowledge, assumptions and language to achieve an effective collaborative decision making process. Moura (2019) defines Geodesign as a method based on the potential of Geographic Information Systems that allows collaborative opinion-making and decision-making processes to be carried out through a codesign environment.

In theory, the concept of Geodesign reflects the idea that a good territorial design depends on the knowledge of existing conditions (constraints and possibilities) and the ability to work with them. This way, the concept of Geodesign is mostly related to the contextualization of proposals in the existing landscape. Miller (2012) argues that the essential aspect of this definition is the idea that design occurs within the context of geographic space as opposed to a conceptual abstract handmade space on pencil and paper or even digitally with the use of computer aided design.

The increasing use of digital technologies, especially Geographic Information Systems (GIS), has made the concept of Geodesign applicable in more complex situations and in larger planning units (Zyngier, Moura, Araújo & Carsalade, 2016). However, the simple data collection and maps overlay do not mean a significant advance in the design methods. This way, Steintz (2012) has proposed a complete methodological scheme, including conceptual framework, design strategies, and procedural techniques. As opposed to returning to the old mapping overlay techniques, it emerges as a new field of research, which integrates and adapts the GIS concept to the design platforms. (Batty, 2013).

Within the Geodesign framework (Steinitz, 2012), design professionals and geographic scientists, information technicians and local stakeholders (people of the place) are supposed to work together to answer key questions about the problem area, integrating data, knowledge and values through planning phases and related models. It suggests the use of six models linked to six key questions whose answers are supposed to describe the whole planning process, as the following:

- **Representation Models:** How should the context be described?
- **Process Models:** How does the context operate?
- **Evaluation Models:** Is the current context working well?
- **Change Models:** How might the context be altered?

- **Impact Models:** What differences might the alterations cause?
- **Decision Models:** Should the context be changed?

The first three models comprise the evaluation process, looking at existing conditions within a geographic context. The second three models comprise the intervention process, looking at how that context might be changed, the potential consequences of those changes, and whether the context should be changed. The author advocates that these questions have to be analyzed and answered three times, according to three different iterations each. The first iteration is aimed at defining the planning goals and objectives to the study area. It is supposed to answer the question "Why study this area?" The second iteration has the function of guiding the planning methods to be used. The question to be answered at this point is "How should this area be studied?" Finally the third round is aimed at the implementation of the chosen scenario, and it involves strategies, policies, plans and projects. It is supposed to answer the questions on "what, where and when interventions in this area"? Figure 4 presents the methodological structure proposed.

Figure 4. Geodesign framework steps for a collaborative planning decision-making process
Source: Adapted from Steintz (2012)

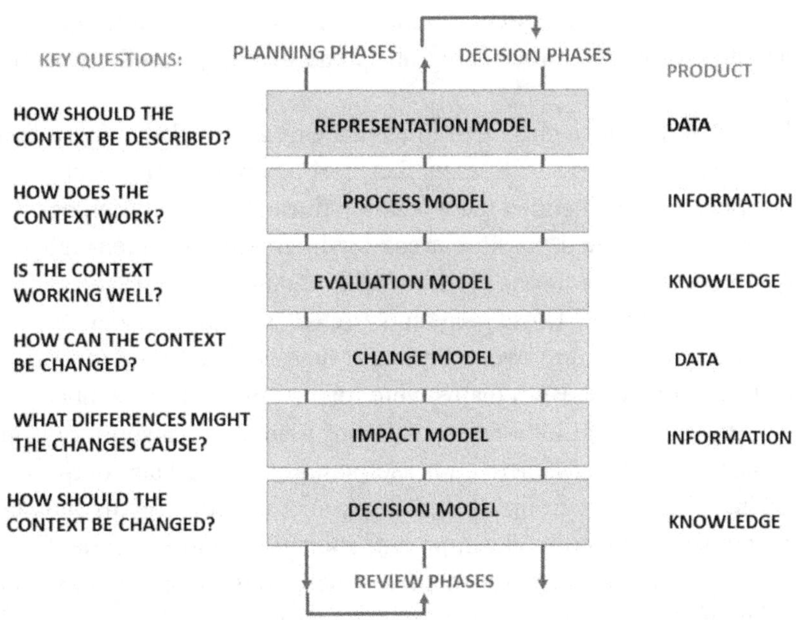

According to this method, the decision model drives the whole process, so that, regardless of the shortage of data and time, a proposed intervention with a reasonable degree of consensus is supposed to come out at the end of the line.

The achievement of an informed decision through collaboration depends, however, on collective knowledge building and for that purpose many different tools and methods have been developed as Planning Support Systems (PSS), in which the use of spatial data visualization has played an important role and produced promising benefits to participatory planning (Harris & Batty, 1993; Geertman, 2008).

IMPROVING URBAN REGULARIZATION PLANS IN BELO HORIZONTE

Aiming at improving citizen's participation in PGE and PRU development processes, the Belo Horizonte Municipality decided to test the Geodesign method, first as a tool to enhance the community involvement, and then as a strategy to reach more democratic decision-making processes. In September 2016 a three-day planning workshop was conducted by faculty members from UFMG GIS Lab, at Belo Horizonte Planning Department, using the Geodesign framework and geovisualization tools to create a collaborative environment and enhance stakeholders' participation for the development of a PRU for the Maria Tereza neighborhood.

The Maria Tereza neighborhood is a fast-growing new settlement located in the northern urban expansion region of Belo Horizonte. It is only ten years old, had a population of 373 inhabitants in 2010, according to the latest census data (IBGE, 2010), which is now assumed to have almost doubled. Its main problems are its isolation with regards to more developed areas, lack of basic infrastructure and public services and occupation of steep slopes and hazard areas.

Existing inventory data and variables on the study area had been previously combined to produce a set of multi criteria thematic maps, named "systems" within the Geodesign framework context. They comprised the representation model, that describes the area's context, and the process model that shows how it is currently working. The two combined resulted in evaluation models that analyze if the area is working well. They included hazard and environmental sensitive areas, sanitary vulnerability, housing conditions, densification possibilities, urban integration and urbanity based on preconditions to form and access centralities.

These systems (Figure 5) were also classified in advance by the workshop coordination team under a methodological supervision. Also, an ICT researcher, responsible for conceiving the Geodesignhub platform[2] followed this process and attended the workshop through videoconference devices. Each map received a classification that indicated possible areas for intervention (green light), areas capable of intervention subjected to specific conditions (yellow light) and areas with no need of intervention (red light), according to the system theme. These maps functioned as change models.

The 60 participants were divided into two main teams that were assigned to work separately on the first and second day of the workshop. Both teams were formed by faculty members, graduate students and public officials from different local authorities, including roads and transportation, green areas, social housing, water and sewers, social assistance, land use planning and urban control departments. Each team was then divided into six interest groups, according to six main focus goals and planning objectives to be pursued. They were socioeconomic development; sociability and urban design, related to public spaces and amenities; existing housing improvement and new housing production; land ownership and regularization; environmental and landscape concerns and mobility. This last topic involved connections within the neighborhood and from the neighborhood to the surrounding areas and to the city, as a whole.

Each group was then stimulated to simulate interest groups and make proposals in terms of projects and policies that were translated into diagrams resulting from points, lines and polygons that were drawn on screen covering the areas where they were supposed to take place. The use of the Geodesign Hub platform allowed for simultaneous visualization and assessment of the proposed diagrams by all groups. At this point, Impact Models were built from a compatible interest matrix so that the groups could see the impacts of their decision in each system, and evaluate the emergence of conflicts caused by each proposal.

Figure 5. Systems and color codes used in the workshop
Source: Material from the authors, 2016.

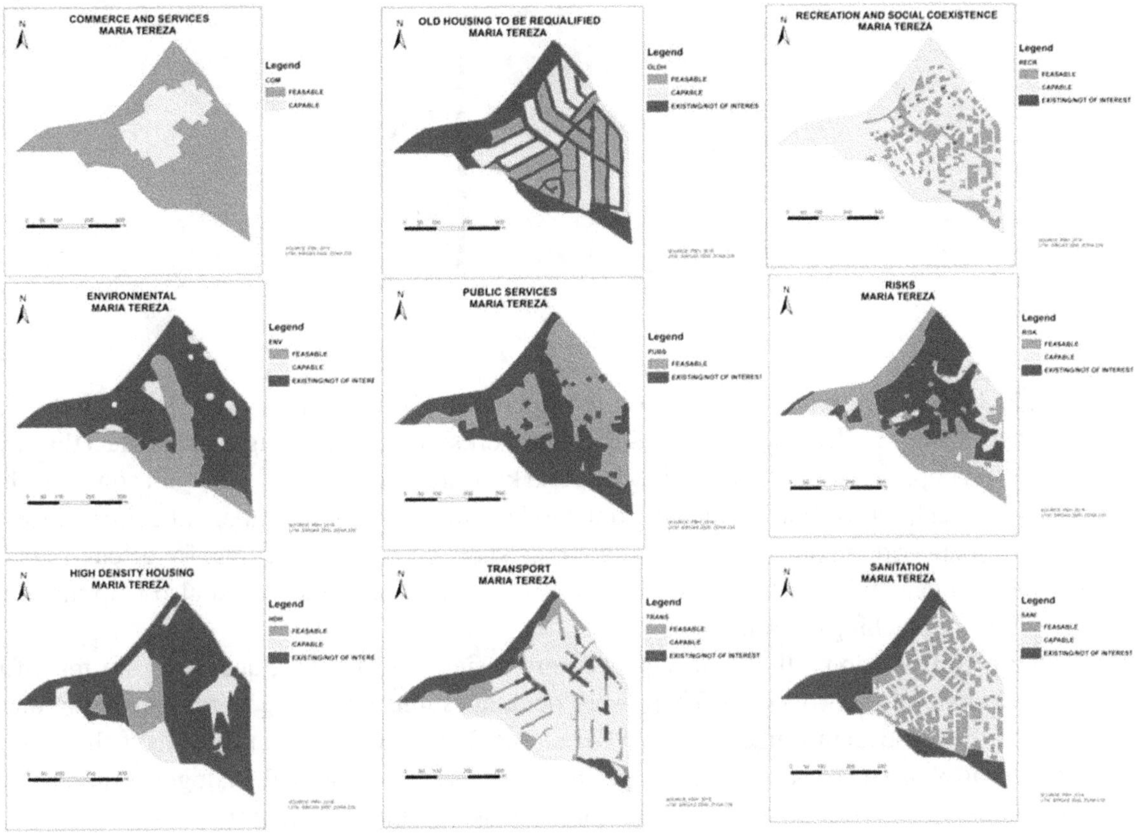

As a second moment in the workshop, participants were regrouped in three teams, and each one was responsible to select the most adequate diagrams designed on the previous step, join them together, and build up a complete master plan for Maria Teresa Settlement. Simultaneously, Geodesign Hub provided the calculation of costs, assessment of impacts and conflicts of the final design, helping in the decision-making process. The participants also used a 3D digital model of the study area to support mental mapping processes.

As a final activity, the three different master plans were compared, and the participants were encouraged to discuss the solutions, merge the proposals and create a single plan. The same procedure was repeated on the following day of the workshop with a second group of 30 participants that also came up with another single plan as a result of a collaborative decision-making process under the same methodology. The comparison of both results showed that although there was a high degree of consensus among the diagrams proposed by the two groups, priorities and final results varied, according to each group's decision model. Figure 6 shows the importance level given to each "system" by each group. It can be noticed, for example, that, although transportation was considered a prime concern by both groups, there are relevant discrepancies with regards to other systems (environmental protection, hazard control and high-density housing) were weighted.

Figure 6. Definition of system weights by workshop participants
Source: Material from the authors, 2016.

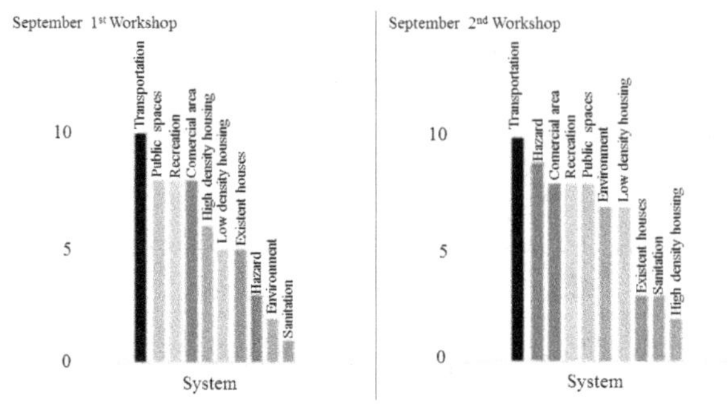

Such results have an important influence on the choice of which diagrams will figure on the final plans that end up having similar but different results. The workshop outcomes highlight the fact that, even though Geodesign improves disciplinary interaction and knowledge exchange, it does not solve conflicts of values, interests and motivations, which continue to be a matter of negotiation and compromise. As a matter of fact, increased previous knowledge about the study area, and more time for discussion were two very important demands pointed out by the workshop participants that could also help maximize the opportunities for reaching consensus.

At the end of that session, all 60 participants were assigned to fill in a questionnaire for a final qualitative evaluation of the workshop. There were open questions on their general perception on the applicability of the Geodesign framework for PGE and PRU development purposes as well as on the positive and negative aspects of using Geodesign in comparison to traditional analogic comprehensive planning methods. Answers to these two sets of questions were summarized in the following tables.

Table 2 shows that, on a general basis, the majority of participants see advantages in using the Geodesign framework as a planning support tool for its transparency, flexibility and economy of time and resources. There were also a significant number of respondents (40% 24 answers out of 60 participants) that relate the method to the early phases of a plan, when a structure base map shall be agreed upon, before it can be developed as a technically defensible final plan. Although many recognize the method's potential to make conflicts clear and promote objective compromises and consensus, there is a significant perception of difficulties low-income communities may face with the use of information technology devices (20%, 12 out of the 60 respondents).

Responses to the question about the positive and negative aspects related to the use of Geodesign compared to traditional comprehensive planning methods were then grouped as shown in Figures 7 and 8. Most mentions of positive aspects were related to the integration of different professional perspectives and the progressive consensus building process provided by the method, followed by the optimization of time and costs.

With regards to the negative aspects, the responses showed that the main concerns about the use of the Geodesign framework are related to the lack of previous knowledge about the study area, the need for more time for preparing and discussing the proposals and difficulties in dealing with the Geodesign Hub platform which was considered not sufficiently friendly by 25%, 15 out of 60 respondents.

Table 2. General perception of workshop participants about the use of the Geodesign framework as a planning support system for PGE and PRU development

Statements	Number of responses
Can provide a structure base for collective planning inputs but needs a technical development follow-up to become a final plan.	24
May face cognitive, graphic, and digital divide, especially with uneducated, low income and elderly population.	12
Makes it easier to establish community priorities and to guide public policy allocation to the study area.	12
Turns conflicts evident and help support opinion forming and decision making.	9
Allows for quick spatial location of proposals and interventions.	8
Allows for simultaneous integration of sectorial proposals, coming from different fields of knowledge.	8
The Geodesign method should be tested on a real plan, involving local stakeholders.	6
Very useful to speed up the diagnosis process.	6
Will reduce costs and optimize results.	6
Improves comparative conditions to justify project financing and decisions on budget allocation	2
Makes it possible for the creation of new proposals based on a quick diagnosis.	1
Geodesign is not an adequate method for PGE and PRU development.	1

Source: Material from the authors, 2016.

Figure 7. Main positive aspects pointed out by workshop participants about the use of Geodesign as a planning support framework for PGE and PRU development
Source: Material from the authors, 2016.

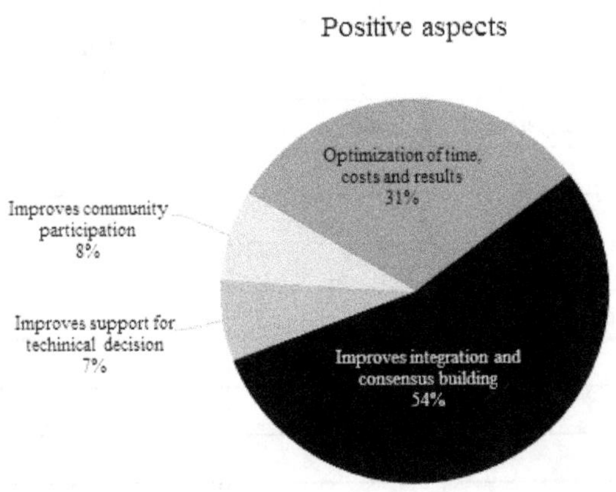

Figure 8. Main negative aspects pointed out by workshop participants about the use of Geodesign as a planning support framework for PGE and PRU development
Source: Material from the authors, 2016.

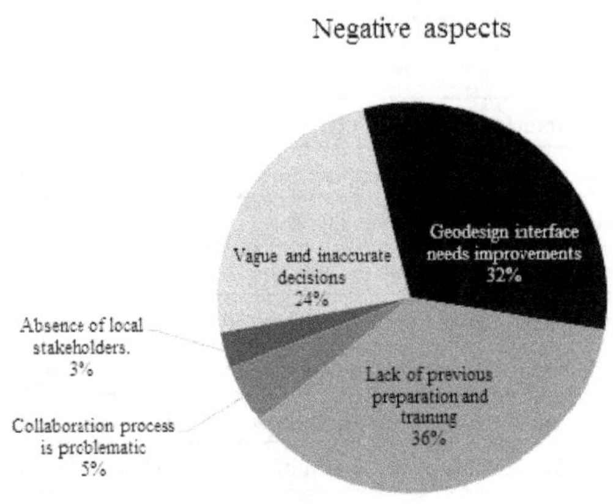

Finally, participants were invited to evaluate the results of the workshop as far as addressing the study area's main planning issues and to make suggestions on the necessary improvements to the method. A ranking of the responses is presented on the following tables.

Responses on Table 3 attest the success of the workshop to introduce the Geodesign framework as a tool to promote participatory planning within a context of little data and money, in a short period of time. Even though many improvements to the interface and to the workshop dynamics were considered necessary as shown on Table 4 the method proved to be a promising support planning tool for more effective and inclusive future PGE and PRU development.

Table 3. Workshop evaluation s as far as far as addressing Maria Tereza Neighborhood's main planning issues

Answers	**N. People**
The spatial allocation of proposals and development of comprehensive scenarios was reached with little data and within a short period of time.	17
Integrated proposals, coming from different planning sectors and fields of knowledge.	17
Optimization of knowledge and resources.	15
Fast consensus on collective proposal and priorities based on successive approximations	12
High consistency among inventory, diagnosis and proposals.	6
Clear picture of conflicts underlying the decision-making process.	5
Fast allocation of basic structure for the plan: main accesses, centralities, community facilities, open spaces and environmental protection areas.	5
Provided visualization of targets, costs and impacts of main proposals.	5
Resulted policies to keep the families in the area and to promote the resettlements.	1
Encouraged objective discussions based on the same methods and criteria.	1

Source: Material from the authors, 2016.

Table 4. Main aspects to be improved for a future use of Geodesign framework for PGE and PRU development

Suggestions	**Number of responses**
Improve interface features: Portuguese version, drawing process, diagram accuracy, system visualization.	15
Provide more information on the study area, mainly social data.	10
Involve community members and local stakeholders (people from the place).	8
Leave more time for group discussions.	7
Involve participants in previous choice of variables and definition of systems and targets.	7
Provide better information on implementation costs and targets.	4
Provide complementary information on surrounding areas, plans and projects.	4
Qualify and detail proposals, including legal, technical and economic feasibility.	4
Consider non spatially allocated proposals.	3
Improve and dedicate more time for training strategies.	2

Source: Material from the authors, 2016.

As a result of this workshop, a decision by the local government to go forward testing the use of Geodesign as a planning tool was taken. The method was then applied in another community: Dandara

(2017) while three other Urban Regularization Plans (Hosana, Novo Santa Cecília e Liberdade) started in 2015, continued to be carried out using the traditional analog planning approach, all involving community residents and other local stakeholders.

The Dandara community results from an organized squatter of a private vacant environmentally sensitive area located on the northern outskirts of the city. The occupation took place in 2009, when 400 homeless families started self-building the settlement, initially guided by a site design plan that provided for a regular street, blocks and lots layout, as well as conservation areas of environmental relevance such as riparian forests and flood hazard areas. In 2017, there were already 1339 dwellings in the area, 210 located in the previously protected areas, 65 of them in the flood hazard zone (Figure 9). The increasing densification of the area, lack of infrastructure and strong community mobilization began to pressure the local government for solutions with regards to both legal recognition of tenure as well as urban improvements.

Figure 9. Dandara Community in 2009 and in 2021
Source: Retrieved from © 2021 Google.

Dandara community in the beginning of the occupation (2009)

Dandara Community in 2021

In 2016, the Dandara Community was granted funds for the development of its Urban Regularization Plan. In 2017 a cooperation agreement between the local planning department and the GIS Lab at UFMG Architecture School was set and the first informal settlement regularization plan using the Geodesign method was put in practice. In 2019, after a long period of legal disputes, with important support from social movements and technical assistance from NGOs and the university, the Dandara Community was finally recognized by local authorities: the land was expropriated by the state government and the area classified as AEIS-2 in the municipal zoning map. For the development of the Dandara Plan, more time was assigned for training community residents on the methodological procedures. Local youngsters, who are usually more receptive to the use of information technologies (Presnky, 2001; Rosen, 2011), were assigned to play a leading role in operating desk top computers for the elderly family members during a three-day workshop. Sessions were held on Saturdays at the neighborhood public school. Between meetings, the municipality also held open follow-up discussions with a larger group of residents in the church hall. Nine young participants (14 to 16 years old) were previously trained to use the Geodesign platform.

In order to make the experience more inclusive and democratic, improvements to the Geodesign interface were also tested. Web Maps were used as a system for visualization and interaction via the Internet. MapServer (an Application Programming Interface) and pMapper (an Internet Map Server), both available for free, were used to develop applications to generate interactive maps from an existing georeferenced database, in order to create a crowd mapping experience. MapServer is basically a tool for displaying geographic data via the web. pMapper works together with MapServer and is an application that facilitates the creation of web maps, making it possible for the user to visualize information and interact with them (Pimenta et al. 2012). This option made the representation and the process Geodesign model steps less complex and more interactive (Paula and Moura, 2018). The option of using open-source software (Figure 10) instead of the Geodesig Hub Platform, as in the previous workshop, provided greater replicability and usability for the target audience.

Figure 10. Dandara WebMap
Source: UFMG GeoLab, 2018

The use of a three-dimensional model of the settlement (Figure 11), built from a drone image captured by UFMG GIS Lab, allowed for dynamic navigation through the area, according to the user's points of interest. The drone model was hosted in the cloud and was easily accessed with the use of a link and internet access, on computer, tablet or cell phone. Interactive representations provide oblique viewing and complement the understanding of the cartographic zenithal view, favoring a better relationship between cartographic representation, mental maps and the reality (Masala and Pensa, 2016; Magalhães and Moura, 2020).

Figure 11. Dandara Community 3D Model
Source: UFMG GeoLab, 2018

Evaluation synthesis maps were produced on the following thematic systems: hazard areas, environmental values, sanitary fragility, transport, economic activities, public facilities, low-density housing, high-density housing, social use and recreation. These maps fed the GeoDesign Hub platform where projects and policies were designed through a collaborative process, followed by the presentation of the proposals and negotiation steps until a final consensus decision was reached for each system.

Fifty-seven people participated in the Geodesign workshops: 41 residents, 8 technical staff (city planning department, utility companies and the University) and 8 civil society representatives (ONGs, social movements and public officials). Following the Geodesign steps, the thematic proposals made by the ten initial groups were aggregated into six integrated plans according to the following approaches:

Plan One: Environment and Landscape
Plan Two: Roads and Transportation
Plan Three: Sanitation and Development
Plan Four: Land Use and Housing
Plan Five: Recreation and Culture
Plan Six: Economy and public Facilities

A presentation of the six design arrangements with all participants was the final activity of this first workshop day.

In the second workshop session, once the positive and negative aspects of the previous week's proposals were discussed, six new integrated plans were prepared by the same groups. According to Steinitz (2012), the first proposal is usually not the best, as the participants are still understanding the process. After listening to the other group's arguments, they feel the need to review their plans. Then, the six groups were regrouped into 3, according to similarity of the plans' proposals, since negotiation is facilitated when groups that think in a similar way are put to work together, leaving major differences to be confronted afterwards (Steinitz, 2012). Three new designs emerged from this new arrangement and once again another break for a broader community meeting took place in the following week.

At the final decision stage, however, when a single design proposal was supposed to emerge from the negotiation of the three previous plans, there was an impasse in the decision-making process involving the displacement of housing units that occupy the protection buffer of watercourses and flood hazard areas and the need for resettlement with higher density buildings somewhere in the community or in a nearby location. Being considered a very sensitive decision, impacting over 150 families, the final plan did not reach consensus within the workshop period.

Due to this controversy, lack of funds and legal warranties regarding the resettlement solution for the population to be removed, the Dandara Plan was only completed in December 2020, presenting two different scenarios with regards to the occupation along the streams and plains: 1. removal of all dwellers to create a conservation area along the watercourse's protection buffer and flood hazard areas; 2. removal of dwellers occupying just the flood hazard area. Both scenarios are still dependent on further technical hydrologic and legal studies to be accomplished.

Meanwhile, the three Urban Regularization Plans developed by the analog approach were concluded in 2018 and two others (Novo Tupi/Novo Lagedo and Jardim Getsêmani) started in 2020 are being carried out under very slow schedules and facing many difficulties due to the Covid-19 pandemic. Consulting planners involved in developing these plans were interviewed about major difficulties and shortcomings of using the traditional analog method and their main remarks are summarized as the following:

Table 5. Major shortcomings of using the traditional analog method for PRU development

Citizen's participation	Residents have difficulty understanding maps and technical issues
	Top-down approach leading to passive participation: people of the place are mostly listeners than decision makers.
	Rigid formats leading to reactive participation. Choice of alternatives restricted to predefined solutions. There is no co-creation.
	Polarization in participation: only major interests are represented in the Reference Group.
Schedule	The traditional methodology presupposes the carrying out of long and detailed diagnoses, which end up making the process very long-lasting (24 to 26 months), expensive and time-consuming.
	The long duration reduces the engagement of the population throughout the process and makes mobilization more difficult.
Complexity	The technical products are very dense and have a difficult vocabulary for the population to understand. Even graphic products have very technical language.

Source: Material from the authors, 2021.

On the other hand, it's been a general perception among interviewees that the pandemic somehow accelerated the use of information and communication technologies such as virtual meetings involving the technical team and people of place; the use of 3D models, virtual presentations and videos to present inventories and other plan steps; surveys accomplished by cell phones rather than in person. They also state that, given the limitations of face-to-face meetings and surveys, requirements by the public authorities who hire those plans have become less rigid, in favor of reducing complexity of the texts and figures, presenting the results in a more graphic way, speeding up the process and enhancing the general public's understanding. However, there were no changes in the general structure of the plans.

Although some people in the settlements have no access to computers nor were trained to use one, most residents could participate in the meetings by cellphone, using the Google Meets Platform. Easy access to the platform ended up contributing to a greater representation of residents in meetings held during the pandemic, than in those held in person in previous plans. At the first meeting, in which the PRU Reference Group of the Hosana settlement was formed, 33 people attended, at Nova Santa Cecília 44 and Jardim Liberdade 125. In Novo Tupi, the same kind of meeting, held on a virtual basis, 128 residents were present.

SOLUTIONS AND RECOMMENDATIONS

The use of the Geodesign Framework allowed for an interactive participatory process involving the simultaneous evaluation of alternative actions as well as a series of possible changes that may be proposed by community members to match local demands and expectations as well as follow up by local social organization groups and leaderships.

Even though the Maria Tereza workshop was conceived as an academic and technical experiment, it involved public officials from different sectoral planning departments, faculty members and private consultants who are usually in charge of developing these plans. The methodology proved to be very effective as an integration tool, enhancing group work and interdisciplinary practice (Moura, Marino, Ballal, Ribeiro & Motta, 2016) and was improved through the Dandara planning experience.

Public participation in PGE and PRU development may be improved even more in all phases but especially at the decision-making process when priorities are supposed to be defined. This planning stage is strongly dependent on community support for fundraising, implementation and maintenance. Not to mention the land regularization component that usually has a very important social, political, legal and compromising component.

As highlighted by Fonseca (2016), the Geodesign methodological framework relates to an open system that is subject to data inputs and outputs throughout the whole process. This allows for changes in both physical designs of proposals as well as redefinition of goals and objectives, which makes the planning process very dynamic.

The method also allows for a systematic and critical assessment of each planning step, integrating inventory, diagnoses and proposal phases as open systems that may receive feedback and alter the initial concepts of a given intervention at any moment (Fonseca, 2016). This way, planners and other stakeholders get involved in the planning process in a more balanced way, even though Steinitz (2012) highlights the need for a more intense participation of planners in the change models generated from the fourth step on.

As a result of the first workshop, the main positive aspect pointed out by the participants was the reduction of costs and optimization of outcomes. This is mostly due to the process speed involving the decision-making and assessment procedures that occur in collaborative and simultaneous ways, avoiding the traditional back and forth movement of paper documents between consultants and public officials responsible for validating each planning step. The Dandara experience proved the use of Geodesign may better balance and fasten decisions on design alternatives but when it comes to involve evictions and displacement of residents, it's very difficult to reach consensus and conflicts must be faced also on political and legal basis.

In Maria Tereza, on the other hand, the short time assigned for discussions was pointed out as a negative aspect of the workshop dynamics according to the participants, which is contradictory to one of the method's main advantages. This indicates the need to save more time for group work in each of the three iterations. Consequently, the methodological shift to the use of the Geodesign framework for the development of PGE and PRU will potentially increase opportunities and save time and resources used in collaborative tasks as opposed to expensive and time-consuming technical diagnosis, as summarized in Table 6. In the Dandara plan, the larger residents meetings held weekly at the local church after each Geodesign workshop provided the expanded discussion and evaluation of alternatives.

The use of a Geodesign platform as in Maria Tereza or WebMaps as in Dandara may also lead to a more equitable participation among community residents and other stakeholders that take part in the planning process.

Table 6. Comparing the traditional comprehensive planning method and the Geodesign framework applied to PGE and PRU development

Comprehensive Planning Method	Geodesign Framework
Lack of fast updating leading to fast obsolescence of plans	Opens system allowing for simultaneous inputs and outputs
Standardized methodology requiring an extensive inventory	Easily adapted method according to the study area reality
Little integration among planning phases and no feedback	Repeating iterations among phases of why, how, when and where to intervene
Time consuming and expensive: too much emphasis on diagnosis, little time for proposals	Simplified diagnosis reducing time and money consumption
Difficult community engagement on a voluntary basis for such a long period	Simultaneous and collaborative decision-making process with immediate evaluation of alternative solutions and their impacts
Little or no knowledge of plans or project goals by the community	Transparency: a democratization of the whole planning process
Final products: technical reports	Simple diagrams and easy to understand color coded maps
Final plan oriented	Decision making process oriented

The Dandara workshop introduced innovations to the representation, process and evaluation stages of the Geodesign method and also allowed the community to better understand the technical and legal constraints related to occupying environmentally sensitive areas (Monteiro et al. 2018). This awareness, however, was not sufficient to overcome basic shelter needs, access to services and job opportunities as well as neighborhood and family ties that are also at stake when it comes to decision making processes.

Both experiences demonstrate that the use of Geodesign brings greater agility, flexibility, and involvement of residents in the planning process, being highly recommended as a powerful tool to enhance citizens participation in planning processes. However, consultant interviewees pointed out that both analog and digital approaches, as well as PGE and PRU methodologies are too much focused on territorial issues such as hazard areas, infrastructure and service gaps, leaving social economic development and land tenure issues unsolved.

This shows the importance of ensuring greater flexibility to the planning process, also making residents aware of the technical and legal motivations for the improvements proposed in the projects. In order to make residents become real decision makers, it is also necessary to improve community cohesion and social development, through capacity building and income improvement programs, towards some degree of autonomy. Only this way, they will be able to take and share greater responsibility for their decisions.

FUTURE RESEARCH DIRECTIONS

Academic research and practice related to the inclusion of social economic themes in plans can be an important contribution to the development of Geodesign Framework to fulfill gaps observed in this chapter.

Also, research projects combining geovisualization strategies, collaborative mapping and participatory planning represent an important direction for futures development of the geodesign framework.

Climate change and local adaptative capacity to face their impacts, especially in poor communities, as well as participation difficulties imposed by the Covid 19 pandemic are issues to be enhanced in both academic and planning practice agendas.

CONCLUSION

According to Campagna (2016), Geodesign provides for a democratic planning environment that is based on the strength and the transparency of information related to choices made during the decision-making process. Collaboration that takes place within a Geodesign framework goes beyond simply information sharing among professionals within a multidisciplinary work. It implies the community's participation in all phases of the planning process, interacting with planners and other stakeholders in very active and diversified ways (Steintz, 2012). This way the quality of the participation process increases even though the quantity of time consumed in participatory events may decrease, also reducing risks of citizens losing interest or becoming absent along the planning process.

Freitas (2020) highlights the importance of Geodesign for macro-scale planning based on existing spatial data available, in order to allocate all the time to prepare and resources used in comprehensive traditional plans for citizens participation and collective construction of proposals and consensus building on priorities. Both experiences reported on this chapter brought evidence of how the use of information and communication technology may enlarge and enhance citizens participation in planning processes.

It is important to improve ICT planning support tools such as the Geodesign framework, but also our traditional planning methods, themes, strategies and approaches, from the citizen's perspectives, creating inclusive and collaborative environments based on their social practices and values, particularly in vulnerable communities.

ACKNOWLEDGMENT

This research was supported by CNPq [project 401066/2016-9] and FAPEMIG [PPM-00368-18] and had the partnership of Prefeitura Municipal de Belo Horizonte and Praxis Projetos e Consultoria Ltda.

The chapter builds upon an article published in the International Journal of E-Planning Research (IJEPR).

REFERENCES

Araujo, R., Moura, A., & Nogueira, T. (2018). Creating Collaborative Environments for the Development of Slum Upgrading and Illegal Settlement Regularization Plans in Brazil: The MariaTereza Neighborhood Case in Belo Horizonte. *International Journal of E-Planning Research, 7*(4), 25–43. doi:10.4018/IJEPR.2018100102

Arnstein, S. (1969). A ladder of citizen participation. *Journal of the American Institute of Planners, 35*, 216-224.

Batty, M. (2013). Defining Geodesign (=GIS + design?). *Environment and Planning. B, Planning & Design, 40*(1), 1–2. doi:10.1068/b4001ed

Bedê, M. C. (2015). Habitação de interesse social e reabilitação de assentamentos precarious. In *Entre o urbano, o social e o ambiental, a Praxis em perspectiva* (pp. 259–277). Gaia Cultural.

Campagna, M. (2016). Sistemas de Suporte ao Planejamento (Planning Support Systems): Retrospectivas e Prospectivas. In *Tecnologias de geoinformação para representar e planejar o território urbano* (pp. 217–248). Interciência.

Conti, A. (2004). A política de intervenção nos assentamentos informais em Belo Horizonte nas décadas de 1980 e 1990 e o "Plano Global Específico". *Cadernos de Arquitetura e Urbanismo, 11*(12), 189–216.

Drumond, W. (2017). *Entendendo o PRU como novo instrumento de planejamento para assentamentos precários de Belo Horizonte. Unpublished Final Undergraduate Research Project. Escola de Arquitetura, Universidade Federal de Minas Gerais*. UFMG.

Edelenbos, J. (1998). Article. In R. Monnikhof (Ed.), Spanning in interactie: een analyse van interactief beleid in de lokale democratie (pp. 11–48). Instituut voor Publiek en Politiek.

Edelenbos, J., & Monnikhof, R. (1977). Lokale interactieve beleidsvorming, Lemma, Utrecht. Cupps, D.S., 'Emerging problems of citizens participation'. *Public Administration Review, 37*, 478–487.

European Commission. (2002). *Guidance document on public participation in relation to the water framework directive. Active involvement, consultation, and public access to information.* Annex I Public participation techniques. Author.

Fonseca, B. (2016). Geodesign: Conceitos e arcabouço Metodológico. In *Tecnologias de geoinformação para representar e planejar o território urbano* (pp. 217–248). Interciência.

Freitas, V. (2020). *Geodesign: uma alternativa para criação de propostas coletivas para assentamentos precários em Belo Horizonte* (Master's thesis). NPGAU, Universidade Federal de Minas Gerais (UFMG), Belo Horizonte.

Geertman, S. (2008). Planning Support Systems (PSS) - a planner's perspective. In *Brail, R. K. Planning Support Systems for Cities and Regions* (pp. 213–274). Lincoln Institute.

Geertman, S., Toppen, F., & Stillwell, J. (Eds.), *Planning Support Systems for Sustainable Development* (pp. 451–467). Springer.

Habitação e Cidades, G. T., & Tonuci, J. (2020). *Casa Verde e Amarela: pontos críticos do novo programa habitacional do Governo Federal.* https://www.observatoriodasmetropoles.net.br/casa-verde-e-amarela-pontos-criticos-do-novo-programa-habitacional-do-governo-federal/

Harris, B., & Batty, M. (1993). Locational models, geographic information and Planning Support Systems. *Journal of Planning Education and Research, 12*(3), 184–198. doi:10.1177/0739456X9301200302

Horizonte, B., & Municipal, C. (2020). Relatório Final do Grupo de Trabalho da Comissão de Direitos Humanos e Defesa do Consumidor sobre Direito à Moradia. Academic Press.

Kapp, S. (2012). *Arquitetos nas favelas: três críticas e uma proposta de atuação.* [Paper presentation]. IV Congresso Brasileiro e III Congresso Ibero-Americano Habitação Social:ciência e tecnologia "Inovação e Responsabilidade". Florianópolis: UFSC.

Lei Federal 10.257 (O Estatuto das Cidades) (2001), Regulamenta os arts. 182 e 183 da Constituição Federal, que estabelece diretrizes gerais da política urbana e dá outras providências. Brasília.

Lei Federal nº 13.465 (Regularização Fundiária) (2017). Dispõe sobre a regularização fundiária rural e urbana, sobre a liquidação de créditos concedidos aos assentados da reforma agrária e sobre a regularização fundiária no âmbito da Amazônia Legal; institui mecanismos para aprimorar a eficiência dos procedimentos de alienação de imóveis da União e dá outras providências. Brasília.

Lei Municipal 7.166 (Lei de parcelamento, ocupação e uso do solo) (1996). Estabelece normas e condições para parcelamento, ocupação e uso do solo urbano no município. Belo Horizonte.

Lei Municipal 9.959 (Altera a Lei n° 7.165, de 27 de agosto de 1996, a Lei n° 7.166, de 27 de agosto de 1996, estabelece normas e condições para a urbanização e a regularização fundiária da Zona de Especial Interesse Social - ZEIS, dispõe sobre parcelamento, ocupação e uso do solo nas Áreas de Especial Interesse Social - AEIS, e dá outras providências.) (2010). Belo Horizonte.

Magalhães, D., & Moura, A. (2020). Aerial Images and Three-Dimensional Models Generated by RPA to Support Geovisualization in Geodesign Workshops. Lecture Notes in Computer Science, 12252, 296-309.

Maguire, D. J., Goodchild, M. F., & Rhind, D. W. (1991). *Geographical Information Systems: Principles and Applications* (Vols. 1–2). Longman Scientific and Technical.

Masala, E., & Pensa, S. (2016). O papel da visualização no Planejamento Urbano: uma abordagem a partir dos conceitos por trás da imagem espacial. In *Tecnologias de geoinformação para representar e planejar o território urbano* (pp. 34–47). Interciência.

Mello, I. (2009). *O espaço da política e as políticas do espaço: Tensões entre o programa de urbanização de favelas "Vila Viva" e as práticas cotidianas no Aglomerado da Serra em Belo Horizonte* (Master's thesis). IGC, Universidade Federal de Minas Gerais (UFMG), Belo Horizonte.

Miller, W. R. (2012). *Introducing Geodesign: The concept Director of Geodesign services*. ESRI Press.

Ministério das Cidades. (2004). *Plano Diretor Participativo: guia para a elaboração pelos municípios e cidadãos*. Ministério das Cidades & Confea.

Ministério das Cidades. (2010). *Guia para o Mapeamento e Caracterização de Assentamentos Precários*.

Ministério do Desenvolvimento Regional. (2020). *Regularização Fundiária e Melhoria Habitacional*. https://www.gov.br/mdr/pt-br/assuntos/habitacao/casa-verde-e-amarela/regularizacao-fundiaria-e-melhoria-habitacional

Monteiro, L., Moura, A., Zyngier, C., Sena, I., & Paula, P. (2018). Geodesign Facing the Urgency of Reducing Poverty: The Cases of Belo Horizonte. *DisegnareCon*, *11*(20), 6.1-6.25. Available in: http://disegnarecon.univaq.it/ojs/index.php/disegnarecon/article/view/362

Moura, A. (2019). O Geodesign como processo de cocriação de acordos coletivos para a paisagem territorial e urbana. In Planejamento e gestão territorial: o papel e os instrumentos do Planejamento Territorial na Interface entre o Urbano e o Rural. Criciúma - SC: UNESC.

Moura, A., Marino, T., Ballal, H., Ribeiro, S., & Motta, S. (2016). Interoperability and visualization as a support for mental maps to face differences in scale in Brazilian Geodesign processes. *Rozwój Regionalny i Polityka Regionalna*, *35*, 89–102.

Nascimento, T. (2011) *Participação institucionalizada e resistência popular na intervenção em favelas: o caso Vila Antena. Belo Horizonte* (Master's thesis). NPGAU, Universidade Federal de Minas Gerais (UFMG), Belo Horizonte.

Paula, P., Moura, A., & Patata, S. (2018). *The application of Geodesign in a Brazilian illegal settlement: Participatory Planning in Dandara Occupation, Case Study* [Paper presentation]. Input. Available at https://www.researchgate.net/publication/327446994_The_application_of_Geodesign_in_a_Brazilian_illegal_settlement_PARTICIPATORY_PLANNING_IN_DANDARA_OCCUPATION_CASE_STUDY

Peets, J., & Leach, B. (2000). *Evaluating methods for public participation: Literature review. R&D technical Report: E135.* Environment Agency.

Salomão, T. (2016) *Linguagem Técnica e (Im)Possibilidades para a Produção Democrática do Espaço Urbano: uma análise a partir de duas experiências participativas em Belo Horizonte* (Master's thesis). UFMG. Belo Horizonte.

Santos, O., & Montandon, D. (2011). *Os Planos Diretores Municipais pós-estatuto da cidade: balanço crítico e perspectivas.* Letra Capital, Observatório das Cidades & Ippur/UFRJ.

Silva, M. (2013) *Água em meio urbano, favelas nas cabeceiras* (Doctoral dissertation). NPGAU, Universidade Federal de Minas Gerais (UFMG), Belo Horizonte.

Steintz, C. (2012). *A Framework for Geodesign.* ESRI Press.

Streich, B. (2004). Information, Wissen, Planungsprozesse und Macht, Die Wissengesllschaft als Herausforderung an die Planungsethik. In Ethik in der Raumplanung, Zagänge und Reflexionen Akademie für Raumforschung und Landesplanung, Forschungs - und Sitzungsberichte (pp. 31-49). Hannover: Band 221.

Urbel. (2021). *PRU: Plano de Regularização Urbanística.* https://prefeitura.pbh.gov.br/urbel/pru-plano-de-regularizacao-urbanistica

Van Den Brink, A. (2007). *Imaging the future geovisualization for participatory spatial planning in Europe.* Wageningen Academic Publishers. doi:10.3920/978-90-8686-625-0

Zyngier, C., Moura, A., Araújo, R., & Carsalade, F. (2016). Geodesign in Pampulha cultural and heritage urban area: Visualization tools to orchestrate urban growth and dynamic transformations. *Rozwój Regionalny i Polityka Regionalna, 35*, 103–117.

KEY TERMS AND DEFINITIONS

Collaborative Planning: Is a participatory approach to design. From governance systems to architectural drafts, collaborative planning seeks to include as many stakeholders as possible in the decision-making process on choosing alternative solutions, defining priorities, and evaluating outcomes. Numerous facilitation techniques are used to encourage consensus and cooperation.

Favelas: Are how slums or shantytowns are called in Brazil. They are informal settlements usually located within or on the outskirts of the country's large cities, occupying steep slopes, flood hazard areas and other inadequate land for urban development. A favela typically comes into being when squatters occupy vacant land and self-construct shanties of salvaged materials. However, they soon become overcrowded and consolidated communities with severe infrastructure and sanitation gaps, poor public services and housing conditions, as well as and economic social problems.

Geodesign: A methodological framework comprising a set of concepts and methods used to involve different stakeholders and professionals in collaboratively designing and achieving consensus on the optimal solution for spatial challenges in the built and natural environments, utilizing all available techniques and data in an integrated process. It includes project conceptualization, analysis, design specification, stakeholder participation and collaboration, design creation, simulation, and evaluation (among other stages). Geodesign builds greatly on a long history of work and research in geographic information science, computer-aided design, landscape architecture, and other environmental design fields. See for instance, the work of Ian McHarg and Carl Steinitz.

Geovisualization: A short for geographic visualization, also known as cartographic visualization. It refers to a set of tools and techniques supporting the analysis of georeferenced data using interactivity and collaboration. Traditional maps have a limited exploratory capability: they are statical, the graphical representations are exclusive linked to the geographical information beneath, and the interpretation depends on a previous cartographic knowledge. Geovisualization allow for more interactive maps; including the ability to show off different layers of the map, to zoom in or out, and to change the symbology, usually on a computer, cellphone, or tablet display.

Illegal Settlements: Areas resulted from an irregular land division. Differently of the favela situation, there are lots and streets defined, and land are sold in the informal market. This means families who occupy that land paid for it and feel they are owners, even though they don't have legal ownership documents. Because the implementations of this settlements are irregular, they do not follow official design criteria (minimum lot area, minimum street width, maximum slope and so on), do not obey environmental constraints (steep slopes, forested areas, springs and water bodies, flood hazard areas, etc.), nor have the basic urban infrastructure that, according to Brazilian urban legislation, is due to real estate developers to implement.

Land Regularization: With regards to urban illegal settlements, land regularization refers to public policies meant to recognize land rights and provide land tenure to dwellers living in informal communities. It is desirable that these actions be part of urban upgrading programs, in order to provide the informal areas with access to municipal services and better living conditions.

Slum Upgrading: Public policies designed to provide or improve urban infrastructure and housing conditions as well as recognize land rights and promote land tenure safety in informal settlements such as favelas and illegal developments. Slum upgrading programs and projects are usually preceded by planning guidelines and actions. It is important to create a collaborative environment where slum dwellers and the government officials and planners can engage in a dialogue about their community needs and possibilities.

ENDNOTES

1 Due to political problems as well as difficulties associated to the Covid 19 Pandemic, the 2020 Demographic Census has been postponed to 2022.

2 Geodesignhub is software for collaborative geodesign. It enables teams to create and share concepts, to design collaboratively, and to receive change-assessments instantly – all in a highly synergetic, efficient, and easy to use environment. It is especially effective in the early stages of a complex study or project, when many alternatives must be rapidly created and considered.

Chapter 6
Bridging the Gap Between Strategic Environmental Assessment and Planning:
A Geodesign Perspective Revised

Michele Campagna
https://orcid.org/0000-0001-7489-7330
University of Cagliari, Italy

Elisabetta Anna Di Cesare
University of Cagliari, Italy

Andrea Matta
University of Cagliari, Italy

Matteo Serra
University of Cagliari, Italy

ABSTRACT

Current planning practices in Europe are affected by regulations aiming at environmental protection and risk reduction; however, planners face difficulties in implementing the norms. This chapter discusses the influence of the introduction of strategic environmental assessment and of spatial data infrastructures in the EU. To address these issues, the geodesign approach is proposed, firstly analyzing its relationships with the current planning regulations at the regional level in Italy and then its opportunities for application in practice with regard to the Sardinian case study (Italy); secondly, examples of Planning Support System are proposed and tested with real-world case studies to implement the core phases of the geodesign approach. In addition, the benefits for SEA of developing state-of-the-art workflows in geodesign collaborative workshops are discussed. The chapter concludes suggesting that geodesign may fruitfully address some of the major issues in the current planning and SEA practice, taking full advantage of the newly available resources of SDI.

DOI: 10.4018/978-1-7998-9090-4.ch006

INTRODUCTION

Over the last two decades, innovations in environmental protection regulations have started to affect the planning practice in Europe, as their principles were transposed into national and regional legislation frameworks in the Member States.

Firstly, the European Directive 2001/42/EC introduced the Strategic Environmental Assessment (SEA) of plans and programs, including regional, urban and sectorial plans. The SEA is a procedure intended to inform the plan-making process since the early stages of elaboration and before plan adoption, with the aim to guide territorial development according to sustainable principles, including the protection of environmental resources, the social and cultural welfare of the communities, the promotion of savvy decision-making and the participation among the stakeholders and the communities involved in the decision process. Secondly, the European Directive 2007/02/EC, currently under implementation, established the INfrastructure for SPatial InfoRmation in Europe (INSPIRE) in order to enable sharing of territorial information among public authorities at all levels, to better facilitate its public access across Europe. This way, while the SEA Directive brought potential innovation into the plan-making process, the INSPIRE Directive contributed to the digital uptake of the planning media.

Nevertheless, while these Directives can be seen as two drivers of innovation and they are indeed starting to substantially affect the way the plan-making process is carried out, difficulties in properly implementing these principles are still widely found. In operational terms, difficulties are found either in developing transparent and democratic informed decision-making processes and in taking advantage of the new digital sources of information which are growingly made available to support data integration, analysis, design, and impact assessment.

To address these current challenges, the authors argue that the geodesign methodological approach may contribute to address some of the current critical issues and pitfalls of SEA in spatial planning, fostering better integration of the principles of recent regulations, and successfully exploiting the benefits introduced by new technologies and digital data in planning. To this end, after a general discussion of the potential of geodesign methods with regard to current planning regulations and practice, examples of the application of Planning Support System (PSS) integrating a range of methods and tools is presented, giving viable examples on how it is possible to provide planners with novel methods and tools for representing and analysing the territory and to improve informed collaborative processes among stakeholders.

The chapter is organized as follows: the first section illustrates in detail the innovation potential introduced in spatial planning by the SEA and the INSPIRE Directives. The second section first gives a brief description of the geodesign methodological approach based on literature review and subsequently highlights current relationships between the principles of the geodesign approach and the normative principles of planning regulations (with reference to European and Italian laws) as well as to current planning practices (with reference to the Sardinia case study in Italy). In the third section, after a brief definition of Planning Support Systems, an operational demonstration of methods and tools addressing major SEA tasks in local land-use planning is given. The overall aim of the chapter is to investigate to what extent the geodesign approach and PSSs may support planners during the decision-making process, bringing a new systematic and technology-based approach to spatial planning, both in terms of process and tools. Bridging the gap between theory and practice is an urgent issue and testing different viable ways to address it may constitute a valuable knowledge base contributing to bring innovation to present and future planning practice.

INNOVATION IN SPATIAL PLANNING

The SEA Directive

The European Directive 2001/42/EC introduced the Strategic Environmental Assessment as a parallel evaluation process to the preparation of plans and programmes that may have significant impacts on the environment. The main goal of SEA is to ensure that environmental considerations and sustainability principles are taken into account since the early stage of plan-making, to foster informed, responsible and transparent decision-making processes (Fundingsland Tetlow & Hanusch, 2012).

Given that the evaluation process is intertwined with the preparation of the plan since the beginning of the procedure up to the choice of its objectives, the strategies to achieve them, their impacts and finally the monitoring of the expected results, SEA deeply altered (at least in its intents) the traditional way to carry out the planning process. Moreover, it promotes approaches addressing such concepts and principles as transparency, participation and collaboration among decision makers, planners and experts from various disciplines, and the prospective involved stake-holders.

However, in spatial planning practice -unfortunately- SEA has not always led to satisfactory results (Parker, 2007; COWI, 2009; Brown & Therivel, 2000), for there is often a lack of shared vision on how to implement its principles in terms of methods and tools. According to Fisher and Gazzola (2006) the effectiveness of SEA is related to the presence of specific context conditions, including a well-structured regulatory apparatus, a satisfactory inter-institutional cooperation, and a strong attitude towards public participation.

As demonstrated by Fisher (2010), shortfalls in demonstrating how the environmental sustainability issues and the participatory processes **inform the design of the plan alternatives** are frequently found. Moreover, considerable deficiencies often persist during the generation of plan alternatives (Fundingsland Tetlow & Hanusch, *ibidem*), which are specifically required by the Directive, and during the post-decisional phase, in which the plan initiatives are implemented (Gachechiladze-Bozhesku & Fischer, 2012). As a matter of fact, the implementation of plan actions should be repeatedly monitored in order to ensure proper achievement of the proposed strategies, but this step is often underestimated due to lack of specific regulatory requirements and of a clear awareness of the benefits that it could bring about (Gachechiladze- Bozhesku & Fischer, *ibidem*).

Another inherent characteristic of SEA, useful to ensure the democratic nature and transparency of the process, is public participation, required in the form of consultations of environmental authorities, stakeholders and citizens, during the elaboration of the plan and of the Environmental Report (ER). Indeed, the territorial knowledge to support decision-making should be not only correct and relevant, but also acceptable to stakeholders, and for this reason it is important that they are called to contribute to its construction (Runhaar, 2009). Currently in planning practices, on the one hand, we assist to a significant increase in cooperation between public authorities, on the other hand, this collaboration is not always followed by an equivalent progress in public consultation that should also involve citizens in the process (Rega and Bonifazi, 2014), due to lack of specific guidelines to better address participatory practices and a lack of public administration willingness (Rega and Baldizzone, 2015).

Moreover, in order to orient the planning process results towards "good governance", the environmental assessment should be deeply consolidated internally (Zoppi, 2012), but such an integration is often difficult because of the many public and private stakeholders involved and because of management complexity that makes the process cumbersome and non-transparent. As a consequence, a strong misalignment between SEA and the plan process is frequent (Zoppi, *ibidem*), with a consequent reduction of SEA effectiveness.

Although many National and Regional Authorities in Europe issued guidelines for a more successful implementation of SEA principles (Partidario, 2012; De Montis *et al*, 2014a), some operative steps of SEA still need more detailed definition in order to address such issues as the adoption of more systematic approaches for the identification of alternative future development scenarios, the identification of methodologies to guide collaboration and participation and the documentation of the reasons which lead to the selection of the preferable scenario (Gonzalez et al, 2015). As will be discussed in this chapter, geodesign offers reliable methods and tools to address the issues above, including informed and transparent decision-making, collaborative design alternative creation and assessment, negotiation to achieve consensus on final decisions.

The INSPIRE Directive

Over the last two decades or so, advances in Information and Communication Technologies enabled broader accessibility to digital spatial data (or geographic information) (Aalders and Moellering, 2001) and Spatial Data Infrastructures (SDI) have flourished worldwide.

The European Directive 2007/02/EC aims at supporting informed and responsible decision-making in domains such as spatial planning, environmental protection and economic development, through the creation of an Infrastructure for Spatial InfoRmation in Europe (INSPIRE) for sharing spatial data among public and private bodies, and with the broader public (Craglia and Annoni, 2007). According to the INSPIRE interoperability principles, public authorities in the Member States should collect and give open online access to 34 spatial data themes, including a variety of territorial themes of relevance to planning. Digital data and metadata are created according to common specifications, in order to facilitate the diffusion and seamless integration of geographic information in transboundary contexts.

The key components for INSPIRE implementation are: (i) metadata, for cataloguing data, (ii) spatial data, spatial data services and technologies, (iii) policies, protocols and agreements on sharing and access, (iv) people (i.e. data providers, service providers and users), (v) procedures and standards (Steiniger and Hunter, 2012). Access to data and services is supported by geoportals, geoweb-technologies able to give free public data access to potential users (Maguire and Longley, 2005).

The INSPIRE Directive contributed to a great extent in promoting the diffusion of SDIs in Europe (Campagna and Craglia, 2012), leading to opportunities and benefits for planners, environmental impact assessment analysts, and local administrations primarily at the local and regional level (Craglia and Campagna, 2009). Indeed, improvement of the quality and quantity of available information may positively affect the analysis and the comprehension of urban and environmental phenomena, offering unprecedented opportunities to enhance informed and responsible decision-making. In Italy, as in many other European Countries, several Regions developed their own Regional SDI and some Italian regional planning laws required their use in planning. The Lombardy Regional Territorial Government Law n. 12/2005, for example, recognized the role of such infrastructures as a reference platform supporting spatial planning processes. In Sardinia the Regional SDI is made up of two interacting elements: the Information System for supporting data management, and the Spatial Data Infrastructure (Vandenbroucke, 2011). It currently collects and gives access to more than 500 spatial data layers, including the geographic, historical, urban and environmental themes. Their accessibility is granted by interoperability Open Geospatial Consortium (OGC) services (i.e. download, Web Map Services and Web Feature Services).

Despite the diffusion of SDIs, their value for spatial planning is still limited due to the widespread lack of skills by professionals in applying advanced spatial analysis techniques. As such, in many contexts, spatial data are used at best to produce thematic maps and the planning documents. The advanced use of Geographic Information Systems (GIS) to inform design and support impact assessment is unfortunately still very limited in practice.

GEODESIGN: NEW OPPORTUNITIES FOR THE PLANNING PRACTICE

Geodesign can be defined as the act of design in the geographic space (Artz *et al.*,2010) and represents a new methodological approach to decision-making. It applies system-thinking and it is supported by spatial information systems and science. Such an approach involves the combination of the analysis of the territorial context with the design of planning proposals and their real-time impact simulation (Flaxman 2010; Artz *et al.*, 2013), using a synergistic iterative approach which integrates not only environmental issues, but also economic and social concerns (Dangermond, 2010). The performance of alternative scenarios can be visualised not only in the geographic space in form of maps, but also through charts, graphs and reports (Ervin, 2011), facilitating knowledge building and communication. Such techniques support the collaborative creation and assessment of spatial development alternatives and the design of improvement measures iteratively when necessary. Geodesign studies are normally carried out by multidisciplinary teams composed by planners and geographic information systems technologists, experts in natural and social sciences, and the local communities, with the purpose of achieving a holistic view of the different problems which may arise during a planning process, in a mutual learning process. Geodesign, therefore, acts as a design approach informed by sustainable development principles: it involves substantive objectives, such as environmental protection and improvement of quality of life of affected communities; procedural objectives, including the democratic nature of the process and its transparency; and instrumental objectives, such as exploiting the power of digital information and technologies to support informed decision-making.

Steinitz (2012) proposed a methodological framework (a.k.a. the Geodesign Framework, GDF) structured in six models and three iterations, in order to guide the practical implementation of the geodesign approach into regional and urban planning and design studies.

The first three models of the Steinitz' framework describe the current state of the territorial system. In their digital implementation, after an accurate analysis of the study area based on a selected set of relevant spatial data layers (Representation Model), the Process Model investigates the future development scenario of the territorial system without the implementation of any planning action (i.e. *do-nothing*). Afterwards, the ongoing dynamics are considered, in order to evaluate risks of and opportunities for change (Evaluation Model). The evaluation model may integrates technical criteria (e.g. environmental constraint) and political objectives (e.g. development in a given system), influencing the change model in an explicit, transparent and trackable way. The assessment phase is followed by the intervention phase in which, starting from the identification of possible alternative plans for the future (Change Model) and their environmental, economic and social impacts (Impact Model), it is possible to choose a shared development alternative as a result of an assessment of their coherence with the broader set of objectives expressed by the decision makers (Decision Model). This is normally done through negotiation, enabling to reach consensus in the face of multiple - often conflicting- objectives. A complete geodesign study should consist of three iterations: during the first one, which can be considered a scoping phase, models are built from the first to the last in order to identify the purpose of the study; in the second iteration, they are detailed in reverse order with the aim to define how to carry out the study in terms of methods and tools, which can be considered as the metaplanning phase (Campagna, 2014a; 2016). Finally, during the third iteration, models are implemented from the first to the sixth again to complete the study.

In operational terms, the GDF entails a rationale which is coherent with the spatial planning process as it is shaped by the SEA Directive, since it guides the study area analysis to inform the design from the early stage of the strategic decision-making process and keeps up with it in a non-linear way, enriching it through iterations (Campagna *et al.*, 2016). Furthermore, this coherence can be also observed comparing the tight relationships among the contents of the Environmental Report, as listed on the Directive's Annex I, and the structure and outputs of the GDF models (Campagna, 2014b). The key assumption is that the geodesign approach can contribute from an operational point of view to deal with many of the current issues and pitfalls encountered in urban and regional planning practices thus contributing to improve SEA.

In the light of the above premises, the GDF seems capable of introducing a new systematic approach to the spatial planning/SEA process, both in terms of workflows and tools.

In order to explore the similarities and differences between the geodesign approach and current planning practice, two analyses were conducted:

- A comparative analysis between geodesign principles and spatial planning laws (case study: Italy)
- A comparative analysis between geodesign methods and tools and planning practices (case study: Sardinia)

These are described in the next sections.

The GDF Principles in Spatial Planning Laws

While the geodesign proposition as an integrated framework is relatively new, many of its key concepts can be traced back in the planning and landscape architecture tradition. Hence, a number of them can be found in existing planning regulations, though often not in an integrated way. In this section a review of spatial planning laws from three Italian regions (Lombardy, Tuscany and Sardinia), that were considered innovative at the time of their adoption is presented, in order to check to what extent similarities between their content and the GDF key concepts can be found.

The regional regulations analysed include:

- Lombardy Regional Territorial Government Law n. 12/2005
- Tuscany Regional Territorial Government Law n. 1/2005
- Sardinian Regional Spatial Planning Law n. 45/1989 and Sardinian Regional Guidelines for SEA of Local Land-Use Plan (LLUP)

The comparative analysis shows that the Representation Model is one of the GDF models whose concepts and contents are more developed in the body of laws. This is not surprising since for the elaboration of urban plans it is commonly required the definition of the study area boundaries and the collection and analysis of spatial data, with the aim of creating a complete local knowledge base. All the spatial planning laws under analysis specify somehow the territorial analysis contents, including the description of geological and hydrogeological characteristics, the selection of natural and vulnerable sites and the description of historical settlements and mobility infrastructures. In addition, Tuscany and Lombardy laws also establish the creation of a Regional SDI as a common platform to be used for collecting and sharing spatial data among all the stakeholders involved in the planning processes. By contrast, the Sardinian regional law does not refer to any specific data management technology, since the establishment of the Sardinian Regional SDI came later the Regional Law, which dates back to 1989. The latter also requires the identification and analysis of other local legislations affecting the territorial context under analysis, and a coherence analysis between their objectives and the objective of the new plan. However, what was not clearly required is the relation between the Representation Model contents and its connection to the other models as expressed in the GDF and its iterations.

In line with the SEA Directive requirements, in the normative body under exam, the Impact Model is specifically required for the assessment of the potential positive or negative effects resulting from alternative plan designs and the identification of mitigation measures, if needed, in order to guarantee the protection of the territorial and environmental resources. However, the regulations do not refer to methods and tools for the assessment and comparison of impacts. The only exception was found in the Sardinian Regional Guidelines for SEA of Local Land-Use Plans, which suggest the use of tools such as impact matrices and overlay mapping. Still, no requirements on the use of advanced spatial analysis methods and tools was found, demonstrating a low awareness of planning professionals of the latest advances in the field. This is not surprising, given that in the Sardinian context, teaching of GIS methods in the academic planning curricula started only in the middle-2005, and a similar situation is found all over Italy; hence senior professionals who developed the guidelines may not be aware of the latest advances.

The identification of the main vulnerabilities and vocations of the territory or, in other words, the sensitivity of each area to hazard and its suitability for given uses, which can be considered an essential part of the Evaluation Model in the GDF, is only partially treated in planning regulations. Reference was found to the necessity to identify those areas which need to be changed or preserved, and to several functional suitabilities with regard to such objectives as environmental requalification, landscape management or ecological preservation. In addition, in this case only the Sardinian Regional Guidelines suggest a methodology to carry out this study, through the SWOT analysis method, with the aim to identify the strengths, weaknesses, opportunities and risks linked to the characteristics of the analysed territory.

It is not easy to find clear links to specific steps of the Process Model either, possibly because the laws do not provide clear guidance on the methodology to be adopted for the preparation of the plan. The only exceptions are generic references to the need to analyse the likely evolution of the environment before plan implementation, through the analysis of the territorial transformation dynamics. More specifically, the Sardinian Regional Spatial Planning Law proposes the analysis of population projections to estimate the housing demand, as a mandatory step for local land-use plan design, but this is clearly not enough to set change targets for all the systems

The Change Model and the Decision Model are certainly the models whose content is least treated within the body of legislation under analysis. Indeed, with the exception of a few references to the need to design a number of possible alternative scenarios during the preparation of the plan, there are not specified methods neither for their design nor to how to make a choice among them in the light of the objectives specified by decision makers. This can be considered one of the possible reasons why the elaboration of planning alternatives is often lacking in practice, as highlighted both by studies in literature and, as a detailed complement, in the second part of this study.

GD in Planning Practices

In this section, the results of the analysis of selected case studies of Sardinian Local Land-Use Plans and Environmental Reports are presented. The analysis aims at identifying which geodesign-like methods and tools, if any, are currently implemented in the planning practice in Sardinia. In this Italian region, Local Land-Use Plans of coastal municipalities have to be in compliance with the Regional Landscape Plan and the Regional Hydrological Risk Assessment Plan. At the same time they are subject to SEA, according to standard procedures. Despite the process of adaptation of the Local Land-Use Plans to the Regional Landscape Plan is a mandatory step for municipalities, only a few municipalities have completed it to date (less than 10% in 2021) since the Regional Landscape Plan adoption in 2006 (De Montis *et al.*, 2014b).

The analysis revealed that in all the case studies analysed in 2017 during the knowledge base construction, which aims to represent the environmental, economic and urban systems state in the municipality (i.e. Representation Model), the territorial context is represented through a set of cartographic representations often built using Geographic Information Systems, combing digital datasets retrieved from the Sardinian Regional SDI by local authorities. Moreover, a set of eleven environmental standard analysis sheets are always given in order to describe environmental components (e.g. air quality, flora and fauna, mobility, energy) through a set of indicators, as required by the Sardinian Guidelines. In the plan-SEA document, the local normative framework is also analysed by coherence matrices showing the consistency of the plan main objectives with the objectives of other overlapping plans.

The analysis of the territorial dynamics before the implementation of the plan (i.e. Process Model) is mainly carried out through predictive models based on the extrapolation of historical data and trends. In practice, in the case studies under exam the most common analysis is a forecast model of the socio-demographic dynamics, which enables the assessment of future housing demand within the municipal boundaries. Some case studies also integrate the quantitative population forecasting with spatial analyses describing processes geographically, in form of thematic maps or reports, depicting phenomena like soil erosion and traffic flows, or a specific hazard occurrence (e.g. landslide risk and potential instability of slopes), as required by the Regional Hydrological Risk Assessment Plan. In spite of all this, the Process Models cannot be considered much developed because not all the main system dynamics within the territory are always analysed systematically.

The assessment of the current territorial conditions (i.e. Evaluation Model) is often carried out through a SWOT analysis, which allows identifying possible Strengths, Weaknesses, Opportunities and Threats of the municipal area in form of textual tables or reports. Seldom, this analysis is supported by maps representing land suitability for selected individual land-uses (e.g. farming or breeding).

Despite the design of alternative development scenarios (i.e. Change Model) represents a key step for a robust SEA process, as required by the European Directive and the Sardinian Regional Guidelines, and as suggested by the technical literature (Partidário, 2012), the formulation of future alternative possible land use scenarios is mostly omitted. Only in a few case studies among those analysed, two plan alternatives are given, including the do-nothing alternative and the final plan. This can be considered a major methodological shortcoming in a SEA process implementation, which generates difficulties in the understanding of how the decision-making process is informed on the basis of the context analysis results.

Usually, also the next two models of the GDF are only partially implemented. Indeed, the aim of the Impact Model is to identify and evaluate the environmental impacts resulting from alternative plan choices, in order to identify the preferred one during the decision phase (i.e. Decision Model). While in all case studies impact assessment analyses are found as required by law, they only address the unique plan alternative, in order to identify its possible negative effects which need to be mitigated. Therefore, the comparison of the environmental impacts of different alternatives is always missing. The results of the impact analysis are usually reported in the form of impact matrices, showing on x-axis the plan actions and on y-axis the environmental components (e.g. air, water, soil). Checklists and indicators showing the expected change of some indicators values as a result of plan choices are also used (e.g. increase of air pollution or energy consumption). Only in few cases are these considerations represented in the geographic space through the overlap of the planned land uses and specific sensitive areas (e.g. protected areas). No reference to the use of advanced spatial analysis for impact assessment was found in any of the case studies analysed.

To summarize, this analysis identified several elements which may recall a geodesign methodological approach. However, not all the environmental analyses, as well as the representation of their results, are modelled in the geographic space as suggested by the geodesign approach. In addition, often analyses and evaluations are implemented in a comprehensive manner, lacking to address the complexity of territorial systems.

If the first phase of our study highlights several existing elements of coherence as well as shortcomings between the GDF models and the main normative principles of selected planning regulations from a theoretical perspective, the second phase identifies the methods and tools relating to a geodesign approach already in use in current planning practices. Such an analysis enabled to identify the potential for geodesign application to improve the practices, but also the current difficulties hindering it. On these premises, it is possible to try to fill the gap with a more advanced application of geodesign principles, methods and tools. The following section presents applied examples in this direction, thanks to the development and application of (geodesign) Planning Support Systems.

PLANNING SUPPORT SYSTEMS IN GEODESIGN PRACTICE

While Planning Support Systems have been proposed since the late 1980s to enable novel models of digital planning, their diffusion up until now has been rather limited in practice considering their potential.

Spatial planning encompasses a range of complex procedures in the attempt to manage territorial system dynamics according to development strategies (Couclelis, 2005). In addition, as Te Brömmelstroet and Schrijnen (2010) argued, the intrinsic complexity of planning processes is affected at the urban and regional level both by the increasing number of actors involved in the process and by their contrasting planning goals that evolve across different spatial scales. In such a context, advances of Information technologies (IT) and computer-aided planning, such as Geographic Information Systems, may help to generate innovative solutions for dealing with this complexity (Te Brömmelstroet *et al., ibidem*). However, despite a certain degree of diffusion of GIS and PSS among planners and practitioners, their efficacy in practice is still limited (Klosterman, 1999; Vonk *et al.*, 2008).

The idea of leveraging planning procedures with technological innovations is not new. In 1989 Harris (1989) gave an early definition of the concept of Planning Support Systems as an integrated system for supporting the planning tasks, coupling a range of digital models and computer-based methods. More operatively PSS were originally conceived as "a user-friendly microcomputer-based planning system, which integrates GIS, sketch tools and spatial models" (Harris, *ibidem*), aiming at dealing with the complexity of planning procedures, so helping to filling the gap in the integration of geo-technologies in planning practices (Geertman and Stillwell, 2004).

A PSS generally gives access to a range of geo-tools through a user friendly interface, in order to support the representation and management of spatial data (e.g. visualization tools), its analysis, to run simulation and forecasts, to design plan alternatives and to assess their impacts on the environment. Many PSS give the possibility to compare alternative scenarios in order to improve the collaboration among experts from different disciplines and the negotiation procedures among stakeholders, in order to produce consensus on design solutions in accordance with the complex system of planning goals. In addition, many PSS also offer reporting tools to present the results of the plan-making phases, supporting the transparent sharing of information about the process and its outputs.

Despite the potential of many existing PSS, their successful integration into planning practices seems to be far from realised (Geertman and Stillwell, *ibidem*; Vonk *et al.*, 2005). Pitfalls in PSS use are related both to the frequently poor flexibility of the PSS addressing different planning processes (Campagna *et al.*, 2006) and their high degree of complexity (Brömmelstroet, 2013).

These problems represent a barrier for the integration of PSS into practices, producing a cause-effect loop that generates a lack of knowledge in the planners community regarding their potential, with a consequent additional underuse of PSS in practice (Vonk, 2006). In an attempt to fill the gap between the planners need for technological support and the available innovations, in the last decades research on PSS produced a range of applications. The most successful examples are software suites such as CommunityViz (Placeways™), What If? (What if? ™) and INDEX (Criterion Planners™).

Among them, the CommunityViz architecture encloses a range of user-friendly dynamic tools to support the implementation of Land Suitability Analysis (LSA), sketch planning, real-time impact assessment through dynamic dashboards, and the comparison of different scenarios with automatic production of reports. These tools were tested by the authors in a workshop aimed at measuring their usability in a planning context that traditionally has a poor level of digital uptake, such as that described in the next section.

GEODESIGN AND PLANNING SUPPORT SYSTEMS: OPERATIONALIZING GEODESIGN TWIGS

A planning process from a geodesign perspective may be seen as an integrated study aimed at reaching consensus on how to improve a territorial system through changes. Although it is grounded on local values and decision models unique in their complexity and locality, a range of methods and tools can be identified which can find broad application in the development of the six models of Steinitz' framework.

In this section, the authors present a practical example on how they integrate a selection of methods in a PSS, aiming at implementing the geodesign methodological approach in a case study of Local Land-Use Planning in Gonnesa (Italy). The main goal of the study was to investigate, through a workshop, how such a PSS could support planners digital-wise during the decision-making process. The following paragraphs describe the technology in action with reference to the six geodesign framework models. Afterwards the results of the technology testing are discussed.

The Representation Model

As mentioned above, the purpose of the Representation model is to represent and visualise the study area from the perspective of the people of the place as this is encoded in the decision model aiming at facilitating them to develop a deeper understanding of the territorial dynamics. Several means can be used for this purpose: from traditional paper maps, which depict static territorial phenomena in a specific moment, to spatio-temporal GIS databases, which enable the interactive visualization of territorial dynamics. Techniques for representing the territory are continuously evolving (Kerski, 2013); nowadays, web-mapping tools allow merging GIS capabilities with user-friendly web interfaces, improving the efficiency of geographic information sharing within collaborative planning processes. This approach to data representation is an intrinsic characteristic of the web-based PSS developed in the Gonnesa case study for communicating and analysing spatial data during the Representation and Process Models development. The web-PSS was developed using ESRI technology (i.e. Story Map and the Web-App builder). It enables to describe and analyse the territorial context through the support of maps and interactive tools.

This web-PSS (Figure 1) can be considered as a local geoportal that creates a wide range of thematic maps combining a variety of spatial datasets retrieved from the regional SDI or from the local municipality. Each thematic map can be complemented by relevant documents and can be progressively enriched with new data, providing increasingly accurate and complete information. The Web-PSS represents the complexity of the territory in a user-friendly way and allows to perform queries or simple spatial analysis operations through automatic pop-up windows. During the workshop, the thematic maps were made available on the internet site of the Gonnesa Municipality, and used by the workshop participants as well as by other interested citizens.

The web-PSS allows managing the spatial information from different sources on a single platform, facilitating its readability by the interested citizens.

Figure 1. User-friendly representation of geospatial data during the plan-making process

While the adopted technologies are not new, their integrated use represents a viable example to give the Representation Model broader access to the involved parties and the broader public.

The Process Model

The Process Model aims at improving the comprehension of environmental and urban processes, investigating territorial dynamics through a range of methods and analyses, such as simple overlay or geoprocessing.

To this end, analytical functions were integrated in the Web-PSS, enabling users to combine different environmental (e.g. protected areas, natural landscape, hydrogeological hazards) and urban data (e.g. land-uses, roads network, historical goods) and to perform queries and spatial analyses on specific issues of interest (e.g. the number of historic sites in a given area) or to compare interactively different data themes. Besides offering geoprocessing functionalities, the Web-PSS displays spatial indicators dynamically calculated (Figure 2). The users can also turn on/off the layers and combine and analyse all or part of them in an easy way, with the aim of producing maps which represent individual phenomena or a combination of them. The interpretation of these phenomena helps participants to improve their awareness of the urban and environmental characteristics of the territory and formulate appropriate and informed strategies during the Change Model.

Indeed, the Web-PSS proved to reduce the difficulty of spatial data representation and sharing and made information available in a more systematic manner, as declared by the participants in the post-workshop survey.

Figure 2. The web-PSS supporting spatial indicator calculation

The Evaluation Model

The aim of the Evaluation Model is to assign a value to the current dynamics in order to recognise strengths and weaknesses of territorial resources and, subsequently, possible opportunities related to the economic, social and/or environmental change. The construction of the Evaluation Model entails the use of a wide range of geo-tools nested in a PSS.

To this end, a customized desktop PSS based on the CommunityViz software enables to build a tool for interactive Land Suitability Analysis (LSA), which was used to dynamically identify the areas suitable for given land uses, in accordance with a chosen set of criteria. More specifically, the LSA operates through two categories of parameters: constraints and factors. A constraint is a Boolean spatial criterion which represents a specific urban/environmental phenomenon that can be considered or not in the analysis (e.g. hydrogeological hazard areas, protected areas), while a factor is a spatial criterion that may assume different weights in the evaluation process according to its relevance (e.g. proximity to main roads, or the slope). When in the PSS a constraint is switched on/off or the weight of a parameter is changed through sliders, it produces a real time response on the evaluation maps and the suitability rate of the different locations for change. In the workshop, this interactive evaluation process fed a discussion about the significance of each parameter affecting the planning goal, promoting a constructive exchange of views among stakeholders with different priorities.

The results of the analysis are depicted in a map in grayscale, where the darker colours represent the most suitable areas and the lighter colours the least suitable one. Such an analysis often requires a large number of parameters and for this reason a GIS-based LSA helps planners to deal with the intrinsic complexity of spatial phenomena in a simpler way (Malczewski, 2004).

The output of these discussions is a set of evaluation maps (Figure 3) which, representing the most suitable areas for specific land uses, represents the input for the design activities and inform decision-making by collaborative discussion.

The Change Model

In order to investigate how the territorial context may be changed, the PSS uses a sketch planning tool, for altering the study area and visualising these changes in the geographic space in real time. This way, it simplifies the evaluation of their consequences (Heppenstall *at al.*, 2001) for supporting the decision-making process (Harris, 2001).

Indeed, the sketch planning tool gives the users the possibility to interactively sketch on a screen, using a digital pen, a set of planning proposals for a specific planning purpose, related both to their own conceptual ideas of future territorial developments (Al-Kodmany, 2002) and to the information coming from the analysis and evaluation of the territorial context. Despite sketch planning is considered an easy-to-use geo-tool, the successful integration into planning practices is influenced by different issues (Al-Kodmany, *ibidem*) such as its tendency to centralize the design activities around a single person in contrast to collaborative processes principles (Ligtenberg *et al*, 201). For this reason, this tool needs to be supported by innovative methods and instruments. In a planning process each designed alternative may represent the proposal of a different stakeholder (or a group of them) with different system of values and interests. This representation method gives the stakeholders the opportunity to compare and edit their own proposal and finally discuss for the most suitable solution to reach the overall agreed planning goals.

Figure 3. The evaluation maps produced through the LSA feeds the sketch planning tool

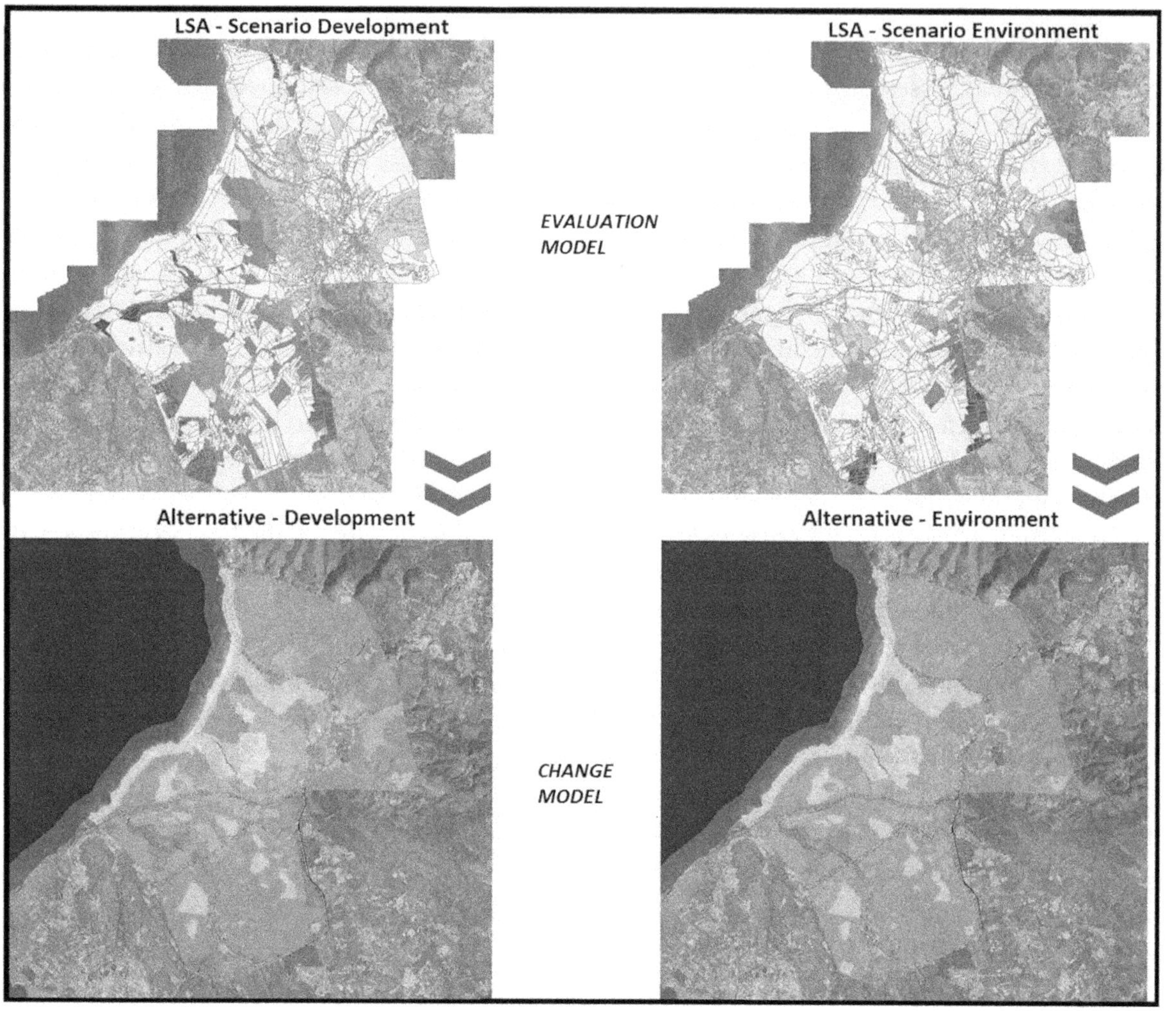

The Impact Model

The Impact Model is used to evaluate the possible impacts of the designed scenarios on the environment and, within the PSS, it operates through dynamic spatial indicators organised on a digital dashboard. The strengthened role of indicators for communicating complex information in an easy way is well-known in the scientific literature (Hammond et al, 1995; Smeets and Wetering, 1999; Niemeijer and de Groot, 2008). The spatial indicators may represent spatial or non-spatial data and may be static or dynamic values related to strategic domains, such as environment, economy or energy. Planners can also change a specific scenario in an effort to improve the performance of a planning solution. In the Local Land-Use Plan case study, in the hypothesis of locating new tourist areas, the indicators represented the impact on natural or anthropic systems (e.g. loss of natural landscapes, environmental resource consumption, costs to support new urbanization services) and their value changed in real-time on the basis of the sketched planning solution. Furthermore, the indicators are linked together in a network integrated in the PSS so that, if a design alterative is edited, the spatial indicators linked with that specific design change. If the planner reduces the size of a touristic area, the number of inhabitants and their consumptions change automatically, as well as the economic costs and the loss of natural values (Figure 4). This process is based on the concept of causal chain, according to which, each operation feeds back the others in a cycle, linking the components in a framework.

During the test workshop the impact model was able to assess the design choices in real time, contributing to the generation of alternatives in accordance with both the planning goals and the sustainable environmental resource consumption.

Figure 4. Real time sketch planning and impact assessment

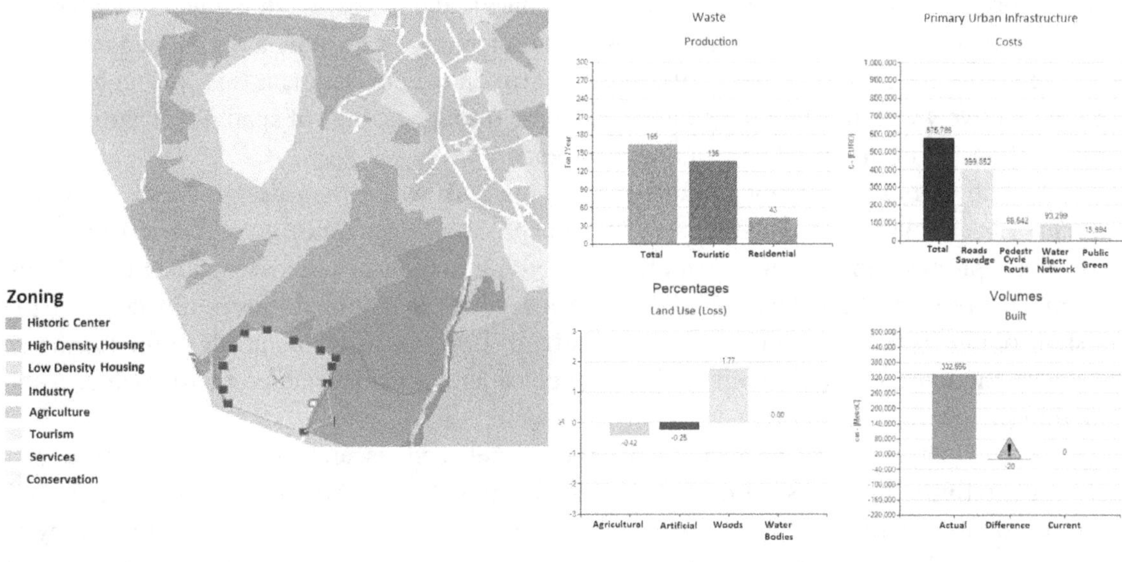

The Decision Model

In order to facilitate the final decision, the PSS supported the comparison of both the maps of design alternatives and their impact indicators. Scenarios could be compared with the aim of evaluating the influence of different alternatives on the environment. Given that a scenario may positively affect certain systems and may have negative effects on others, the comparison tool showed the stakeholders the impacts of their design choices fostering collaboration. Planners, during the negotiation phase, could edit scenarios in order to generate a final decision which achieve the planning goal with regard to sustainable resource consumption. Furthermore, the final decision can be accompanied by the production of reports, which include maps, documents and indicators, which can be used to enhance the transparency of the decision-making process.

The PSS described above was tested in a planning study conducted in the municipality of Gonnesa (Sardinia), during an experimental workshop carried out by the authors, with the purpose of investigating how it may support planners in a decision-making process. The case study concerned the process of adaptation of the Gonnesa Local Land-Use Plan to the Sardinian Regional Landscape Plan. The goal of the workshop, organised in collaboration with the local administration through a set of preliminary meetings and interviews, was to identify the most suitable area to assign to new tourism development.

A range of participants such as planners, local administrators, researchers and other professionals, was selected to contribute to the workshop activities, which were based on the experiential learning cycle of Kolb and Fry (1974). They were asked to analyse the territorial characteristics through the web-PSS (i.e. Representation and Process models) and, in the context of the Evaluation Model phase, to discuss about criteria and weights that could be dynamically changed through sliders, producing a real time feedback as suitability maps. The output of this discussion was a set of two evaluation maps corresponding to two different development scenarios: the "environmental protection oriented map" which gives the highest priority to the locations not compromising natural protected areas, and the "development-oriented map" which gives the highest priority to proximity to roads and services. The two evaluation maps represented the input for the design activities carried out during the next workshop phase, in which users were called to sketch a set of planning proposals on an interactive screen by a digital pen (i.e. Change Model). Thereafter, the impact of every different scenario on the environment (i.e. Impact Model) was assessed interactively with the digital dynamic dashboard through a range of spatial indicators changing real-time on the basis of the sketched planning solutions.

The workshop users were finally asked to discuss the reasons for their design choices, comparing the impacts of each solution on the environment. This procedure generated a collaborative process, in which the users provided their own contributions according to their own expertise and their role in the decision-making process. After the discussion phase, a negotiation among the users allowed them to edit their design solutions and eventually reach a shared final scenario fitting the needs of the local authority to develop the tourism sector while taking into consideration sustainable environmental resource consumption (i.e. Decision Model).

The workshop represented an endeavour to fill the gap between research and planning practices, with the aim of integrating innovations in Local Land-Use Planning. It contributed to investigate how the PSS architecture enabled, on the one hand, to support planners for dealing with the range of planning tasks related to SEA and Planning and, on the other hand, to demonstrate a possible viable way for integrating geodesign workflows in planning process, as synthetized in Table 1.

These results show possible ways to improve single planning tasks within the plan-making-SEA process. It is worth noting that other tools can be used to implement the workflow described above. Notably, the ESRI ArcGis GeoPlanner web-based PSS enables the interactive development of evaluation maps, sketch-design, and dynamic impact indicators dashboard creation.

In the next section, the geodesign workshop with Geodesignhub is presented to demonstrate how the dynamic workflows can successfully enable, by the use of digital data and technology, to contribute to develop design alternatives expressing different sets of values, to assess their performance iteratively, and to support negotiations towards consensus while monitoring the process unfolding. All these analyses can be considered essential requirements of the SEA, and are currently poorly addressed in the planning practice.

Table 1. Relationships between Environmental Report contents (ex Annex I of Directive 2001/42/EC), the geodesign framework models and the PSS features

Environmental Report	Geodesign Framework	PSS features
(b) the relevant aspects of the current state of the environment […]; (c) the environmental characteristics of areas likely to be significantly affected; (e) the environmental protection objectives, established at international, Community or Member State level, which are relevant to the plan or programme […];	Representation Model	web-based PSS: web-mapping tools, merging GIS capabilities with easy-to-use Web-interfaces, for representing the complexity of the territory.
(b) the relevant aspects of the current state of the environment and the likely evolution thereof without implementation of the plan or programme; d) any existing environmental problems which are relevant to the plan or programme including, in particular, those relating to any areas of a particular environmental importance, such as areas designated pursuant to Directives 79/409/EEC and 92/43/EEC;	Process Model	web-based PSS: web-analysis tools for interactively combining and comparing different data themes and for performing queries and spatial analyses on specific issues of interest.
(f) the likely significant effects on the environment, including on issues such as biodiversity, population, human health, fauna, flora, soil, water, air, climatic factors, material assets, cultural heritage including architectural and archaeological heritage, landscape and the interrelationship between the above factors;	Evaluation Model	Desktop PSS: Land Suitability Analysis for dynamically identifying the areas suitable for given land uses, in accordance with a chosen set of criteria.
(e) the environmental protection objectives, established at international, Community or Member State level, which are relevant to the plan or programme and the way those objectives and any environmental considerations have been taken into account during its preparation; (g) the measures envisaged to prevent, reduce and as fully as possible offset any significant adverse effects on the environment of implementing the plan or programme;	Change Model	Desktop PSS: real time sketch planning tool for altering the study area and visualising the alternative designs in the geographic space.
(f) the likely significant effects on the environment, including on issues such as biodiversity, population, human health, fauna, flora, soil, water, air, climatic factors, material assets, cultural heritage including architectural and archaeological heritage, landscape and the interrelationship between the above factors;	Impact Model	Desktop PSS: spatial impact indicators on a digital dashboard for evaluating the possible impacts of the designed alternatives on the environment.
(h) an outline of the reasons for selecting the alternatives dealt with, and a description of how the assessment was undertaken including any difficulties (such as technical deficiencies or lack of know-how) encountered in compiling the required information;	Decision Model	Desktop PSS: scenario comparison tool for assessing the influence of different alternatives on the environment in order to support a final decision.

GEODESIGN WORKSHOPS

Several recent studies which addressed more specifically the relationships between the Evaluation, Change, Impact, and Decision Models (Rivero *et al*, 2015; Nyerges *et al, 2016;* Campagna et al, 2016, Pettit et al, 2019)were recently developed relying on the novel Geodesignhub PSS (www.geodesignhub.com). Geodesignhub is a web-based PSS which was originally developed (Ballal, 2015) to implement the Steinitz framework for geodesign (Steinitz, 2012). It supports collaborative design by taking as input evaluation maps externally developed with GIS software, which are used as informative base to crowd-source projects and policies as geographic diagrams by a potentially unlimited number of workshop participants (Fig. 5). Using the system evaluation maps allows to inform locational choices for design in order to avoid negative impacts and select best locations for projects and policies.

Figure 5. Projects and policies crowdsourcing with geodesignhub

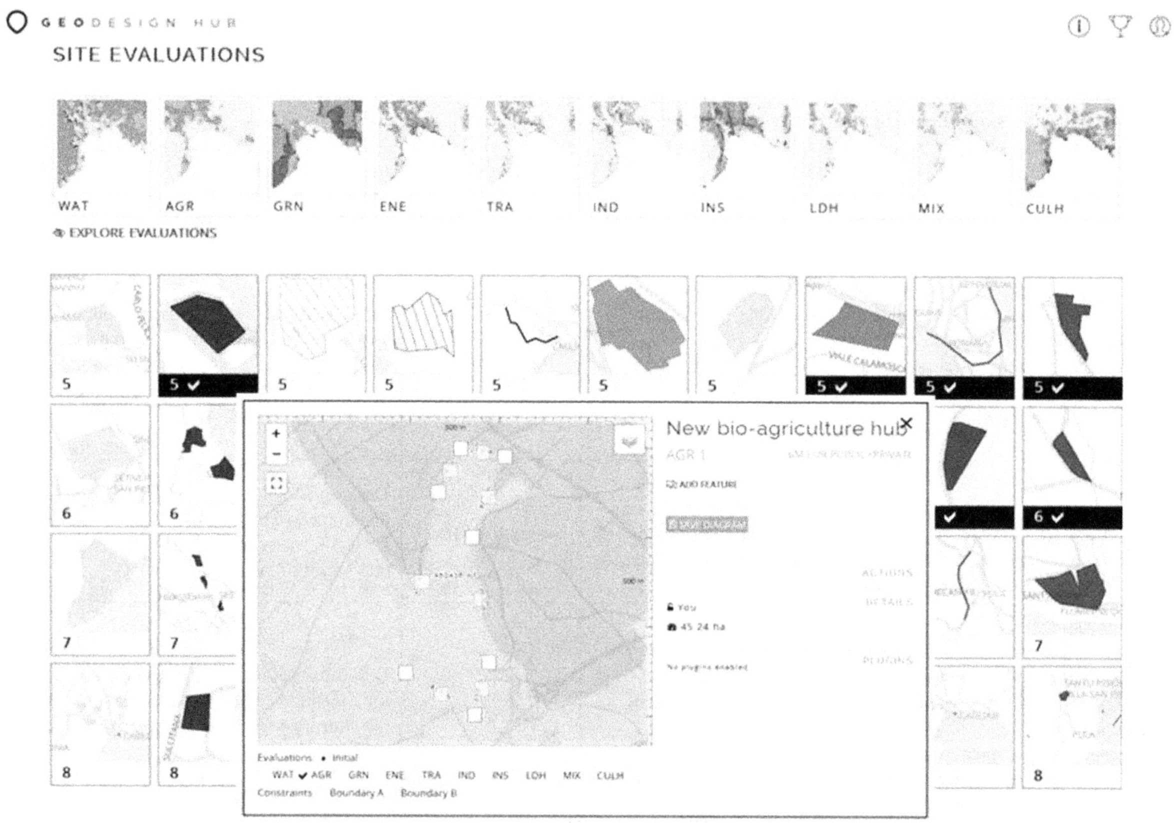

Individual project and policies diagrams are shared among participants and then available for selection by teams with diverse interests and objectives to compose complex integrated design, or *syntheses*. Syntheses are assessed for further impacts dynamically in real-time, so that the impact model can support their iterative improvement. The syntheses of the different teams represent alternatives design proposals, which can be compared with the support of performance and impact indicators, and mediated toward consensus through negotiation. In fact, Geodesignhub offers a set of reliable tools to support negotiation. Furthermore, Geodesignhub was built with interoperability standard technology, so that external models can be linked via API.

Geodesign workshops with Geodesignhub are particularly effective in supporting collaborative strategic planning for they allow collaboration among a wide number of stake-holders in a reasonably short time (i.e. normally a workshop last between 12 and 15 hours). Furthermore, the Geodesignhub interface is user-friendly, enabling the seamless participation of participants with little or no technical background and skills. Regular post-workshop satisfaction surveys run by the authors in several research, teaching, and planning practice case studies confirmed a high level of satisfaction by workshop participants.

- The major advantages of using geodesign workshops in the SEA-planning process can be summarized as follows:Planning alternatives are generated on the base of explicit and transparent evaluation, impact, and decision models;

- Final decisions are made through informed negotiation towards consensus;
- Planning alternatives are iteratively improved with the support of the impact model;
- A high number of stakeholders can take part to the design, enriching the consultation with the possibility for them to directly contribute to the final design;
- Design results are immediately available for further processing and reporting, improving design evolution traceability and transparency.

Last, but not least, the latest experiences by the authors in coordinating geodesign workshops with Geodesignhub in real world planning process at the time of COVID-19 pandemic, proved workshops can be effectively run fully digital (i.e. using Geodesignhub in combination with a virtual conferencing platform such as Zoom) simplifying organization and logistic and facilitating wider participation.

CONCLUSION

Notwithstanding the introduction of Strategic Environmental Assessment and the growing availability of Spatial Data Infrastructure in Europe, contemporary planning practice often shows difficulties in up-taking their innovation potential. The geodesign approach may offer the missing methods and tools needed to address some of the most critical parts of the planning process which often determine the failure of a correct SEA application.

The analyses of current practices in Sardinia and in Italy shows that the adoption of the digital media as a base for plan-making is still very limited. A similar situation may be expected in many other regions in Europe looking at existing literature. In the best cases, spatial data and GIS are used to create maps, and very seldom advanced spatial analyses techniques are used to support environmental savvy informed design and decision-making in SEA-based planning. Likewise, very seldom digital communication technology are used to support collaboration and participation.

However, as it happened in the past in other periods, we are now living a wave of progress due to rapid advances in Information Communication Technology, and this may represent an opportunity to bring innovation in spatial planning and design further. While geodesign techniques can in principle be applied without computer support, the use of digital data and technology can bring new dynamics in the analysis, the design, the impact assessment, and the decision-making steps of planning processes. Currently, unlike in the past, this opportunity appear more mature and should not be missed. The growing interest by academia in the use of new technology in planning research, and in the education of the future (digital native) planners, will possibly, eventually, enable to cope with many challenges which were abandoned at time in the past, such as the use of advanced operational models, visualization tools, and planning support systems towards more sustainable planning as auspicated in Agenda 21.

With this chapter, the authors aim at giving a contribution in explaining the relationships among normative, methodology, and technology frameworks and the planning process in practice. The proposed case studies, if may show only a possible way to bridge the existing gap, may also contribute to demonstrate the feasibility of the geodesign approach and its potential to address some of the most critical SEA pitfall in planning.

ACKNOWLEDGMENT

The original work presented in this chapter was developed by the authors within the research project "Efficacia ed efficienza della governance paesaggistica e territoriale in Sardegna: il ruolo della VAS e delle IDT" [Efficacy and efficiency of landscape and environmental management in Sardinia: the role of SEA and of SDI] CUP: J81J11001420007 funded by the Autonomous Region of Sardinia under the Regional Law n° 7/2007 "Promozione della ricerca scientifica e dell'innovazione tecnologica in Sardegna". The revised version was developed by the authors within the research project "Investigating the relationships between knowledge-building and design and decision-making in spatial planning with geodesign, funded by Fondazione di Sardegna CUP F74I19001040007

The chapter builds upon an article published in the International Journal of E-Planning Research (IJEPR).

REFERENCES

Aalders, H. J., & Moellering, H. (2001). Spatial data infrastructure. *Proceedings of the 20th International Cartographic Conference*, 4.

Artz, M., Batty, M., Dangermond, J., Fisher, T., McElvaney, S., Richardson, K., & Steinitz, C. (2013). Geodesign: Past, Present, and Future. Esri Press.

Artz, M., Dangermond, J., Ball, M., & Abukhater, A. (2010). *Changing Geography by Design. Selected readings in GeoDesign*. Esri Press.

Ballal, H. (2015). *Collaborative Planning with Digital Design Synthesis* (Ph.D. Thesis). UCL (University College London), London, UK.

Brown, A. L., & Thérivel, R. (2000). Principles to guide the development of strategic environmental assessment methodology. *Impact Assessment and Project Appraisal, 18*(3), 183–189. doi:10.3152/147154600781767385

Campagna, M. (2014a). Geodesign from Theory to Practice: From Metaplanning to 2nd Generation of Planning Support Systems. *Tema. Journal of Land Use, Mobility and Environment.*

Campagna, M. (2014b). Geodesign: dai contenuti metodologici all'innovazione nelle pratiche. In Atti della XVII Conferenza Nazionale SIU – Società Italiana degli Urbanisti. Planum Publisher.

Campagna, M. (2016). Metaplanning: About designing the Geodesign process. *Landscape and Urban Planning, 156*, 118–128. Advance online publication. doi:10.1016/j.landurbplan.2015.08.019

Campagna, M., & Craglia, M. (2012). The socioeconomic impact of the spatial data infrastructure of Lombardy. *Environment and Planning. B, Planning & Design, 39*(6), 1069–1083. doi:10.1068/b38006

Campagna, M., De Montis, A., & Deplano, G. (2006). PSS design: A general framework perspective. *International Journal of Environmental Technology and Management, 6*(1-2), 163–179. doi:10.1504/IJETM.2006.008259

Campagna, M., & Di Cesare, E. A. (2016). Geodesign: Lost in Regulations (and in Practice). In Smart Energy in the Smart City (pp. 307-327). Springer International Publishing.

Campagna, M., Steinitz, C., Di Cesare, E. A., Cocco, C., Ballal, H., & Canfield, T. (2016). *Collaboration in Planning: the Geodesign approach.* Rozwój Regionalny I Polityka Regionalna.

Couclelis, H. (2005). "Where has the future gone?" Rethinking the role of integrated land-use models in spatial planning. *Environment & Planning A, 37*(8), 1353–1371. doi:10.1068/a3785

COWI. (2009). *Study concerning the report on the application and effectiveness of the SEA Directive* (2001/42/EC). DG Environment European Commission, Report no. p-67683-a, Issue no. 2.

Craglia, M., & Annoni, A. (2007). INSPIRE: An innovative approach to the development of spatial data infrastructures in Europe. *Research and Theory in Advancing Spatial Data Infrastructure Concepts*, 93-105.

Craglia, M., & Campagna, M. (2009). *Advanced regional spatial data infrastructures in Europe.* Luxembourg: Office for official publications of the European Communities.

Dangermond, J. (2010). *GeoDesign and GIS–Designing our Futures. Peer Reviewed Proceedings of Digital Landscape Architecture.* Anhalt University of Applied Science.

De Montis, A., Ledda, A., Caschili, S., Ganciu, A., & Barra, M. (2014). a. SEA effectiveness for landscape and master planning: An investigation in Sardinia. *Environmental Impact Assessment Review, 47,* 1–13. doi:10.1016/j.eiar.2014.03.002

De Montis, A., Ledda, A., Caschili, S., Ganciu, A., & Barra, M. (2014). b. SEA effectiveness for landscape and master planning: An investigation in Sardinia. *Environmental Impact Assessment Review, 47,* 1–13. doi:10.1016/j.eiar.2014.03.002

Ervin, S. (2011). A system for GeoDesign. *Proceedings of Digital Landscape Architecture, Anhalt University of Applied Science*, 145-154.

Fischer, T. B. (2010). Reviewing the quality of strategic environmental assessment reports for English spatial plan core strategies. *Environmental Impact Assessment Review, 30*(1), 62–69. doi:10.1016/j.eiar.2009.04.002

Fischer, T. B., & Gazzola, P. (2006). SEA effectiveness criteria—Equally valid in all countries? The case of Italy. *Environmental Impact Assessment Review, 26*(4), 396–409. doi:10.1016/j.eiar.2005.11.006

Flaxman, M. (2010). Fundamentals of geodesign. *Proceedings of Digital Landscape Architecture, Anhalt University of Applied Science.*

Fundingsland Tetlow, M., & Hanusch, M. (2012). Strategic environmental assessment: The state of the art. *Impact Assessment and Project Appraisal, 30*(1), 15–24. doi:10.1080/14615517.2012.666400

Geertman, S., & Stillwell, J. (2004). Planning support systems: An inventory of current practice. *Computers, Environment and Urban Systems, 28*(4), 291–310. doi:10.1016/S0198-9715(03)00024-3

González, A., Thérivel, R., Fry, J., & Foley, W. (2015). Advancing practice relating to SEA alternatives. *Environmental Impact Assessment Review, 53*, 52–63. doi:10.1016/j.eiar.2015.04.003

Hammond, A., Adriaanse, A., Rodenburg, E., Bryant, D., & Woodward, R. (1995). *Environmental Indicators: A Systematic Approach to Measuring and Reporting on Environmental Policy Performance in the Context of Sustainable Development.* World Resources Institute.

Harris, B. (2001). Sketch planning: Systematic methods in planning and its support. *Planning Support Systems: Integrating Geographic Information Systems, Models, and Visualization Tools*, 59-80.

Heppenstall, A. J., Crooks, A. T., See, L. M., & Batty, M. (Eds.). (2011). *Agent-based models of geographical systems.* Springer Science & Business Media.

Klosterman, R. E. (1999). The what if? Collaborative planning support system. *Environment and Planning. B, Planning & Design*, *26*(3), 393–408. doi:10.1068/b260393

Kolb, D. A., & Fry, R. E. (1974). *Toward an applied theory of experiential learning. MIT Alfred P.* Sloan School of Management.

Maguire, D. J., & Longley, P. A. (2005). The emergence of geoportals and their role in spatial data infrastructures. *Computers, Environment and Urban Systems*, *29*(1), 3–14. doi:10.1016/S0198-9715(04)00045-6

Malczewski, J. (2004). GIS-based land-use suitability analysis: A critical overview. *Progress in Planning*, *62*(1), 3–65. doi:10.1016/j.progress.2003.09.002

Niemeijer, D., & de Groot, R. S. (2008). Framing environmental indicators: Moving from causal chains to causal networks. *Environment, Development and Sustainability*, *10*(1), 89–106. doi:10.100710668-006-9040-9

Nyerges, T., Ballal, H., Steinitz, C., Canfield, T., Roderick, M., Ritzman, J., & Thanatemaneerat, W. (2016). Geodesign dynamics for sustainable urban watershed development. *Sustainable Cities and Society*, *25*(6), 13–24. doi:10.1016/j.scs.2016.04.016

Parker, J. (2007). *Strategic Environmental Assessment and SFs Operational Programmes: An assessment.* DG Environment European Commission presentation at the Aarhus Workshop on Public Participation in Strategic Decision- Making (PPSD), Sofia, Bulgaria.

Partidário, M. R. (2012). *Strategic Environmental Assessment Better Practice Guide.* Agência Portuguesa do Ambiente e Redes Energéticas Nacionais.

Pettit, C., Hawken, S., Zarpelon, S., Ticzon, C., Afrooz, A., Steinitz, C., Ballal, H., Canfield, T., & Lieske, S. (2019). Breaking Down the Silos Through Geodesign—Envisioning Sydney's Urban future. *Environment and Planning. B, Urban Analytics and City Science*, *46*(8), 1387–1404. doi:10.1177/2399808318812887

Rega, C., & Baldizzone, G. (2015). Public participation in Strategic Environmental Assessment: A practitioners' perspective. *Environmental Impact Assessment Review*, *50*, 105–115. doi:10.1016/j.eiar.2014.09.007

Rega, C., & Bonifazi, A. (2014). Strategic Environmental Assessment and spatial planning in Italy: Sustainability, integration and democracy. *Journal of Environmental Planning and Management*, *57*(9), 1333–1358. doi:10.1080/09640568.2013.804404

Rivero, R., Smith, A., Ballal, H., & Steinitz, C. (2015) Promoting Collaborative Geodesign in a Multidisciplinary and Multiscale Environment: Coastal Georgia 2050, USA. *Proceedings of the 2015 Digital Landscape Architecture Conference.*

Runhaar, H. (2009). Putting SEA in context: A discourse perspective on how SEA contributes to decision-making. *Environmental Impact Assessment Review*, *29*(3), 200–209. doi:10.1016/j.eiar.2008.09.003

Smeets, E., & Weterings, R. (1999). *Environmental indicators: Typology and overview*. Copenhagen: European Environment Agency, EEA Technical report No 25.

Steiniger, S., & Hunter, A. J. (2012). Free and open source GIS software for building a spatial data infrastructure. *Geospatial free and open source software in the 21st century*, 247-261.

Steinitz, C. (2012). *A Framework for Geodesign. Changing Geography by Design.* Esri Press.

te Brömmelstroet, M. (2013). Performance of planning support systems: What is it, and how do we report on it? *Computers, Environment and Urban Systems*, *41*, 299–308. doi:10.1016/j.compenvurbsys.2012.07.004

te Brömmelstroet, M., & Schrijnen, P. M. (2010). From planning support systems to mediated planning support: A structured dialogue to overcome the implementation gap. *Environment and Planning. B, Planning & Design*, *37*(1), 3–20. doi:10.1068/b35019

Vandenbroucke, D., & Biliouris, D. (2011). Spatial Data Infrastructures in Italy: State of play spring 2011. K.U. Leuven (SADL + ICRI).

Vonk, G., & Geertman, S. (2008). Improving the adoption and use of planning support systems in practice. *Applied Spatial Analysis and Policy*, *1*(3), 153–173. doi:10.100712061-008-9011-7

Vonk, G., Geertman, S., & Schot, P. P. (2005). Bottlenecks blocking widespread usage of planning support systems. *Environment & Planning A*, *37*(5), 909–924. doi:10.1068/a3712

Vonk, G. A. (2006). *Improving planning support: the use of planning support systems for spatial planning*. KNAG/Netherlands Geographical Studies.

Zoppi, C. (2012). Valutazione come sostegno all'efficacia del piano. In Valutazione e pianificazione delle trasformazioni territoriali nei processi di governance ed e-governance–Sostenibilità ed e-governance nella pianificazione del territorio. Milano: FrancoAngeli.

Chapter 7
Responding to Future Climatic Conditions of Heat and Flood: A Design Decision Support System for Street Trees

Nano Langenheim
University of Melbourne, Australia

Marcus White
Swinburne University of Technology, Australia

ABSTRACT

The escalating intensity and duration of heat and flood events in cities increase the importance of green infrastructure design that responds to climate change challenges. The aim of this chapter is to demonstrate the potential of a rapid precinct scaled design approach for street tree planting choices that enables simultaneous assessment of visual impact and environmental performance factors including shade quantification and flood responsiveness. This chapter describes a street tree design decision support system (DDSS) drawing from advances in computational botany, entertainment industries, light engineering, and hydrology modelling. The DDSS is tested at both street and precinct scale using two case studies with results demonstrating the feasibility of rapid scenario assessment of tree placement and species selection. The DDSS allows the streetscape street tree design process to integrate the consideration of changing climatic challenges alongside community aesthetic preferences and provides an extensible framework for designing more resilient and comfortable urban spaces.

INTRODUCTION: FROM AESTHETICS TO GREEN INFRASTRUCTURE

Visual Benefits of Trees

Trees are often an integral part of urban design in new or renewed cities of the 19th century, profoundly influencing the quality of streetscapes. Traditional urban street tree planting is arranged in evenly spaced

DOI: 10.4018/978-1-7998-9090-4.ch007

symmetrical rows, reflecting the influence of French Formal compositional ideas on urban design during the expansion of the British Colonial Empire (Lawrence, 2008). These historic, formal tree species, placement and spacing choices were driven by visual and symbolic principles of the time, rather than environmental concerns. Visually, formal street tree arrangements 'provide a sense of streetscape enclosure', 'define the street as an outdoor room', and 'distinguish a boundary between different street uses' (Arnold, 1980; Ewing & Handy, 2009). However, over the past century, urban trees have become increasingly valuable to cities for reasons beyond the visual, expanding to include their functional attributes in moderating environmental problems related to the impacts of climate change.

Environmental Benefits of Trees

City streets are complicated, spatially competitive, and climatically diverse environments. Trees, as opposed to low shrubs, groundcover and lawn, are particularly important for spatially constrained city streets, as they require comparatively small amounts of high value land for substantial environmental and human health returns or a high 'green plot ratio' (Ong, 2003; Taylor et al., 2015). The environmental returns of urban trees are increasingly understood to be extensive, including noise attenuation and air pollution filtration through particulate interception (Maher et al., 2013), urban heat moderation through surface shading (Lindberg & Grimmond, 2011), air temperature cooling through evapotranspiration (Shashua-Bar et al., 2010), reduction of pedestrian UV exposure (Parisi et al., 2000, 2001), and storm water management, through water absorption or the 'sponge' actions of the root system, and canopy rainfall interception (Carlyle-Moses et al., 2020; Szota et al., 2018).

Melbourne Context

Melbourne, Australia, is an illustrative example of a heat and flood prone city, seeking to harness urban forestry to improve climatic resilience. Melbourne and its local government areas (LGAs) have large proportions of low diversity, over mature tree stock planted during colonial settlement, that struggled to withstand the impacts of urban expansion, development intensity, and climate change, particularly the prolonged heat waves and flash flooding of the past two decades (BOM, 2013; Rizmal, 2019). Extensive amounts of these historically significant existing trees now require urgent replacement.

With this major tree-stock renewal challenge, comes a unique opportunity to reconsider not only street tree pallets, more suitable for the local climate, but also the design and decision-making processes for tree selection to respond to spatial and temporal conditions and implications (Dobbs et al., 2014; Langenheim et al., 2020; White & Langenheim, 2018b). For instance, shading the impermeable, dark asphalt and bluestone surfaces of Melbourne's wide exposed streets with strategically placed trees would substantially contribute to heat mitigation, particularly in areas of low and mid-rise density (Coutts, 2014; Mohd Sanusi, 2015), while the speed of storm water flow in flash flood events could be reduced through increased used of dense canopied trees to intercept rainfall (Carlyle-Moses et al., 2020).

METHODS FOR MAKING CLIMATICALLY RESPONSIVE TREE DECISIONS

Making climatically responsive tree replacement decisions in Melbourne, will require departures from traditional street tree species selections and arrangements. Likewise, communicating the benefits of these departures with local residents and community groups, will require more sophisticated decision support material than the traditional 2D formats of plans, sections and elevations to avoid tree replacement that is simply 'like for like'. The dual values that urban trees are now recognised to hold, in equal measure performance-based environmental function and qualitatively-based visual impact, require street tree decision-making to evolve (Escobedo et al., 2011; Oke, 1988). New tree decision-making methods and decision-support tools, need to integrate and work with the constrained complexities of urban environments and quantitative performance-based tree modelling, to maximise both the visual and environmental benefits of new planting choices in urban streets (Beatley, 2011; Dobbs et al., 2014; Langenheim et al., 2020; Norton et al., 2015).

Tree Modelling Used by Landscape Architects and Urban Designers

In past decades, selection of trees for streets in Australian cities, involved referring to a Local Government Area (LGA) preferred species list nominating tree planting for each street. These lists comprise an amalgamation of species, both observed to grow well in the local area in the past, and to reflect the specific civic image of that LGA (Ely, 2009). For example, the preferred species list of the LGA of Moreland, has a high proportion of native and indigenous species (Moreland, 2009), while the City of Melbourne's list, (prior to a 2011 re-evaluation of this method of tree selection), had a high proportion of exotic deciduous boulevard species in keeping with the historical character of the inner city (City of Melbourne, 2011).

For community consultation forums, in which changes to street tree plantings are discussed, proposed trees might be 'modelled' using 2D cross sections and plan drawings, or post-production 'collage' methods. Trees are inserted as photographic montage over either flattened views of modelled environments or site photos using Adobe Photoshop™ (White & Langenheim, 2018b). These analogue methods of modelling tree decisions, are still the predominant 'norm' for presentation of landscape designs as evidenced in a number of recent local reports (COM, 2016; King et al., 2017; O'Shea, 2020).

While these 2D modelling methods are rapid, convey evocative visual qualities, and are traditionally prefered by community groups, who find them more 'open for comment' than digitally modelled photo realistic represntations (MacEachren, 1995), they do not convey complex 3D, spatial or climatic interactions. Spatial interactions between trees and the urban environment they inhabit, often demand substantial canopy pruning for building facades, power line, vehicle and pedestrian clearances while asymetrical climatic conditions of light, soil and water common in urban streets have substantial impacts on tree performance and growth (McPherson & Peper, 2012; Měch & Prusinkiewicz, 1996).

Recent methods that use 2.5D tree modelling (GIS tree points geospatially located on a map with height, species and other data) loosely coupled to visual 'billboard' libraries of drag and drop landscape elements such as those developed for the Victorian Resources Online tool, are begining to work with the specifics of space and climate (Grêt-Regamey et al., 2017). Recent updates to ESRI City Engine™ allows users to replace billboard trees with more visually compelling Lumen RT™ 3D tree Libraries at render time. While these advancements are a significant step forward in linking geospatial data with tree design visualisation, these more sophisticated approaches still have limitations.

They do not work with the geometric proportions of tree canopies. The coupled 3D Lumen RT library is only conjured at render time for visualisation purposes, the trees do not respond to spatial clashes such as overhead powerlines. The library is also currently limited to eighty plant species including ground covers and shrubs, most of which are common only in the US.

CURRENT LIMITATIONS – WHY AREN'T ACCURATE 3D TREE MODELS USED IN PRECINCT SCALE MODELLING?

While it has been possible to model complex tree geometry, shrubs and grass with highly realistic results for over two decades (Weber & Penn, 1995) and while public acceptance of digital visualisation grows with increasing exposure to platforms such as Google Earth, there have been problems with balancing the degree of realism with the speed of the model to regenerate (Rebollo et al., 2014). Difficulties arise in relation to levels of detail (LOD), between the close streetscape view and the precinct view. Tree models used for visualisation at a close streetscape scale, with five or ten trees, are often too detailed and computationally intensive for precinct or larger scale visualisations that can contain many thousands of trees (Kwon et al., 2018). Visualisations of precincts created through digital modelling have not had wide spread use due in part to the time investment needed, the expense of computer hardware and software and high-level skills required for their construction (Durdyev et al., 2021; Jones, 2020).

Tree Modelling used by Disciplines Outside of Urban Design

Ecology and Computational Botany - Functional Structural Applications

Since the late 1960's, researchers have developed computational tree and plant modelling for different purposes in a range of disciplines (Prusinkiewicz et al., 1990). In botany, forestry, agronomy and ecohydrology two types of plant model evolved, Geometric Models (GM's) representing plants in realistic detail at close range with high fidelity to botanical structure and Process Based Models (PBM's) representing effects of interactions between plant and environment, impact on aspects such as yield, partitioning of biomass or survival in hostile conditions (de Reffye et al., 1988). These two approaches later merged into Functional-Structural Modelling Applications (FSMA), to produce botanically accurate topological plant models, to a cellular level with growth parameters responsive to their environment (Benes et al., 2016; Pirk et al., 2012).

Process based plant models (PBMs) have been included in microclimate modelling platforms such as Envi MET™, to study the impact plants have on an environment. Within these microclimate modelling platforms, trees are represented by abstract voxels, or sets of cubes with a similar visual quality to Minecraft or Lego block trees (Y. Yang et al., 2019). Interoperability between microclimate and city design modelling software platforms is also quite limited, with only recent development of a plugin for a one way conversion of Rhino Grasshopper urban form models to the Envi MET required format (Di Nunzio, 2019). Without bi-directional interoperability, any use of high quality FSMA tree models lose capacity for interactive plant growth governing controls.

Forestry - Empirical Modelling Applications

Empirical models that build on increasingly available GIS tree inventory data, utilised by the Forestry and Urban Forestry industries, use similar base technology to the aforementioned 2.5D tree GIS database modelling. Applications such as iTree™ released in 2006 and Urban Forest Effects (UFORE), are open-source GIS database programs developed by the USDA Forest service which can be used to record complex urban growth parameters and output predictive height over time as abstract lollypop form for specific tree species under differing climatic, management and environmental regimes (McPherson & Peper, 2012).

Defence and Entertainment Industry - Pure Geometric (Aesthetic) Modelling

Research in the fields of defence and computer visualisation undertaken in the last two decades has focused on the morphology of plants, similar to the structural models described above, but with a focus on computational efficiency, allowing for realistic looking vegetation for real-time rendering in flight simulators, film, gaming and other virtual outdoor environments (Rebollo et al., 2006). Most of these applications are designed for use by 'non-botanists' and to that end plant morphology terms are sometimes translated into geometric parameters.

Tree libraries such as Xfrog™, Laubworks™, Speed Tree™, Onyx Tree™ and LumenRT™ contain highly realistic modelled 3D trees but provide limited variations of form and tree health representing growth in idealised conditions, may contain a selection of species unsuitable for the given environment (primarily European and North American species), or levels of detail that make the model prohibitively computationally costly.

Software developers have recently focused upon the development of tree modelling applications that are integrated *within* spatial design modelling environment platforms as 'plug-ins', allowing editing and manipulation of tree models directly within their spatial context. This integration will be discussed later in the chapter [see Figure 2].

Processing power and graphics cards developed primarily towards the gaming industry have substantially improved in the past fifteen years. Advances in software that take advantage of processor and graphics card improvements by storing the geometry in the memory or on the graphics card's Video Random Access Memory (VRAM). The graphics card can store the rendering calculations of one modelled polygon mesh object and use algorithms to calculate the rest of the 'proxy-objects' based on that first piece of geometry. Proxy-object models can be displayed as simplified point clouds on screen to further improve the efficiency of on-screen display, substituting the point cloud version for full detailed mesh version when rendered. This method of proxy-object and point cloud modelling is now commonly used in the animation industry due to its significant benefits in reduced pressure on both graphics card (GPU) and processor (CPU) within the modelling environment providing substantial time-saving benefits with dramatically reduced render times.

Illumination Engineering nd Architectural Sciences - Lighting Modelling

It is now possible to include grids of light meters to calculate lux levels in a digital model based on daylight expected at a given time of day, for a given day of the year that considers direct sunlight, indirect ambient light from the sky and reflected light from surrounding buildings. These light meter grids were originally designed for internal use (for example to assess task lighting) but have recently been shown to be useful assessment for external scenarios, in conjunction with accurate geographic sun systems at an urban scale to assess the shade impact of individual trees or use at a larger precinct (White & Langenheim, 2018a).

SELECTING MODELLING PLATFORMS AND DEVELOPING A DECISION SUPPORT SYSTEM

The Aim

The aim of this chapter is to demonstrate the potential of a new approach to streetscape and precinct scale streetscape design modelling that utilises detailed 3D trees. The approach is aimed at providing landscape architects and urban designers a rapid method for assessing both environmental performance and visual impact of streetscape tree planting design, in particular, quantitative shade impact, and flood responsive species selection. The purpose of the approach is to make tree planting considerations a more integral part of the street design process, resulting in higher quality, resilient and more comfortable urban spaces.

METHOD

In this chapter we present a three-part study. Firstly, we provide a review of current methods of tree modelling in the design profession, alongside alternatives evolving in a range of other disciplines, that have potential for adoption and integration into street tree decision-making processes.

Secondly, based on the findings of this review and the classic definition of a *decision support system as*: an architecture that accommodates 'a system of systems' facilitating customisation and flexibility fit for a specific purpose (Batty, 2016; Geertman & Stillwell, 2009), we outline a prototype urban street tree *design decision support system* (DDSS). The prototype DDSS brings together disparate data sources, recent advances in software and hardware, with digital tree modelling techniques to enable a clearer, more quantifiable method of assessing the spatio-temporal and ecosystem service implications of design decision making involving trees. The prototype DDSS utilises computational 'geometric' plant modelling, traditionally focused on the architectural form of plants and developed around the needs of agronomy, forestry and defence (Cournede et al., 2007; de Reffye et al., 1988; Weber & Penn, 1995), together with 'parametric proxy object' modelling developed for the entertainment industry (Rebollo, 2007; Rebollo et al., 2014) and photometric daylight analysis modelling (Kota et al., 2014).

Thirdly, we demonstrate the application of the DDSS through two case studies in the north-west of Melbourne. We use the DDSS to output both informative and compelling visualisations that demonstrate the visual impact of tree renewal strategies, in both the short and long terms coupled with measurable ecosystem services, in one case, quantification of shade and in the other, growing tree numbers associated with increasingly saline soils due to sea water intrusion into ground water.

REVIEW OF TREE MODELLING METHODS

The Choice of Modelling Platform and Decision Support System

Designing for continued inclusion of trees in cities, is important but complex, requiring equal visual and performative considerations. Approaching these considerations with a decision support system that allows iterative testing of street tree design options with geometric, spatial and visual accuracy, responsive to underlying soil conditions or green infrastructure associations would be highly beneficial.

Commonly used software in architecture and urban design for visualisation, can be customized to include modelling for these performative considerations through addition of low cost or free/open-source tools. The benefits of utilising existing standard industry software as the base platform for a decision support system is twofold. It increases likelihood of uptake by practitioners who currently have low awareness and little interest in using DSS developed by academics (Hussnain et al., 2020), and also builds industry capacity to address environmental performance criteria (De los Rios & Charnley, 2017). In addition, the outputs of this integrated modelling method, can be fed back into data driven microclimatic and hydrological models particularly those that analyse heat and flood performance.

Decision support systems (DSS), as broadly defined by Sprague are 'interactive computer-based systems, which help decision makers utilise data and [digital] models to solve unstructured problems'. More specifically, the DSS described here will respond to urban design professional's analytical, forecasting and design tasks, dealing with outcomes of multiple time periods and alternative scenarios (1980). DSS are often brought together under a simplified user interface, dashboard or standalone platform, however, for our design focused DSS (DDSS), we choose to keep the components loosely coupled, allowing for component alternatives and substitution as they become available and as hardware and software develops.

For this part of the study the authors put forward their prototype DDSS which brings together urban design modelling software with botanically-accurate geometric tree modelling from computational botany and military and entertainment industries; efficient parametric 'proxy-object' modelling from the entertainment industry, and photometric daylight analysis from architectural and lighting engineering disciplines.

The system has been set up with a range of inputs such as street tree data, street networks, building geometry, services data and geolocation, appearing on the left-hand side of the diagram shown in Figure 1. This data feeds into the algorithmic botany modelling, urban morphology modelling, storm water and solar modelling. These models are then used in conjunction with design proposals to output as visual analysis (visualisation renders), 3D clash detection (geometry overlap) and solar analysis (shadow studies) and storm water modelling. The results are interpreted by the designer and further design iterations are fed back into the system in an iterative fashion (Figure 1).

Urban form modelling can be performed in a wide range of applications such as ESRI City Engine™, McNeal's Rhino 3D, Trimble's SketchUp Pro™, Blender (open source), Cinema 4D™ and Autodesk's Infraworks™, Maya™ and 3ds Max™, as well as game and real-time engines such as Unity™, Lumion™, Epic Game's Unreal Engine™ and Twin Motion™. Each of these programs have the potential for inclusion of 3D trees to some extent, but for this study, the authors have chosen Autodesk's 3ds Max™ as the central software for a variety of reasons. 3ds Max™ is common in architecture, landscape architecture and urban design offices (used for visualisation) integrating with civil engineering software (Civil CAD™), CAD (AutoCAD™) and GIS software. It has inbuilt animation capabilities allowing for demonstration of change over time. The program has physically accurate material shaders allowing different levels of material (leaf) reflectivity and emissivity using sophisticated environmental simulation (render) engines such as the inbuilt Arnold™ (previously NVIDIA Mental Ray™) or plugins such as Chaos Group's V-Ray™ that can perform daylight assessment for different global locations and times. It has 'proxy-object' modelling and rendering features for efficient rendering and parametric and scripted procedural modelling capabilities that allow rapid iterative design changes. A substantial number of tree libraries currently exist which can be dynamically linked into urban models into 3ds Max™ and in addition, is open to inexpensive third-party plugins such as the user-friendly algorithmic tree modelling GrowFX™ by Exlevel™, allowing for tree modelling directly within the urban design modelling environment.

Figure 1. This diagram describes the conceptual layout of the design decision support system

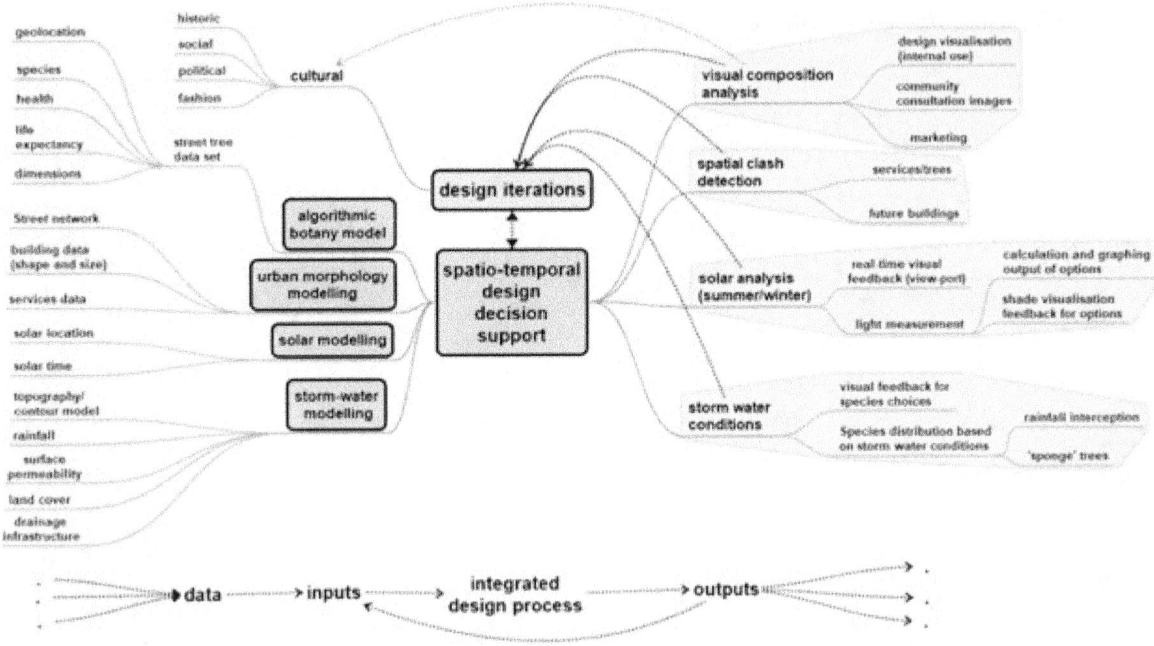

With 3ds Max™ as the base modelling package (Figure 2), the authors modelled trees using Xfrog™ and where simple pruning and temporal changes (e.g. tree growth) within the urban model were required, the authors used the integrated GrowFX™ plugin. These detailed polygon mesh tree models were converted to 'Proxy-Objects'[1] for efficient display and rapid rendering.

For the placement of the proxy-object trees, the authors used a scripted procedural approach allowing thousands of trees to be distributed in a scene. The tree placement was managed by procedural distribution polylines – placing trees based on a set centre to centre distance along vector shapes derived from block and footpath data-sets. This data was sourced from either AURIN's data portal or Open Street Map (OSM), geo-processed in QGIS and then cleaned in Autodesk's AutoCAD™ (Figure 2). Other urban morphology was generated using a mix of 2D (extruded in 3ds Max) and 3D data collected from Google Earth™ and the Klaas Nienhuis's Maproom script, which can import data and textures from Open Street map (OSM), and other aerial photographic textures.

For the first study (Test Case 1) the authors used 'Mental Ray Proxy-Objects' with the now discontinued Mental Ray™ render engine, and on the second study (Test Case 2), the authors used 'V-Ray Proxy-Objects' which used the Chaos Group's V-Ray render engine. These render engines were chosen due to their efficient Proxy-Object feature but also due their inclusion of the aforementioned daylighting analysis capabilities (Figure 2 and Figure 8).

The first study (Test Case 1) was conducted on a Metabox P150EM laptop, running an i7 3720QM processor, NVIDIA GTX 675M GPU and 16GB DDR3 RAM. The second study (Test Case 2) was conducted on a Razer 15 (2018) laptop running an i7-8750H processor, NVDIA GTX 1070 Max-Q, and 16GB DDR4 RAM.

Figure 2. This diagram shows the technical make-up of the decision support system with data sources, geo-processing and modelling software used to feed into the central model, as well as the analysis technology for assessing the user's different design iterations

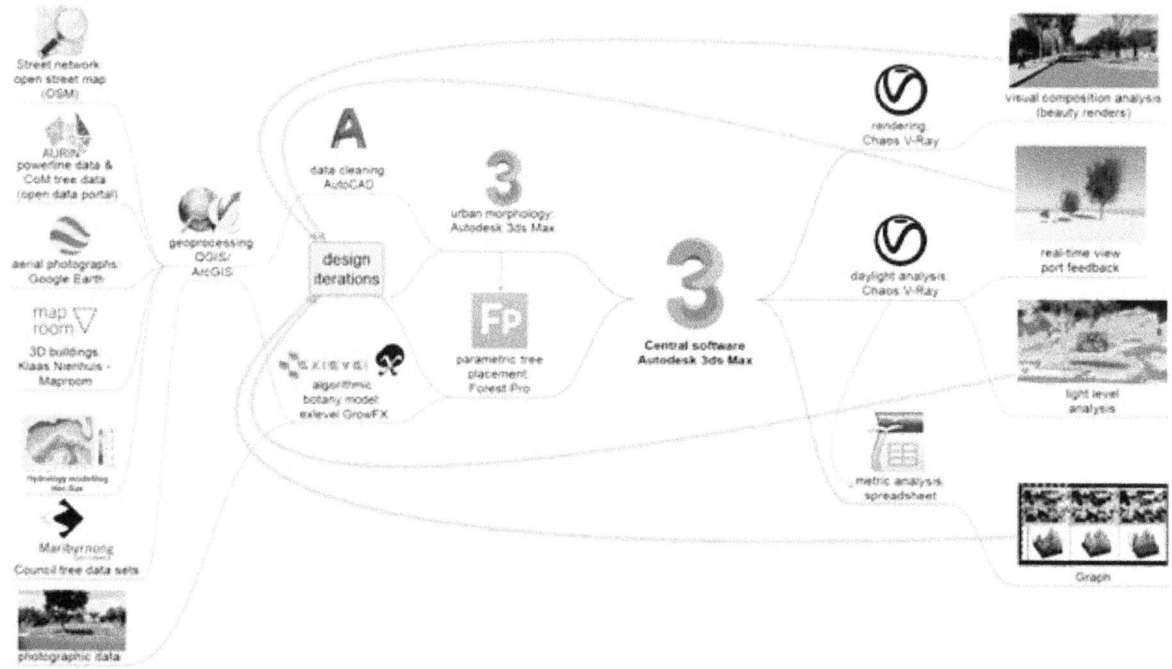

TEST CASE STUDY APPLICATION OF DECISION SUPPORT SYSTEM

Test Case 1: Beevers Street, Maribyrnong - Modelling change over time and different species options

The small suburban test case study of Beevers Street is located in Footscray, in the City of Maribyrnong, Melbourne Australia. The existing Beevers Street trees, consisted of species poorly suited to changing climatic conditions, have been removed due to poor health, while those remaining, are at the end of their 'useful life expectancy'. As a tree replacement strategy, it was important to test the visual impact at street level of different options at different periods of growth.

To test the prototype design decision support system, a variety of species at various stages of tree development were tested. For demonstration purposes, this section of the chapter will show the visual representation over time of the first option (Option A) using the deciduous Japanese zelkova (*Zelkova serrata*) species [Figure 3 to Figure 6], then compare shade impact of Option A at maturity against a second option (Option B), an evergreen gum tree *Corymbia citriodora* in both summer and winter [Table 1].

The model was quick to produce taking only three hours to generate the street geometry and place trees. As the tree species geometry was dynamically linked, testing alternative species was extremely rapid – taking only seconds once appropriate tree models had been selected and entered into the scene. Rendering was also fast, with full HD resolution (1080p) renders being produced in approximately 25 seconds each. The renders give a clear visual image of the look and feel of the streets at various stages of plant growth [Figure 3 to Figure 5] as well as in winter where the deciduous Option A would lose its leaves [Figure 6]. Perspective views of the two options were produced at maturity allowing a visual comparison of species choice in the long term [between Figure 5 and Figure 7].

Figure 3. Digital model of Beevers Street Footscray, Japanese Zelkova (Option A) planting scheme at inception showing footpath and road surface shading in summer

Figure 4. Digital model of Beevers Street Footscray, Japanese Zelkova (Option A) planting scheme at year 5 showing footpath and road surface shading in summer

Though a visual indication of the level of shade provided by the proposed trees can be obtained by looking at the perspective renders, a more detailed level of analysis was also tested to obtain numeric readings for a quantifiable comparison between the two options. A light meter grid was setup to take readings 0.5-meter intervals across full 16-meter width of the street for a length of 80 meters [Figure 8]. Readings were taken for both Option A and Option B in both winter and summer. These readings were tabled [Table 1] showing the total number of light meter readings listed as "# readings", the number of readings considered to be not in shade (above 20,000 lx) are listed as "# unshaded", average total light levels readings (lux), and percentages of area in shade and or full exposure. From this table, it is clear that the amount of shade provided by these two species is not dramatically different in summer with Option B providing just 1.8% more shade cover, but as expected, is dramatically different in winter with Option A casting significantly less shade than Option B due to the loss of leaves [Table 1]. It is worth noting that the average total light reading is slightly higher in summer for Option B than Option A, despite Option B providing a higher percentage of shade cover. Though this may appear to be a contradictory result, it is due to the different tree geometries and their impact on indirect light levels as well as different leaf density (canopy gap fraction) of the trees.

Figure 5. Digital model of Beevers Street Footscray, Japanese Zelkova (Option A) planting scheme at maturity showing footpath and road surface shading in summer

Figure 6. Digital model of Beevers Street Footscray, Japanese Zelkova (Option A) planting scheme at maturity showing reduced footpath and road surface shading in winter due to species deciduous nature, and visually grey 'dead' appearance

Figure 7. Digital model of Beevers Street Footscray, Corymbia citriodora (Option B) planting scheme at maturity showing footpath and road surface shading in summer. Note the visual appearance of this species in winter is almost identical to its appearance in summer

Figure 8. Screen grab the showing a grid of light meter readings covering the street and footpath area measuring total illuminance (combining Direct Illuminance and Indirect Illuminance) (Option B)

Table 1. Table showing simple comparison of shadow impact of Option A – Japanese Zelkova and Option B – Corymbia citriodora at maturity with total illuminance readings taken at 0.5-meter intervals over a 16m x 80m grid in both winter and summer

Summer	# readings	# unshaded	Average	% in Shade	% Exposure
A) Zelkova-22Dec_2pm	5120	4577	82955 lx	**10.6%**	**89.4%**
B) Corymbia-22Dec_2pm	5120	4486	87553 lx	**12.4%**	**87.6%**
Winter					
A) Zelkova-22Dec_2pm	5120	4944	43876 lx	**3.4%**	**96.6%**
B) Corymbia-22Dec_2pm	5120	4025	37636 lx	**21.4%**	**78.6%**

The time taken to set up the model and render of the first perspective image for this study was more than what a skilled user of Adobe Photoshop™ may have taken to produce a similar post-production image, scaling trees by eye (approximately 30-45 minutes if using pre-cut-out tree library images). The level of accuracy of scale and shadows was significantly higher however, and allowed for extremely rapid testing of different scenarios, taking a few seconds to adjust the tree growth parameter (age of tree) or substitute species and time of year. Though not necessarily considered to be photo-realism, these images [Figure 3 to Figure 7] were considered to be of a similar level to an Adobe Photoshop™ post-production collage that is commonly used by landscape architects for design visualisation and advocacy. This assessment was supported by feedback from by Maribyrnong City Council staff who deemed the images as being of high quality and were 'excited by the prospect of using the images in public consultation'.

Test Case 2 Elwood: Modelling Species Selections Responding to Flood Conditions and Topography

Context of Study: Elwood, a Low Lying Area of Melbourne with Major Future Flood Challenges

Elwood is a coastal suburb to the southeast of Melbourne, at the lowest elevation point of the Elster Creek catchment, within the Local Government Area (LGA) of Port Philip in Victoria, Australia. Prior to European settlement, it was a swampy flood plain that drained into localised sub-catchments and only occasionally out to Port Philip Bay. The swamp land was partially filled, and the Elster Creek channelized in 1890, but some areas are still below high tide levels, meaning Elwood can still flood even if it is not raining but the tide is exceptionally high (MMBW, 1979).

In 2012 Melbourne Water released the document, *Planning for Sea Level Rise* (2012), discussing the impact of the International Panel on Climate Change (IPCC) predicted increase of .8m in the 100-year flood level by the year 2100, on flood prone areas such as Elwood in the Port Phillip and Western Port Region of Victoria. Under this sea level rise scenario, the mouth of Elwood's channelized Elster creek would be below sea level at high tide, and would thus, no longer allow overland flow to drain into Port Philip Bay, but would instead, work as a conduit for sea water to flow inland. Research conducted by the CRC for Water Sensitive Cities in 2014-16, developed an urban storm water modelling approach to analyse and simulate, detailed iterative urban flooding scenarios for this complex hydraulic future of Elwood (Rogers et al., 2016; Urich & Rauch, 2014). In this test case study, we investigate the integration of the outputs of their flood modelling, into our street tree decision-making approach.

The existing urban forest of Elwood is considered iconic, with 61% (2,400) of its street trees consisting of historically significant, large healthy 1930-1940's *Platanus* × *acerifolia* (London Plane) and *Ulmus procera* (Dutch Elm) plantings [Figure 9]. Under the sea level rise scenario described above, the associated levels of soil salinity would be too high for more than half of these traditional plantings, which would thus need replacement with species tolerant to coastal flooding or saline water tables. These species, adapted to niche saline and water table conditions, known as halophytes and phreatophytes are few, and their appearance or visual characteristics are 'unconventional' in terms of formal tree-scape norms in common use in Melbourne. Using halophytes and phreatophytes in streetscapes would, be in keeping with an 'ecological design approach' aesthetic, which, has been met with mixed responses when compared to 'conventional' public space design in visual preference studies (B. Yang et al., 2013). It is therefore important that in the process of tree renewal, the aesthetic 'visual impact' of changing tree planting strategy be considered alongside the ecological design response.

Figure 9. Shows a modelled visualisation of the traditional street tree plantings of Elwood, dominated by plantings of Platanus × acerifolia (London Plane) and Ulmus procera (Dutch Elm) planted with a generally symmetrical approach to placement. Building footprints in grey are left 2-dimensional for clarity

Hydrological modelling

Modelling future urban floods, requires integration of existing conditions such as the capacity of the drainage infrastructure, alongside future conditions, such as changed rainfall intensity and duration patterns, changed tide levels due to sea level rise, and changed land cover conditions such as increased or decreased canopy cover or urban development (Urich et al., 2013). While flood modelling has multiple outputs such as extent, water depths, flow velocity and risk and while it is critically intertwined with 2.5 and 3D topographic Digital Elevation Models (DEM), the outputs of flood models, particularly extent and depth, can be successfully visualised in a 2D format, as shown in Figure 10 (1 in 100-year flood with sea level rise of 2m).

How the DDSS was Applied (Inputs into the Model)

Urban Morphology and Streetscape Model

In this case the inputs consisted of a simple suburban morphology model, generated through two data sets; building footprint outlines (maintained by Melbourne Water Corporation) and street centre lines (maintained by the Public Sector Mapping Agency (PSMA)), accessed through the Australian Urban Infrastructure Research Network (AURIN). Using the inbuilt capabilities of the base platform, building footprint data, was given a height extrusion and hipped roof structures, while the street centreline data was given an offset by width and extrusion to generate accurate streets with footpaths.

Figure 10. 1: 10,000 Overlay of CRCWSC 1% AEP and 2m sea level rise flood impact 2100 for Elwood (shown in pink). Drawing created by authors using QGIS) using data from (Rogers et al., 2016; Urich & Rauch, 2014)

The suburban morphology model was then overlayed with a georeferenced map of future flood conditions and categorised into two conditions accordingly: **Condition 01: Flood impacted,** includes low lying areas where trees could assist in clearing water from detention systems (trees as pumps); and **Condition 02: High ground,** which includes higher areas, unaffected by salinity or flood and where tree canopy rainfall interception could assist in slowing overland flow [Figure 11].

Tree Species Selection for Environmental Condition

The primary input into the model is the selection of tree species suitable for each condition and the matching of these species to an associated 3D tree model. Species were selected for their capacity to survive and perform optimally in site specific storm water management infrastructure. For **Condition 02: High ground,** tree species that assist in reducing the speed of overland flow, through canopy interception were selected that consisted primarily of large evergreen, densely foliaged, well-known species such as pines and cedars [Table 2]. 3D models of these species were readily available in the extensive Xfrog tree libraries.

Figure 11. Overlay of the modelled 2100 flood extent and Elwood's existing urban forest, showing trees within Condition 01: Flood impacted in red that will need to be replaced with salt tolerant species that assist in removing storm water from detention systems, and trees within Condition 02: High ground in green, unaffected by flood, but where future species selections may need to support integrated water management infrastructure, slowing stormwater overland flow through canopy interception of rainfall

For **Condition 01: Flood impacted,** tree species tolerant of saline, boggy conditions that can assist in clearing water holding capacity in bio retention systems (pumps), were selected. These species are uncommon in urban settings, have specialised unfamiliar growth habits and only occur in niche environments. These species were matched to visually similar 3D high polygon tree models from the extensive Xfrog and ExLevel tree libraries and modified using the algorithmic tree modelling GrowFX™ by Exlevel™ tools to represent the species selections more accurately as seen in Table 3.

Table 2. Condition 01: Flood impacted (pumps): Salt tolerant high-water use species for low lying sites matched to Xfrog model variation (visually similar)

Selected species	*Casuarina glauca*	*Acacia ampliceps*	*Melaleuca halmaturorum*	*Melaleuca ericifolia*	*Eucalyptus occidentalis*	*Banksia attenuata or menziesii or prionotes*
Rendered model						
Based on Xfrog/ ExLevel models	*OC07_var 9 Casuarina equisetifolia*	OC22_var 2 *Acacia harpophylla*	*OC49 var 7 Eucalyptus dumosa*	*OC57_var 9 Melaleuca alternifolia*	*OC47 var 3 Corymbia calophylla*	*OC26 var 6 Banksia serrata*

Table 3. Condition 02: High ground flood management support: High rainfall interception species for non-flooding areas matched to Xfrog model (exact species often available)

Species selected from Xiao et al.	*Picea pungens Pinus ponderosa*	*Tsuga canadensis*	Picea engelmannii
Rendered model			
Xfrog model	*USA conifers Var01_Pinus muricata*	*USA conifers Var03_ Tsuga canadensis*	*OC49 var 7 Eucalyptus dumosa*

Flood Extent Map Applied to the Site

Bringing the tree models into the site base model used a combination of raster-based distribution and vector-based controls. From the GIS base file refer [Figure 11], a flood extent map was exported as a black and white (binary) geoTIFF, (GDA94/MGA94 zone 55 projection) in which black represents **Condition 01: Flood impacted** areas and white represents **Condition 02: Higher ground** areas. This map was brought into the AutoCAD environment, from where it was then 'passed' to the base modelling platform with a specified insertion point, representing a known latitude, longitude value. The map was then accurately 'draped' (or projected) over the terrain mesh of the site (created from contour data).

Collections of Tree Species Models Applied to the Site

Two collections of tree species models (one for **Flood impacted** areas and one for **Higher ground** areas) were created, with allocation of species controlled by percentage values. These collections were then applied to the site topographic polygon mesh. The distribution of each collection of tree models was controlled by the underlying conditions over which they occur (i.e. black areas of the map suppress the **High ground** area species collection and distribute **Flood impacted** species [see Figure 13], white areas of the map supress the **Flood impacted** area species collection and distribute the **High ground** species, [see Figure 12]. Further to this, to arrange tree models into lines of street trees, additional vector control was used. Vectors representing alignment of street trees on footpath edges, generated from offsets of the street centreline geometry, constrained occurrence of tree models to either side of streets with both species collections [see Figure 14].

Resulting Outputs of the Model

The outputs of the model were able to be used in multiple ways. Firstly, broad scale suburb wide visualisation of the future of Elwood's urban forest was achieved, [see Figure 14, Figure 15, and Figure 16]. Secondly, the model can be used to calculate total numbers of required replacement trees, also important given the departures this plan presents from the traditional species selections grown by commercial contract nurseries that supply councils with street trees, refer Table 4.

Thirdly, the model can also be used for microscale or streetscape visualisation for community consultation on localised tree replacement projects, [refer Figure 17, Figure 18, and Figure 19]. This is particularly important given the substantial visual departure from the traditional, symmetrical European planting implementation this plan would entail. The advantage of having both the broadscale and the microscale visualisation producible from the single model, means that connections between the smaller and larger scale plans can be demonstrated to residents. In addition, any adjustments made at the micro scale projects can also be fed back into the suburb wide model, concurrently adjusting broadscale visual outcomes and numbers of species requiring contract growing.

Table 4. Total count of required replacement trees

Collection	Species percentage value	Total number of specimens
Flood area species	*Acacia ampliceps 45%*	984
(2187) total	*Melaleuca halmaturorum 25%*	547
	Melaleuca ericifolia 18%	394
	Banksia attenuate 12%	262
High area species	*Pinus ponderosa 78%*	1181
(1515 total)	*Tsuga canadensis 12%*	181
	Picea engelmannii 10%	151

Figure 12. Tree model species collection for High ground (non-impacted by flood) areas occurrence controlled by underlying colour of the draped flooding extent map

Figure 13. Tree model species collection for Flood impacted areas, occurrence controlled by underlying colour of the draped flooding extent map

DISCUSSION

In these studies, the authors set out to develop and test a rapid, integrated approach for precinct scaled streetscape modelling with highly detailed 3D trees to provide landscape architects and urban designers a prototype decision support system to enhance spatio-temporal design decision making for street trees.

The authors provided a review of current practice for tree modelling concluding with a proposed design decision support system drawing from both existing conventional and emerging modelling technologies in urban design alongside developments in computational botany, military and entertainment industries, hydrological modelling, as well as architectural and lighting engineering disciplines. The approach allows thousands of detailed trees to be included in a scene without making the model untenably slow to regenerate or render. The adoption of entertainment industry use of proxy-object modelling provided significant improvement on previous modelling approaches, reducing the level of abstraction required and therefore increasing the visual and spatial accuracy of the model.

To test the flexibility and capacity to address multiscale, multi-criteria tree replacement questions, the authors applied the approach at both suburb and street scale to two studies. In the case of the Beevers Street study, the approach was used to quantify tree shade provision, and in the case of Elwood, the approach was used to respond to a future flood scenario.

Figure 14. Broadscale visualisation of the future of Elwood's urban forest showing both High ground and Flood impacted areas combined

Figure 15. Broadscale aerial view from above Port Phillip Bay looking North-East showing a visualisation of the proposed future of Elwood's urban forest

Figure 16. Broadscale visualisation aerial view from above Elster Creak looking towards the East showing the proposed future of Elwood's urban forest

Figure 17. Intersection model of Flood impacted lower catchment streetscape using functional species selection. Housing has been elevated to accommodate periodic flooding

In the Beevers Street study application of the DDSS resulted in rapid modelling and testing of potential design scenarios with a clear graphic indication of visual impact of each option. The approach combined a mixture of subjective visual and cultural preferences and objective shade measurements to inform both debate and decision. The results of the shadow impact analysis [Table 1] suggest that the deciduous Option A was the preferable option in terms of ecosystem services as this option provides similar levels of shade to Option B, but allows considerably more light into the street in winter. This however does not necessarily match the visual assessment as Option B might be considered to appear 'more lush and green' throughout winter as it keeps its leaves. This ability to assess visual impact concurrently with the ecosystem service potential (in this case shade) of tree selections is a key point of difference between the DDSS presented here, and other tree modelling methods currently in use.

Figure 18. Flood impacted area visualisation to show impact of in-flood scenario

Figure 19. Higher ground area visualisation to show impact of high rainfall interception canopied trees

In the Elwood study, application of the DDSS demonstrated a method to build relationships between future street tree choices and future flood conditions, that could inform both residents and contract tree growers of impending changes. In the case of representing change to residents, the concerns are dominantly visual, while for contract tree growers, change might be the impact on their current processes and species growing lists. The method of visualisation has potential to demonstrate to residents the relationships between change at the micro and macro levels, or the bigger picture impacts of climate change and what can be done to moderate these impacts through street tree choices. The representations of alternative tree selections demonstrate the potential for raising community awareness and engagement with alternative courses of action that must be taken over the next decade to moderate impacts of climate change, while also informing more technical aspects such as changes to growing methods.

LIMITATIONS AND FUTURE RESEARCH

Once the process for building the models was developed, the models themselves were reasonably quick to generate, including the street geometry and placement of trees taking around four hours. Though the setup time for the models and initial rendering tests were greater than a post-production Adobe Photoshop™ approach, that additional upfront time would be recouped if more than two or three options were to be considered or if change over time was required for the visualisation.

Due to the system building upon commonly used software, there is strong likelihood of the DDSS overcoming issues of poor uptake in industry and be widely adopted by urban designers and landscape architects to analyse existing urban environments and test proposed design scenarios. This belief does, however, remain untested at this stage and would require further research including user-group testing.

In the future, the authors would expect to be able to integrate even more site-specific data such as soil conditions, pavement replacement timelines and some of the complexities of process-based tree modelling to further inform the design, bringing together the effects of urban environments *on tree growth* and the effect of tree growth on *urban environments* simultaneously. For example, it may be possible for growth parameters to be adjusted to reflect potential impact of available root zones. A tree species that is planted in a street with limited cubic metres of soil available can be modelled to represent stunted growth, whereas a tree that has a more generous amount of soil can be modelled with higher growth and leaf density parameters.

Another limitation is that the proposed system has been developed and tested by, and for urban designers and landscape architects with mid to high-level of operational skill in using these common industry programs. This limitation means that the uptake of the system is more likely to occur in design practices that are comfortable using the base software than in organisations that have less skills in this area. For example, for the system to be used by a council planning officer not familiar with the base software, the planning officer would require considerable training considerable work, or the substantial work would be needed to develop a user-friendly graphic user interface. The system may also need to be simplified in this process to improve the beginner user experience. Depending upon the extent of simplification required, the resulting system may lose its flexibility and power to such an extent that it does not offer any improvement on other existing tools.

As the DDSS has been set up as a flexible framework, as modules may be improved or may stagnate or may even be discontinued, this approach to the DDSS can continue to evolve. The adaptability of the system allows users to swap out modules or add new modules.

For example, in the time between the authors conducting Case 1 and Case 2, the render engine 'module' in the system was swapped out for an improved render engine to speed up and improve the quality of rendering. In the rapidly evolving field of render engines and real-time renderers, we will continue to explore other options including GPU based engines (such as Octane™ and Chaos Vantage™) and potential direct links with real-time game/render engines such as Unity™ and Twinmotion™ so that it may include a virtual reality capability. It may also be possible to swap out other components, for example, the base software could conceivably be replaced with Autodesk Maya™, Rhino 3D™, or even the open-source Blender. Alternative approaches to tree placement might be found such as using particle systems, or Grasshopper scripting.

It is also worth noting that while the base software is common in design practice and only requires the addition of cheap (around $500 US) plug-in software, in practices or local governments that do not currently use this software, total cost may be prohibitive.

CONCLUSION

This research allows further analysis of the impact of large-scale tree planting or replacements. It demonstrates to an iterative street design process where multiple scenarios and multiple species can be assessed quickly at various stages of growth with a reasonable level of accuracy, using detailed three-dimensional representations of trees in a spatial environment which provides accurate measurement of shade qualities.

This approach to modelling street trees could potentially be used by urban designers and landscape architects working with councils to supplement GIS/database decision making, allowing solar impact, flood condition response, and visual assessment alongside other less three-dimensional aspects of tree data. Current urban forest research is enabling more quantifiable strategic street tree selection and the proposed design decision support system will allow these choices to be made in a more transparent way.

Being able to communicate change involving trees in an iterative manner which expresses their visual, shade impact and flood condition responsiveness in very realistic terms may be beneficial to the study of urban forestry and presentation of the strategic decisions it makes possible and potentially be a powerful advocacy tool.

This method raises potential further investigation of the three-dimensional street qualities with the additional dimension of time – looking at how the movement of the sun over various times of the day and different angles throughout the year might influence the design of north-south or east-west streets.

The initial results of this study demonstrate that it is becoming feasible to integrate algorithmic botany modelling into street design models at precinct scale and this flexible design decision support system allows rapid scenario testing of tree placement, species choice, tree age and size, and solar impact.

In this study, the authors were looking at quantitative shade impact assessment; visual assessment; impact of tree development over time; impact of species selection based on flooding conditions; spacing; and positioning. There is scope to develop the method to output eco-system benefit, micro climate and thermal comfort mapping. There is also potential to combine the shade modelling with designated walking paths and bicycle routes to improve the human thermal comfort whilst on the route, or even reconsider the routes base on shade. The system may potentially integrate with pedestrian modelling tools such as WalkScore.com and PedCatch.com, or, just as walkonomics.com calculates the 'most beautiful route', perhaps a tool to calculate the shadiest (or sunniest) route, based on the current time of day, could be developed building upon early prototypes by White & Kimm, (2015) and more recent explorations (BRUDA, 2020; Xia et al., 2020).

The prototype design decision support system put forward here promotes consideration of the spatio-temporal characteristics of trees and provides a greater level of tree integration into climate responsive urban design than previously possible.

ACKNOWLEDGMENT

The research here builds upon work originally funded in part by the Graham Treloar Fellowship (MSD), the James Cox PhD Scholarship in Architecture (MADA), and University of Melbourne Early Career Research Grant, "Getting to yes for Nature-based solutions".

Initial findings have been presented at the following conferences 7th International Urban Design Conference Designing Productive Cities, in Adelaide, Australia; 21st International Seminar on Urban Form, in Porto, Portugal; and the 7th International Conference on Urban Tree Diversity, in Melbourne, Australia. The chapter builds upon an article published in the International Journal of E-Planning Research (IJEPR) (White & Langenheim, 2018a).

REFERENCES

Arnold, H. F. (1980). *Trees in Urban Design* (1st ed.). Van Nostrand Reinhold Company.

Batty, M. (2016). Urban Resilience: How Cities Need to Adapt to Unanticipated and Sudden Change. *Perspectives on Complex Global Challenges: Education, Energy, Healthcare, Security and Resilience: Education, Energy, Healthcare, Security and Resilience*, 169–171.

Beatley, T. (2011). *Biophilic Cities: Integrating Nature Into Urban Design and Planning*. Island Press. doi:10.5822/978-1-59726-986-5

Benes, B., Deussen, O., Pirk, S., Chen, B., Mech, R., & Ijiri, T. (2016). Modeling plant life in computer graphics. *ACM SIGGRAPH 2016 Courses on - SIGGRAPH '16*, 1–180. doi:10.1145/2897826.2927332

BOM. (2013). *2013 shaping up to be one of Australia's hottest years on record* [Climate Update]. Bureau of Meteorology. http://www.bom.gov.au/climate/updates/articles/a003-2013-temperature.shtml#july-update

BRUDA. (2020). *Cool walks*. https://www2.bcnregional.com/

Carlyle-Moses, D. E., Livesley, S., Baptista, M. D., Thom, J., & Szota, C. (2020). Urban Trees as Green Infrastructure for Stormwater Mitigation and Use. In D. F. Levia, D. E. Carlyle-Moses, S. Iida, B. Michalzik, K. Nanko, & A. Tischer (Eds.), *Forest-Water Interactions* (pp. 397–432). Springer International Publishing. doi:10.1007/978-3-030-26086-6_17

City of Melbourne. (2011). *Urban Forest Diversity Guidelines: 2011 tree species selection strategy for the City of Melbourne*. Author.

COM. (2016). *Draft Arden Vision Framework*. City of Melbourne.

Cournede, P.-H., Mathieu, A., Houllier, F., Barthelemy, D., & de Reffye, P. (2007). Computing Competition for Light in the GREENLAB Model of Plant Growth: A Contribution to the Study of the Effects of Density on Resource Acquisition and Architectural Development. *Annals of Botany*, *101*(8), 1207–1219. doi:10.1093/aob/mcm272 PMID:18037666

Coutts, Loughnan, Tapper, White, Thom, Broadbent, & Harris. (2014). *The Impacts of WSUD solutions on human thermal comfort.* Green Cities and Micro-Cilmate B3.1 Cooperatice Research Centre for Water Sensitive Cities.

De los Rios, I. C., & Charnley, F. J. S. (2017). Skills and capabilities for a sustainable and circular economy: The changing role of design. *Journal of Cleaner Production*, *160*, 109–122. doi:10.1016/j.jclepro.2016.10.130

de Reffye, P., Edelin, C., Francon, J., Jaeger, M., & Puech, C. (1988). Plant models faithful to botanical structure and development p151-de_reffye 1988.pdf. *Computer Graphics*, *22*(4), 151–158. doi:10.1145/378456.378505

Di Nunzio, A. (2019). *Envi_MET.* https://www.envi-met.com/learning-support/plugins/

Dobbs, C., Kendal, D., & Nitschke, C. R. (2014). Multiple ecosystem services and disservices of the urban forest establishing their connections with landscape structure and sociodemographics. *Ecological Indicators*, *43*, 44–55. doi:10.1016/j.ecolind.2014.02.007

Durdyev, S., Mbachu, J., Thurnell, D., Zhao, L., & Hosseini, M. R. (2021). BIM Adoption in the Cambodian Construction Industry: Key Drivers and Barriers. *ISPRS International Journal of Geo-Information*, *10*(4), 215. doi:10.3390/ijgi10040215

Ely, M. (2009). Planning for Trees in Urban Environments. *TREENET Proceedings of the 10th National Street Tree Symposium 3rd and 4th September*, 87–98.

Escobedo, F. J., Kroeger, T., & Wagner, J. E. (2011). Urban Forests and Pollution Mitigation: Analyzing Ecosystem Services and Disservices. *Environmental Pollution*, *159*(8), 2078–2087. doi:10.1016/j.envpol.2011.01.010 PMID:21316130

Ewing, R., & Handy, S. (2009). Measuring the Unmeasurable: Urban Design Qualities Related to Walkability. *Journal of Urban Design*, *14*(1), 65–84. doi:10.1080/13574800802451155

Geertman, S., & Stillwell, J. (2009). Planning Support Systems: Content, Issues and Trends. In S. Geertman & J. Stillwell (Eds.), *Planning Support Systems Best Practice and New Methods* (Vol. 95, pp. 1–26). Springer Netherlands. doi:10.1007/978-1-4020-8952-7_1

Grêt-Regamey, A., Sirén, E., Brunner, S. H., & Weibel, B. (2017). Review of decision support tools to operationalize the ecosystem services concept. *Ecosystem Services*, *26*, 306–315. doi:10.1016/j.ecoser.2016.10.012

Hussnain, M. Q., Waheed, A., Anjum, G. A., Naeem, M. A., Hussain, E., Wakil, K., & Pettit, C. J. (2020). A framework to bridge digital planning tools' utilization gap in peri-urban spatial planning; lessons from Pakistan. *Computers, Environment and Urban Systems*, *80*, 101451. doi:10.1016/j.compenvurbsys.2019.101451

Jones, B. I. (2020). A study of Building Information Modeling (BIM) uptake and proposed evaluation framework. *Journal of Information Technology in Construction*, *25*, 452–468. doi:10.36680/j.itcon.2020.026

King, C. (2017). *Armidale Regional Council.* Armidale CBD Revitalisation Project. https://yoursay.armidale.nsw.gov.au/32201/widgets/185587/documents/68266

Kota, S., Haberl, J. S., Clayton, M. J., & Yan, W. (2014). Building Information Modeling (BIM)-based daylighting simulation and analysis. *Energy & Buildings. Complete*, *81*, 391–403. doi:10.1016/j.enbuild.2014.06.043

Kwon, S., Kim, B. C., Mun, D., & Han, S. (2018). User-assisted integrated method for controlling level of detail of large-scale B-rep assembly models. *International Journal of Computer Integrated Manufacturing*, *31*(9), 881–892. doi:10.1080/0951192X.2018.1448114

Langenheim, N., White, M., Tapper, N., Livesley, S. J., & Ramirez-Lovering, D. (2020). Right tree, right place, right time: A visual-functional design approach to select and place trees for optimal shade benefit to commuting pedestrians. *Sustainable Cities and Society*, *52*, 101816. doi:10.1016/j.scs.2019.101816

Lawrence, H. W. (2008). *City trees: A historical geography from the Renaissance through the nineteenth century*. University of Virginia Press.

Lindberg, F., & Grimmond, C. (2011). The influence of vegetation and building morphology on shadow patterns and mean radiant temperatures in urban areas: Model development and evaluation. *Theoretical and Applied Climatology*, *105*(3–4), 311–323. doi:10.100700704-010-0382-8

MacEachren, A. M. (1995). *How Maps Work: Representation, Visualization, and Design*. Guilford Press.

Maher, B. A., Ahmed, I. A., Davison, B., Karloukovski, V., & Clarke, R. (2013). Impact of Roadside Tree Lines on Indoor Concentrations of Traffic-Derived Particulate Matter. *Environmental Science & Technology*, *47*(23), 13737–13744. doi:10.1021/es404363m PMID:24215538

McPherson, E. G., & Peper, P. J. (2012). Urban Tree Growth Modeling. *Arboriculture & Urban Forestry*, *38*, 175–183.

Měch, R., & Prusinkiewicz, P. (1996). Visual models of plants interacting with their environment. *Proceedings of the 23rd Annual Conference on Computer Graphics and Interactive Techniques - SIGGRAPH '96*, 397–410. 10.1145/237170.237279

Melbourne Water Corporation. (2012). *Planning for sea level rise guidelines*. Author.

MMBW. (1979). *Elster Creek Drainage System Development* (No. D0022). Melbourne Meltropolitan Board of Works. https://www.portphillip.vic.gov.au/Elster%20Creek%20Drainage%20System%20Development_MMBW_Dec%201979%20(1).pdf

Mohd Sanusi, A. R. (2015). *The importance of street trees for human thermal comfort in a warming climate*. Academic Press.

Moreland. (2009). *Moreland Landscape Guidelines*. MCC.

Norton, B., Coutts, A., Livesley, S., Harris, R., Hunter, A., & Williams, N. (2015). Planning for cooler cities: A framework to prioritise green infrastructure to mitigate high temperatures in urban landscapes. *Landscape and Urban Planning*, *134*, 127–138. doi:10.1016/j.landurbplan.2014.10.018

O'Shea, J. (2020, September 2). *First phase of work to transform MacCurtain Street starts tomorrow*. Cork Beo. https://www.corkbeo.ie/news/local-news/work-transform-maccurtain-street-victorian-18865812

Oke, T. R. (1988). Street design and urban canopy layer climate. *Energy and Building*, *11*(1), 103–113. doi:10.1016/0378-7788(88)90026-6

Ong, B. L. (2003). Green plot ratio: An ecological measure for architecture and urban planning. *Landscape and Urban Planning*, *63*(4), 197–211. doi:10.1016/S0169-2046(02)00191-3

Parisi, A. V., Kimlin, M. G., Wong, J., & Wilson, M. (2000). Personal Exposure Distribution of Solar Erythemal Ultraviolet Radiation in Tree Shade over Summer. *Physics in Medicine and Biology*, *45*(2), 349–356. doi:10.1088/0031-9155/45/2/307 PMID:10701508

Parisi, A. V., Kimlin, M. G., Wong, J. C., & Wilson, M. (2001). Solar ultraviolet exposures at ground level in tree shade during summer in south east Queensland. *International Journal of Environmental Health Research*, *11*(2), 117–127. doi:10.1080/09603120020047500 PMID:11382345

Pirk, S., Stava, O., Kratt, J., Said, M. A. M., Neubert, B., Měch, R., Benes, B., & Deussen, O. (2012). Plastic trees: Interactive self-adapting botanical tree models. *ACM Transactions on Graphics*, *31*(4), 1–10. doi:10.1145/2185520.2185546

Prusinkiewicz, P., Lindenmayer, A., & Hanan, J. (1990). *The Algorithmic Beauty of Plants*. The Virtual Laboratory. doi:10.1007/978-1-4613-8476-2

Rebollo, C. (2007). *An efficient continuous level of detail model for foliage*. Academic Press.

Rebollo, C., Remolar, I., Chover, M., & Ripollés, O. (2006). An Efficient Continuous Level of Detail Model for Foliage. *WSCG'2006*.

Rebollo, C., Remolar, I., Gumbau, J., & Chover, M. (2014). Three-dimensional trees for virtual globes. *International Journal of Digital Earth*, *7*(10), 789–810. doi:10.1080/17538947.2013.783126

Rizmal, Z. (2019, October 24). *Hate Melbourne's plane trees? Relief is coming, but don't hold your breath*. ABC News. https://www.abc.net.au/news/2019-10-24/plane-trees-to-be-culled-in-melbourne-city/11635846

Rogers, B., Bertram, N., Gunn, A., Löwe, R., Murphy, C., Pasman, R., Radhakrishnan, M., Urich, C., Wong, T., & Arnbjerg-Nielsen, K. (2016). An interdisciplinary approach to identify adaptation strategies that enhance flood resilience and urban liveability. *IWA World Water Congress & Exhibition*.

Shashua-Bar, L., Tsiros, I. X., & Hoffman, M. E. (2010). A Modeling Study for Evaluating Passive Cooling Scenarios in Urban Streets with Trees. Case Study: Athens, Greece. *Building and Environment*, *45*(12), 2798–2807. doi:10.1016/j.buildenv.2010.06.008

Sprague, R. H. (1980). A Framework for the Development of Decision Support Systems. *Management Information Systems Quarterly*, *4*(4), 1. doi:10.2307/248957

Szota, C., McCarthy, M. J., Sanders, G. J., Farrell, C., Fletcher, T. D., Arndt, S. K., & Livesley, S. J. (2018). Tree water-use strategies to improve stormwater retention performance of biofiltration systems. *Water Research*, *144*, 285–295. doi:10.1016/j.watres.2018.07.044 PMID:30048867

Taylor, M. S., Wheeler, B. W., White, M. P., Economou, T., & Osborne, N. J. (2015). Research note: Urban street tree density and antidepressant prescription rates—A cross-sectional study in London, UK. *Landscape and Urban Planning*, *136*, 174–179. doi:10.1016/j.landurbplan.2014.12.005

Urich, C., Bach, P. M., Sitzenfrei, R., Kleidorfer, M., McCarthy, D. T., Deletic, A., & Rauch, W. (2013). Modelling cities and water infrastructure dynamics. *Proceedings of the Institution of Civil Engineers. Engineering Sustainability*, *166*(5), 301–308. doi:10.1680/ensu.12.00037

Urich, C., & Rauch, W. (2014). Exploring critical pathways for urban water management to identify robust strategies under deep uncertainties. *Water Research*, *66*, 374–389. doi:10.1016/j.watres.2014.08.020 PMID:25240118

Weber, J., & Penn, J. (1995). Creation and rendering of realistic trees. *Proceedings of the 22nd Annual Conference on Computer Graphics and Interactive Techniques*, 119–128.

White, M., & Kimm, G. (2015). Measuring sky view factor of urban canyons using hacked Gopro hemispheric video processing. In R. H. Crawford & A. Stephan (Eds.), *Research for a Better Built Environment: 49th International Conference of the Architectural Science Association 2015* (pp. 525–535). Faculty of Architecture, Building and Planning (UniMelb). http://apo.org.au/node/61762

White, M., & Langenheim, N. (2018). A Spatio-Temporal Decision Support System for Designing with Street Trees. *International Journal of E-Planning Research*, *7*(4), 1–24. doi:10.4018/IJEPR.2018100101

Xia, T., Jatowt, A., Wang, Z., Si, R., Zhang, H., Liu, X., Shibasaki, R., Song, X., & Kim, K. (2020). CoolPath: An Application for Recommending Pedestrian Routes with Reduced Heatstroke Risk. In S. Di Martino, Z. Fang, & K.-J. Li (Eds.), *Web and Wireless Geographical Information Systems* (pp. 14–23). Springer International Publishing. doi:10.1007/978-3-030-60952-8_2

Yang, B., Li, S., Elder, B. R., & Wang, Z. (2013). Community-planning approaches and residents' perceived safety: A landscape analysis of park design in the Woodlands, Texas. *Journal of Architectural and Planning Research*, *30*(3), 311–327.

Yang, Y., Gatto, E., Gao, Z., Buccolieri, R., Morakinyo, T. E., & Lan, H. (2019). The "plant evaluation model" for the assessment of the impact of vegetation on outdoor microclimate in the urban environment. *Building and Environment*, *159*, 106151. doi:10.1016/j.buildenv.2019.05.029

ENDNOTE

1 For Test Case 1, authors used 'Mental Ray Proxy-Objects', and 'V-Ray Proxy-Objects' for Test Case 2.

Chapter 8
Health Mobility, Telemedicine, and Local Areas:
Exploring the Relationships in the Piedmont Region

Sylvie Occelli
Istituto di Ricerche Economico Sociali del Piemonte, Italy

Marco Dalmasso
Servizio Sovrazonale di Epidemiologia, ASL TO3, Italy

ABSTRACT

Telemedicine solutions help improve the delivery of and access to healthcare services. The recent pandemic has given a burst to the use of telemedicine services, allowing to overcome or set aside, at least temporarily, some of the difficulties which usually hamper the implementation of innovative services. It has also spurred a new attention to the issues of access to healthcare services, and to the role telemedicine services could play in overcoming or reducing the physical distance between patients and healthcare premises. This study makes a preliminary step to sharpen the topics for the Piedmont region. Data retrieved from the Regional Health Information System are used to profile the health mobility pattern in the regional local areas, through a set of spatial performance indicators. A few hypotheses of the possible impact of telemedicine on health mobility are formulated and their impact is investigated in an exploratory analysis of the regional health flows.

INTRODUCTION

Telemedicine solutions improve the delivery of and access to healthcare services (WHO, 2010). Their importance has increased significantly during the past decade (European Commission, 2018; Joint Action to support the eHealth Network, 2017) and the recent pandemic had given them a further burst.[1] It has spurred government response to accelerate the deployment of TM services, providing new evidence of the benefits of digital health to professionals and the public, such as making access easier to healthcare

DOI: 10.4018/978-1-7998-9090-4.ch008

services (Bashur et al., 2020; Esper et al., 2020; Ferrara, 2021; HIMSS, 2021; National Academies of Sciences, Engineering, and Medicine, 2021).

The pandemic experience has also revealed that the level of implementation of TM services varies considerably across areas and that in Europe connectivity and integration gaps exist in several Member States (HISSM, 2021a). This is the case of Italy, where a survey of TM initiatives, carried out before the pandemic (Ministero della Salute, 2018) clearly showed a patchy pattern in terms of the type of services and implementation level across the regions (see Box 1).

Latest reflections about the TM experiences shed new light on the fact that:

- TM initiatives are developed in situated complex socio-technical contexts which include many different stakeholders, legal, organisational and financial conditions as well as a set of behavioural and ethical concepts (Occelli & Scelfo, 2020);
- implementation of these initiatives proceeds along a path of continuous improvement based on a scale of maturity stages. By questioning the models' feasibility to assess telemedicine initiatives proactively developers can have a practical support to guide them in the process (Otto, Whitehouse & Schlieter, 2019; Van Dyk, 2014);
- to put in practice TM projects needs a supportive policy environment which facilitates, enables and promotes the spread of best practices through knowledge transfer and dissemination (Cravo & Hashiguchi, 2020). To reap the benefits of telemedicine and limit its risks, actually, care systems should engage in continuous learning and improvement, taking advantage of new and innovative care models, and exploiting all sources of data (OECD, 2019b; European Commission, 2020).

The last remarks, in particular, suggest that enquiring TM benefits and risks should embrace a perspective which acknowledges issues of access to healthcare without neglecting the latest societal concerns about sustainability.

Access is a complex notion entailing a few dimensions of health services which are fundamental if universal health coverage is to be attained, including: the definition of healthcare needs, the strategy to seek healthcare services, the possibility to reach, obtain or use health care services and to actually have the need for services fulfilled (Levesque, Harris & Russell, 2013). Measuring these aspects of access also requires to take into account their socio-organizational and/or geographical attributes (Aday & Andersen, 1974). The former makes reference to all those characteristics of the resources, other than spatial attributes, that either facilitate or hinder the efforts of users to obtain care. The latter refers to the fact that to get care users must cover a distance (use time) and this involves a cost. In addition, their willingness to seek care depends on their attitudes toward and knowledge about health care and the social and cultural definitions of illness they have learned.

Discussion about health access has emphasized that (Aday & Andersen, 1974; Levesque, Harris & Russell, 2013 OECD, 2019a; Penchansky & Thomas, 1981):

- evidence of access is use of service, not simply the presence of a facility;
- access can be measured by the level of use in relation to need;
- users and professionals evaluate need differently;
- barriers to access are not only financial but also psychological, informational, social, organizational, spatial, temporal, etc.;

- access has a pivotal role for equity as health systems are equitable if they provide access to services based on need.

Access to healthcare is also at the core of the present study for the Piedmont region although it mainly focuses on the spatial dimension of access.

Rationales of the study are manifold but the following ones are worth recalling.

First, travel restrictions imposed by the Covid-19 made it difficult for many Piedmont citizens to reach healthcare services. The situation drew attention to the role TM services could play to overcome the distance barrier separating patients and healthcare premises and spurred Piedmont regional government to issue regulations for telemedicine delivery.

Second, earlier studies of the 2013-17 change in Piedmont health mobility revealed that the length of journeys to get to the healthcare premises increased in the period, in spite of a reduction in the number of journeys (Landini et al., 2017; Occelli & Dalmasso, 2019). The fact that this twofold phenomenon was particularly significant for specialist care explains the interest for this category of healthcare services.

Finally, mobilising a knowledge which can bridge summative and formative research for designing and evaluating telemedicine services (Ekeland, Bowesc, & Flottorpd, 2011) would be a valuable contribution to regional telemedicine strategy. The approach taken in the study in fact, complements some of the analytic steps for assessing the impact of TM services (Kidholm et al., 2012). More specifically, it provides evidence for better understanding the regional healthcare context and reasoning about consequences of the introduction of the new healthcare services. Questions it helps address include: how many patients would benefit of an easier access to healthcare services in the case telemedicine is used for reaching those services located beyond a certain distance threshold from their residence?; how many journeys could be saved and travelled distances reduced?; which regional sub-areas would most profit from these changes?; to what extent the number of estimated telemedicine users might be served, depending on the available resources allotted to telemedicine services by healthcare providers?.

Answering these questions can be a part of the continuous learning activity which is required to make TM service design and delivery more capable to meet population's health needs in the regional sub-areas.

Telemedicine Projects in Italy in 2018

A survey of TM projects was conducted by the Italian Health Ministry just before the pandemic broke out in the late 2019 (Ministero della Salute, 2018).

269 active TM projects were reported which concentrated in the more populous regions of the country. The survey provided an overview of the efforts undertaken in the country to implement these innovative health services. It also made clearly apparent that TM experiences do vary markedly across regions. This is not surprising as Italian regions have a certain autonomy at seeing to the provision of health services, including innovative ones such as TM. In this respect it is worth reminding that Italy's National Health Service (NHS) is decentralised and regionally based. The central government allocates general tax revenues for financing public health care, defines the benefit package and exerts overall stewardship. Each region is responsible for organizing and delivering health services through local health units and via public and accredited private hospitals (OECD, 2021).

As for the types of TM services, the main ones were specialist tele-consultations (30%), tele-reporting (24%) and tele-monitoring (21%). The shares vary appreciably within regions Fig.1a, with Piedmont and Veneto having the most similar values to the national ones.

At national level, 1 out of 4 patients treated with TM services were outpatients suffering multi-chronic diseases, Fig. 1b. The proportion is almost threefold higher in regions with richer TM experiences, such as Lombardia, Piemonte and Sicilia; it is however much lower (10%) in Emilia Romagna, the Italian region with the greatest number of TM initiatives.

Differences among regions also exist when considering the efforts made to evaluate the (expected) consequences of TM projects, regarding a few dimensions, such as the clinical effectiveness, the patients' perception, the impact on the budget and the organization and the legal and ethical aspects of the new services (Kidholm et al., 2012). Overall, the organizational impact, the clinical effectiveness and ethical and legal aspects are the dimensions most frequently assessed by more than 30% of the surveyed projects, Fig. 1c. Both assessment rate and prioritization of dimensions vary across regions. In Lazio, for example, the impact of ethical and legal aspects is given priority by more than 60% of the projects which have been evaluated. In Piedmont assessment rates for all the dimensions is higher than the average ones, with clinical effectiveness being the prioritized dimension.

The remainder of this paper is organized in three parts. The first presents the approach strategy which guided the study of the TM-access in the Piedmont region. Concepts of health access are recalled and their relationships with mobility highlighted. Then the health mobility data to specialist care used in the study are introduced and the information sources illustrated. The second part explains the methodological underpinning of the study. Building on the existing literature, a set of Spatial Information Indicators are implemented. They help profile the Piedmont health mobility at sub-regional level and appreciate its changes as TM services may reduce or reshape patients' needs to reach healthcare facilities. A descriptive framework of health mobility is also proposed which gives ground to the formulation of a few hypotheses for investigating the impact of TM on the regional health mobility. To show the information potential of the exploratory analysis, a selection of results is discussed in the third part which progressively zoom from regional to zone levels. Lastly, conclusions summarize the main findings of study and provides suggestions for future research.

THE STUDY APPROACH

The investigation of TM-access relationships belongs to a broad multi-disciplinary approach to health systems which aims to give practical ground to concepts related to effective coverage and technical efficiency of healthcare (Durán et al., 2012).

Recently, concerns about care access by an increasing number of people in frail conditions and attention to the role of active living in healthy-ageing are fostering research interest in people's daily mobility. Greater consideration is given to how geographical, spatial and environmental aspects related to mobility directly or indirectly bear upon people's health and ultimately affect their quality of life (Feuillet et al., 2019; Perchoux et al., 2014).

Figure 1. An overview of (active) telemedicine initiatives in the Italian regions, 2018
(*) Total number of initiatives is shown at the bottom of the bars. Regions are ordered by decreasing number of initiatives in the North, Centre and South areas of the country. Regions with missing patient data are not shown.
Source:(Ministero della Salute)

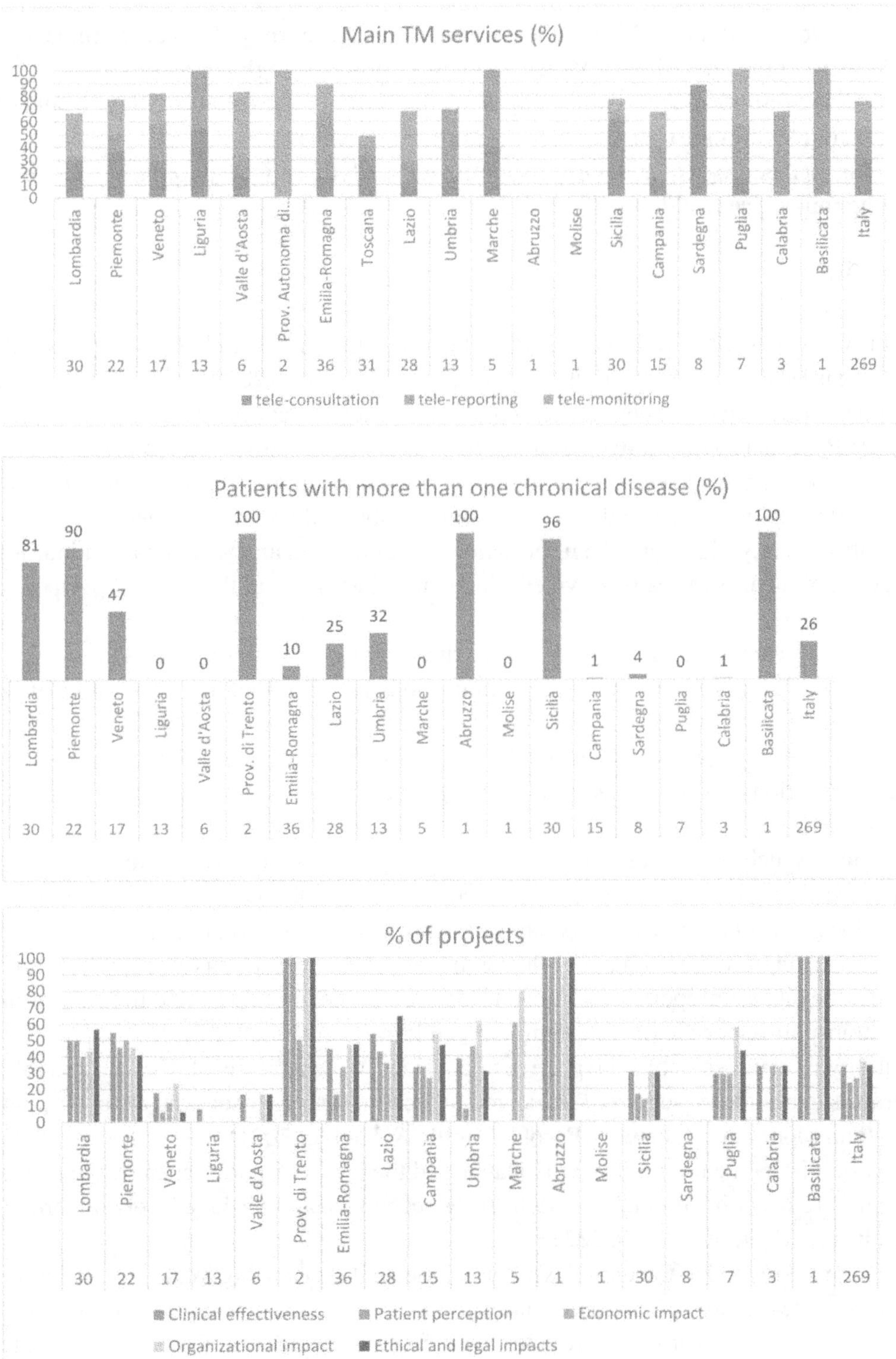

While sharing some of these issues, this study is part of an ongoing activity, undertaken in Piedmont to monitor and analyse the functioning of the regional health system (Costa et al., 2017; Occelli & Dalmasso, 2019). It is motivated by three facts:

- the first is substantial and holds that health mobility, e.g. journeys to healthcare, is a main reference domain for the analysis of TM-access relationships at a regional level;
- the second is operational and acknowledges the opportunity to exploit existing (health) data regularly gathered for administrative and/or institutional purposes;
- the third is methodological and purports to provide analytic support approaches for designing and managing innovative health services,

Health Mobility

Health mobility is the primary variable considered in the study.[2] The number of journeys to health services can be regarded as a measure of revealed accessibility (Gatrell & Wood, 2012) and represent the actual use of health services. Journeys are a main correlate of physical accessibility which, according to Evans, Hsua & Boerma (2013, p.546), "is understood as the availability of good health services within reasonable reach of those who need them and of opening hours, appointment systems and other aspects of service organization and delivery that allow people to obtain the services when they need them".

This definition clearly highlights the information potential of an analysis of patient journeys to draw a spatial account of the use of health services which takes into account the twofold perspective:

- of population's care needs in the areas where the health journeys are generated;
- of the delivering capacity of healthcare organisations in the destination areas of health journeys.

Both these perspectives play a role in developing TM services as their joint articulation would help prioritizing interventions for those areas where health needs are high but services are poor and/or difficult to reach.

Spatial attributes such as distance, travel times and transportation means are main correlated attributes of physical accessibility to reach healthcare. A recent study by OECD (2019b), for example, provides evidence that long travelling or distance is likely to delay or forego care and that this affects to a greater extent people living in rural area or having a lower income. According to this study, Italy is the European country having the highest proportion of unmet health needs due to distance and the only country where distance and transport issues are greater in urban than in rural areas (OECD, 2019b, p. 142).

Sustainability is an unavoidable perspective to consider in the analysis of TM - access relationships. By saving travels to healthcare service, TM can reduce access costs for users and the externalities caused by transport (emissions and accidents) (Purohit, Smith, & Hibble, 2021).

This study builds upon earlier works which analysed the journeys undertaken by the residents of the Piedmont region to reach health facilities delivering secondary and tertiary levels of care· (Landini et al., 2019; Dalmasso & Occelli, 2019, 2021)

Levels of care are related to the complexity of the medical cases being treated as well as to the skills and specialities of the providers. They are usually divided into the categories of primary care, secondary care, and tertiary care: a primary care provides the essentials medical care and is responsible for coordinating care among specialists and other levels of care. Primary care providers (PCP) may be doctors, nurse practitioners, or physician assistants; b. secondary care delivers specialist care. Specialists focus either on a specific system of the body or a specific disease or condition; c. tertiary care (Hospitalization) requires highly specialized equipment and expertise and involves complex treatments or procedures.

The present study focuses on the 2019 journeys for specialist care, about 15 million, which represent by far the largest share of regional health mobility (85%). It is worth reminding that between 2013 and 2019 regional access to these services dropped by about 4%, more than the 3% fall observed in the regional population, whilst the total travelled distance for inter-municipality journeys increased by about 8%. The study area consists of 47 zones which served as a reference in regional transport analysis (see Box 2).

An Outlook at the Piedmont Area

Situated in the North-west of the country, Piedmont has a population of 4.3 million inhabitants, most of which (85%) live in the 21 urban zones of the study area. The other 26 non urban (rural) zones include the mountainous and less populated municipalities of the region, Figure 2. Population density in the urban zones, 809 inhabitants for 100 ha, is more than 5 times higher than that in rural areas (non urban zones).

Piedmont is an area where population ageing is putting pressure on the regional developmental path and on healthcare delivery as well. In 2019, the old-age dependency ratiowas as high as 203, with the highest values in the Biella and Alessandria provinces (262 and 243, respectively) (Osservatorio demografico territoriale del Piemonte, 2022). The regional indicator value is expected to grow by 20% in 2038, with the most noticeable increases (above 25%), in the Novara province and Turin metropolitan area which concentrates more than half of the regional population.

Figure 2. The study area

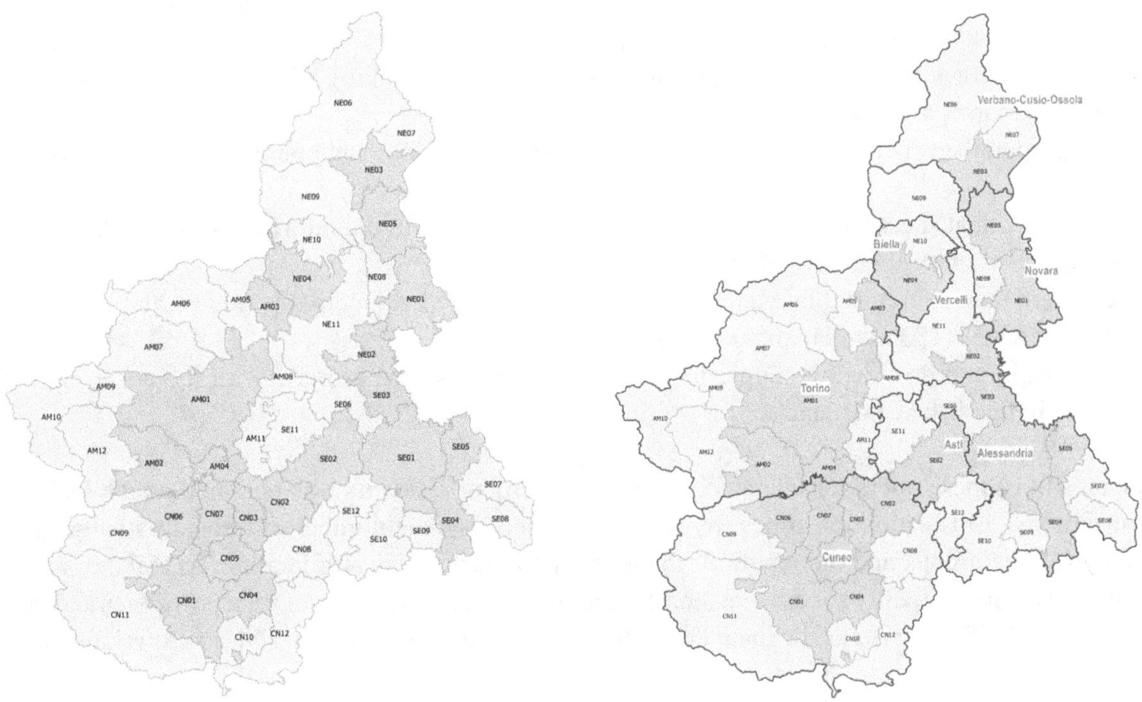

a. The 47 zones (urban areas are grey coloured)

b. The study area

Information Sources

The study exploits (health) data regularly gathered for administrative and/or institutional purposes which fuel the Regional Health Information System. This is a comprehensive and heterogeneous data platform and consists of several information sources. In this study the following ones are used:

- information about the population and patients by residence zone and age group;
- the yearly number of access to specialist care services (laboratories and medical cabinets for consultation or diagnostic exams) by the resident population. These data are collected for administrative purpose according to protocols defined by national and regional regulations;
- the location of healthcare facilities delivering specialist care services (only those which are enrolled in the National and Regional Health Service are considered);
- general information about socio-economic and geographical characteristics of municipalities and regional sub-areas (provinces, planning local areas, health districts, etc.) which belong to the institutional database of the Piedmont Region. Included in this group is a 1181x1181 matrix table of travel distances between the centroids of the 1181 municipality areas. Distances are measured on the regional road network and are an output of the regional transportation model managed by 5T mobility service company (www.5t.torino.it).

Access to most of the above information sources is restricted to healthcare personnel or authorized users and submitted to GDPR privacy regulations. Data treatment has been performed using anonymized information according to the GDPR guidelines.

A number of operations were carried out in order to fix the databases at the municipality level as this is regarded as a convenient spatial unit for applying a meso level approach to health mobility. It enables an observation lens which is between a macro regional approach as applied in most health mobility studies, and a micro approach, as used in epidemiological study to investigate individuals' health determinants (Costa et al. 2017). It also permits a straightforward linking of health mobility data with other socio-economic and geographical data.

Building upon the experience gained in a previous work, the following operations were undertaken:

- 2019 individual access data to specialist care were merged with the information about the population enrolled in local health and classified by 6 age groups;
- data were then aggregated by patients' residence zone (1181 municipality codes) and health care delivery zones (150 municipality codes);
- each origin and destination couple in the access data was then merged with the corresponding couple in the distance database.

As a result of these operations a database for the Piedmont residents was obtained which includes: 1181 origin and 150 destination zones (by municipality codes), the number of treated patients and of access by (6) age groups and the length of the journey to reach the health facility. To make easier the computation the municipality zones were then complemented with the codes of transport zones and provinces (see Fig. 2). This serves as the main data reference for the analysis of the TM-access relationships.

METODOLOGICAL UNDERPINNINGS

Measuring Health Journeys

To investigate health mobility in the Piedmont region, reference is made to the Spatial Performance Indicators (SPI), an approach which became popular in the eighties as government strategies (mainly in the UK) called for adequate indicators to measure efficiency and performance in public services and nationalized industries (Clarke & Wilson, 1994).

A distinctive feature of these indicators is that they take into account the spatial relationships existing among zones, and make it possible to assess (health) services from the twofold point of view of users (patients), in their residence zone, and of (healthcare) providers, in their facility location.

The information sources described in the previous section have been used to implement a number of SPI indicators, and more specifically:

a. Indicators accounting for the spatial pattern of patients' health treatment:

a1. proportion of patients in age group k treated within a zone, wz^k_{ii}.

$$wz^k_{ii} = T^k_{ii}/\Sigma_j T^k_{ij} \quad (1a)$$

a2. a complementary indicator is the proportion of patients treated in a surrounding zone ws^k_{ij},

$$ws^k_{ij} = T^k_{ij}/\Sigma_j T^k_{ij} \quad (1b)$$

where T_{ij} is the flow of patients from residence in i to health facilities in j. In this study the type of health service is not distinguished. Indicator (1a), also known as self-containment indicator, is commonly used for describing zone spatial profiles. In this case, a higher indicator value would mean that the services delivered by the health facilities located within the zone are more capable to treat the patients living within the zone.

Note that the term "flow of patients" represents both the functional and spatial relationships between demand and supply of healthcare. On a functional ground it accounts for health access, that is the number (and type) of patients using health services in a certain time interval. On a spatial ground, it measures the number of journeys made by patients to get a care. Introducing TM services is expected to decouple the two relationships, reinforcing the functional and reconfiguring (reducing) the spatial.

b. Indicators measuring the distance covered by patients in their journeys to reach health facilities (here the articulation by k population group is dropped):

b1. total distance travelled by residents in zone i to get to health services in zone j:

$$D_i = \Sigma_j T_{ij}\, d_{ij} \quad (2a)$$

b2. average length of a journey (kms) for residents in zone *i* to reach health services:

$$AD_i = \Sigma_j T_{ij} d_{ij}/\Sigma_j T_i \quad (2b)$$

b3. average distance (kms) travelled by residents needing care in zone *i* to reach health services:

$$ADP_i = \Sigma_j T_{ij} d_{ij} / PH_i \quad (2c)$$

where d_{ij} is the distance from residence in *i* to health facilities in *j* and PH_i is the population having a care need;

b4. journey average distance (kms) per resident needing care in zone *i* to reach health services:

$$ADPF_i = \Sigma_j T_{ij} d_{ij} / f_{i*} MH_i \quad (2d)$$

where f_i is frequency of health journeys for patients moving outside their residence zone (MH_i).

c. Indicators measuring the health burden in an area. A general indicator can be expressed as the ratio between the population having health needs PH_i (patients) and the residential population P_i:

$$B_i = PH_i / P_i \quad (3)$$

d. Indicators measuring the treatment rate in an area. They can be distinguished by considering;

d1. the total number of health flows generated in a residential zone i, divided by the total number of patients:

$$R_i = \Sigma_j T_{ij} / PH_i \quad (4a)$$

Indicator (4a), therefore, is the number of health journeys averagely made by patients in zones i;

d2. the number of health flows moving outside residential zones i, divided by the respective number of patients (MH_i):

$$Q_i = \Sigma_j T_{lj} / MH_i$$

with l not equal i (4b)Indicator (4b) measures the frequency of health journeys made by patients to reach healthcare premises outside their residence zone i.

By properly articulating the indicators by population groups and/or types of services, and summing over i or j, a whole set of measurements of the use of health services can be obtained at different sub regional spatial levels, such as the (47) transport zones and the (8) provinces.

It is worth noting, furthermore, that most of these indicators can be easily updated as they use data which are regularly gathered for institutional purposes.

A Framework of Analysis

The scheme in Fig. 3 shows the main variables underlying health mobility and how they are linked as a result of the social, functional and spatial relationships existing among them. It provides a viewpoint to observe the regional health system.

Three types of variables are distinguished: a. context variables which represent those components of a regional health system which structure and affect the health mobility in the area (green coloured in

Fig. 3); b. primary (level) variables which account for the generation of patient flows and characterize their spatial profile (grey coloured in Fig.3); c. secondary variables associated with attributes of flows which are targeted in investigating regional health mobility (yellow coloured in Fig. 3). These are typically expressed through (SPI) indicators.

In addition to the connections between variables, the scheme draws attention to a few feed-back relationships, which may affect some context variables under the control of policy decision-makers and healthcare organizations.

The scheme also serves as a reference framework to guide an investigation of the TM-access relationships. It helps disentangle the questions about the type of impacts TM may have on the health mobility for specialist care in the regional sub-areas. More importantly, it prompts us develop a system view of the issue at stake, that is to see how things are connected to each other within some notion of a whole entity (Peters, 2014).

Figure 3. A framework for the analysis of health mobility

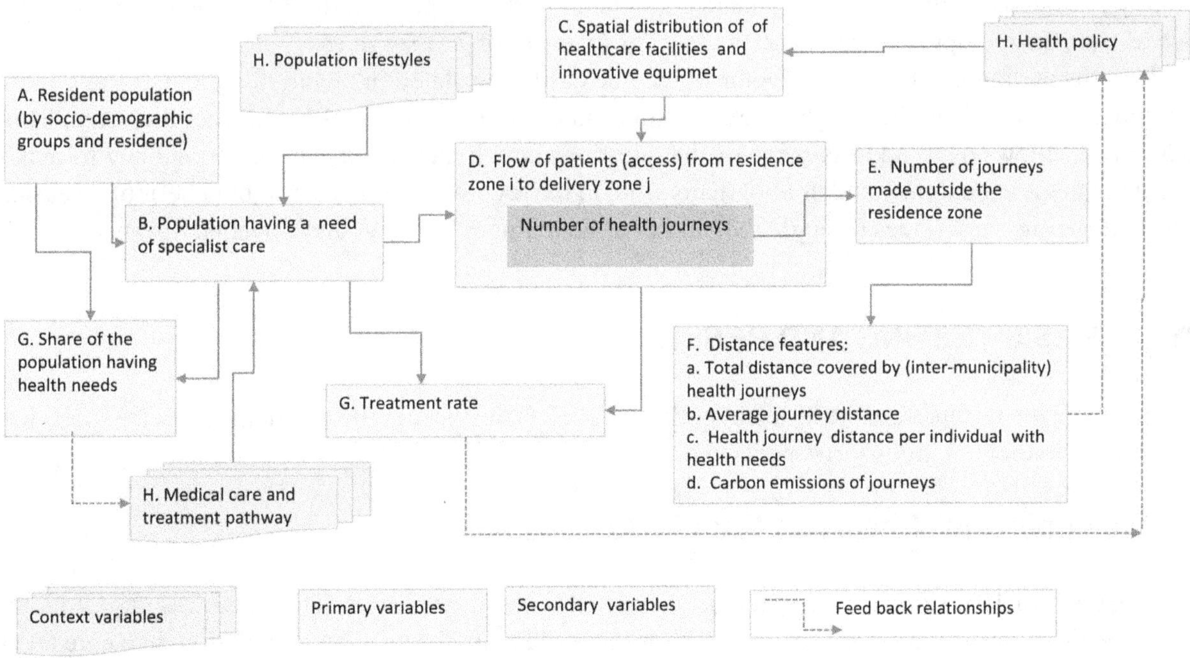

For example, it reminds us that, by enabling care at a distance, TM services can eliminate or reduce the need to travel to get care (see box D-E-F in Fig. 3), thus saving time and costs for patients and caregivers. In addition, although the topic is neglected in this study, TM can lead to a reduction of carbon emissions; as pointed out by the recent literature (Oliveira, Barlow & Gonçalves, 2013), these reductions may be not negligible and should be considered when exploring actions to reduce carbon footprints of health care systems.

Second, it suggests that by making available innovative ways of providing medical care (see Box H in Fig. 3), TM can offer patients a wider range and higher level of healthcare services, which would positively influence the outcome of their accessing. Finally, it gives cues that, through more timely disease prevention services, TM may have beneficial effects on population's health and wellbeing and on the long term, by changing lifestyle, reduce the share of population needing care (see Box H in Fig. 3).

In this study only the first type of impacts is addressed. The analysis, therefore, assumes that TM services enable patients to get care at their residence place and avoid travelling to the places where specialist healthcare services are delivered.

Here, travel avoidance is purported on the basis of three hypotheses:

- HP1. TM is used for reaching those services which are located beyond a certain distance threshold from patients' residence;
- HP2. TM is used for reaching care services which are delivered by those healthcare facilities providing telemedicine services;
- HP3. TM is used by those patients who have to travel frequently to get medical care.

The formulation of these hypotheses is posited taking into account the features of health mobility database, discussed earlier. It however neglects some important aspects in the TM-access relationships, such as the clinical aspects of care and the level of maturity of TM services.

Notwithstanding this lack, the hypotheses can be easily translated in "what if" conditions and their consequences investigated through an exploratory analysis of the regional health mobility data. This undertaking allows us to address some of the questions which have been raised, and namely to make educated guesses about the number of patients who, in the regional sub-areas, might benefit of an easier access to healthcare services through TM, and of a reduction in journeys and travelled distances.

HYPOTHESIS TESTING AND RESULTS

To appreciate the results from the tested hypotheses, a brief overview of the health mobility for specialist care in the Piedmont region is first presented.

Health Mobility for Specialist Care in 2019

In 2019, 6 out of 10 residents used specialist care at least one time, Tab. 1. The share was largest in the VCO (see Fig.2) and smallest in the Turin metropolitan area. On the average, population with care need made 5.4 journeys to get healthcare, with the VCO province having the lowest value.

Inter-municipality journeys accounted for 60% of the regional mobility, Fig. 4. The proportion varies significantly across provinces, as it depends on the socio-demographic size of the municipalities as well as on the availability of healthcare services: it is lowest in the Turin province (52%) which concentrates nearly half of the population of the region and 4 out of 10 of the places where healthcare services are delivered. At sub-regional level, inter-municipality journeys are more frequent in the less urbanized zones, situated in mountainous areas of the region, or in more remote areas with lower availability of healthcare services, Fig. 4.

Table 1. Health mobility for specialist care, in the Piedmont provinces, 2019 (*)

	Resident population	% of the population needing specialist care (see eq. 3)	Number of journeys	Average number of journeys per patient (see eq. 4a)	% of inter-municipality journeys (see eq. 1b)	Average journey length (kms) (see eq. 2b)	Distance covered per inhabitant to reach healthcare (kms) (see eq. 2c)
Torino (AM)	2230944	59	7346759	5,5	52	19	120
Cuneo (CN)	586064	61	1808616	5,1	70	25	137
Vercelli (NE)	169387	61	573253	5,5	66	30	182
Novara (NE)	364980	60	1188566	5,4	67	21	125
Biella (NE)	174170	62	563310	5,2	94	17	97
VCO (NE)	156320	63	495355	5,0	65	26	148
Asti (SE)	212010	62	707125	5,4	66	26	152
Alessandria (SE)	417283	60	1315787	5,3	63	28	171
Piedmont	4311158	60	13998771	5,4	60	22	132

(*) AM is the Turin metropolitan area and corresponds to the North-western part of the region; CN, the Cuneo province represents the South-western area; NE, the North-eastern area includes Biella, Vercelli, VCO and Novara provinces; SE the South-eastern area consists of Asti and Alessandria provinces.

Source: (Piedmont Health Registry, ISTAT, 5T)

Figure 4. Share of inter-municipality flows in the study zones, 2019
Source: Piedmont Health Registry

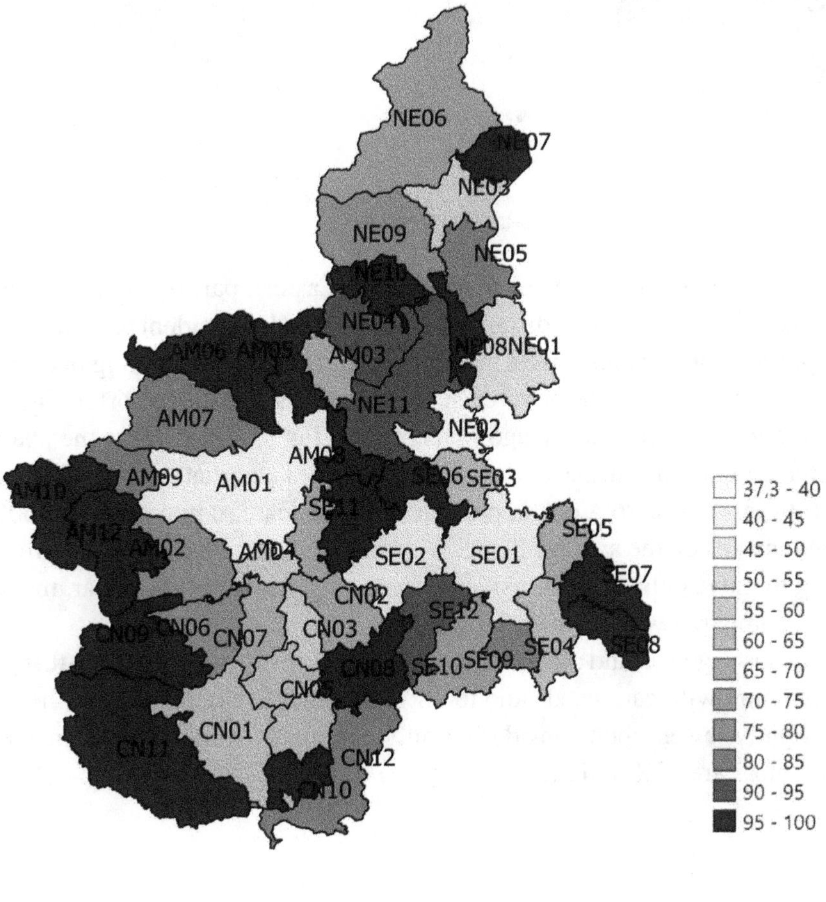

To reach specialist healthcare facilities situated outside his/her place of residence, in 2019, a resident made about 132 kms (eq. 2c), say 61 kms per journey on the average (eq. 2d) (the average length of journey was 22 kms, eq. 2b). The distance covered by all patients in 2019 was as high as 184.5 million kms, and increased by about 8% since 2013.

Figure 5. Self-containment value of health journeys (see eq. 1a) and share of the total covered distances within a zone (see eq. 2a) in the study zones, 2019
Source: (Piedmont Health Registry, 5T)

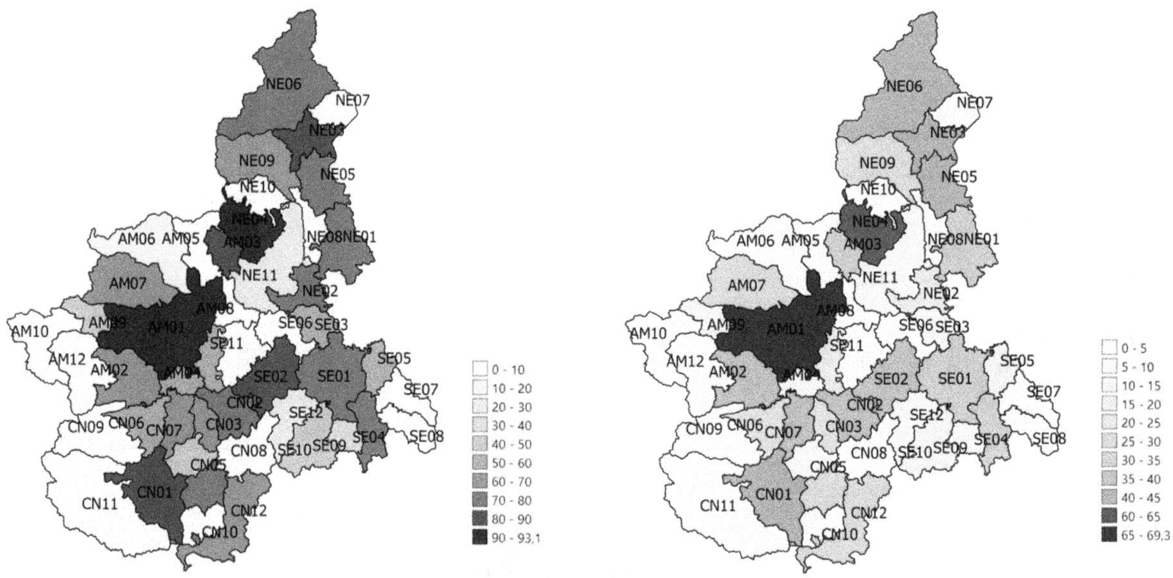

a. Journeys (%) b. Distance Covered (%)

At sub-regional level, Vercelli and Alessandria, in the Eastern part of the region are the provinces where the average journey lengths and the distances yearly covered by resident patients are longest (Tab. 1).

As could be expected, zones which have a great proportion of journeys taking place within them, Fig. 5a, are also the zones where the journeys to reach healthcare services are shorter, Fig. 6.

An outlook at the mobility profile by age brackets for, Fig. 7, shows that the share of population needing care and the number of journeys raise as age increases and that the highest values occur in the 75-84 age bracket. In this class, 86% of the population needs care, compared with 60% for the whole population; a patient makes on the average 8 health journeys a year compared with a regional average of 5.4; patients having to move outside their residence zone to get care, in particular, make on the average 3.3 journeys a year, compared with an average of 2.8.

In 2019, seniors, those aged 65 and older, are the 26% of the resident population; this proportion grows to 35% for the population with care needs and to 48% for patients having to reach healthcare facilities. These values are slightly lower when considering only patients having to move outside their residence zone: in this case seniors are 33% and their journeys 47%.

Figure 6. Average journey length (see eq. 2b) and distance yearly covered by a patient (see eq. 2c) in the study zones, 2019
Source: (Piedmont Health Registry, 5T)

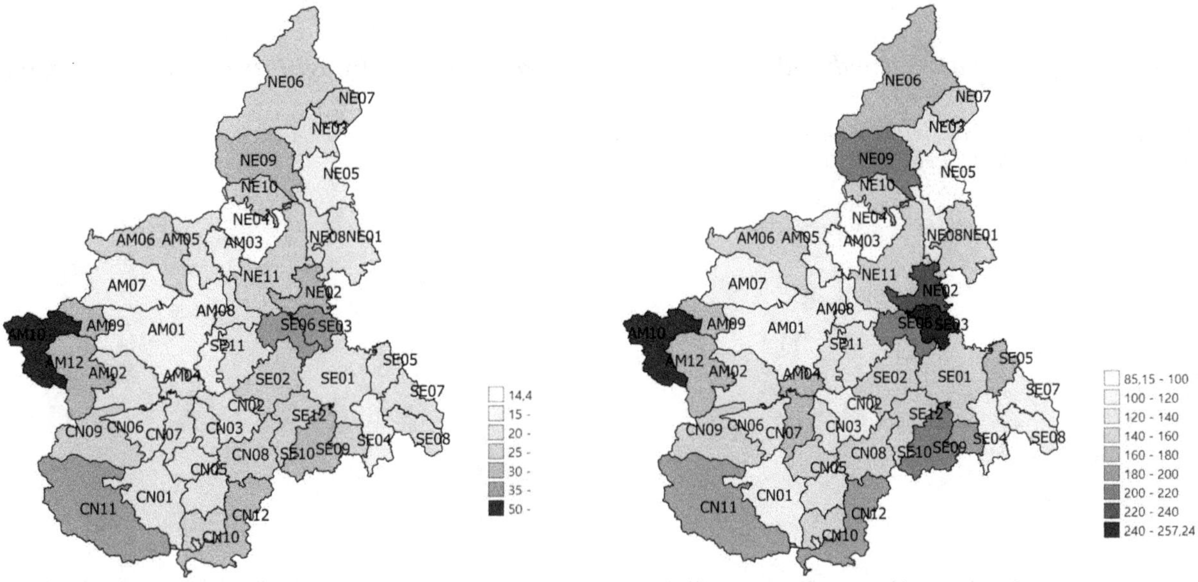

a. Average journey length (kms) b. Distance yearly covered by a patient (kms)

Figure 7. Share of Piedmont population with care needs and average number of journeys by age brackets, 2019
Source: (Piedmont Health Registry)

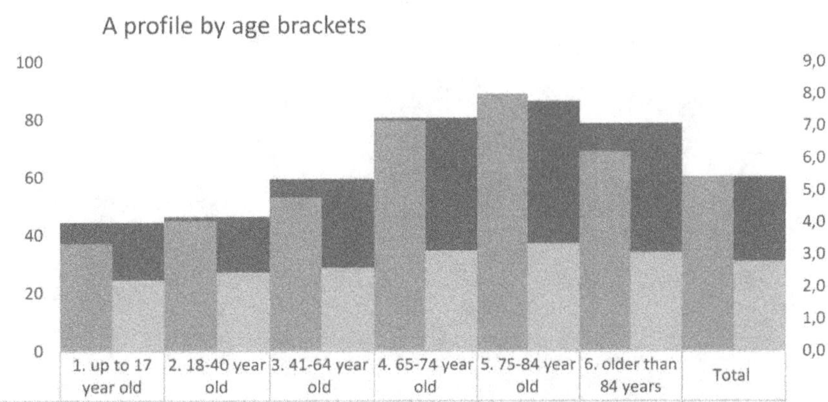

	1. up to 17 year old	2. 18-40 year old	3. 41-64 year old	4. 65-74 year old	5. 75-84 year old	6. older than 84 years	Total
% of the resident population needing care (eq.3) (left axis)	44	46	59	80	86	79	60
Number of health journeys per patient (eq. 4a)	3,4	4,1	4,8	7,1	8,0	6,2	5,4
Av. number of journeys per patient moving outside the residence zone (eq. 4b)	2,2	2,4	2,6	3,1	3,3	3,1	2,8

Although seniors travel more frequently than younger patients, they use healthcare services which are closer to home than younger patients do, Fig. 8a. On yearly basis, however, seniors' travelled distance is longer, Fig. 8b.

Figure 8. Average journey length and distance yearly covered by patient by age brackets in the Piedmont provinces, 2019 (*)
(*) In both graphs the x axis intersects the y axis at the regional value.
Source: (Piedmont Health Registry, 5T)

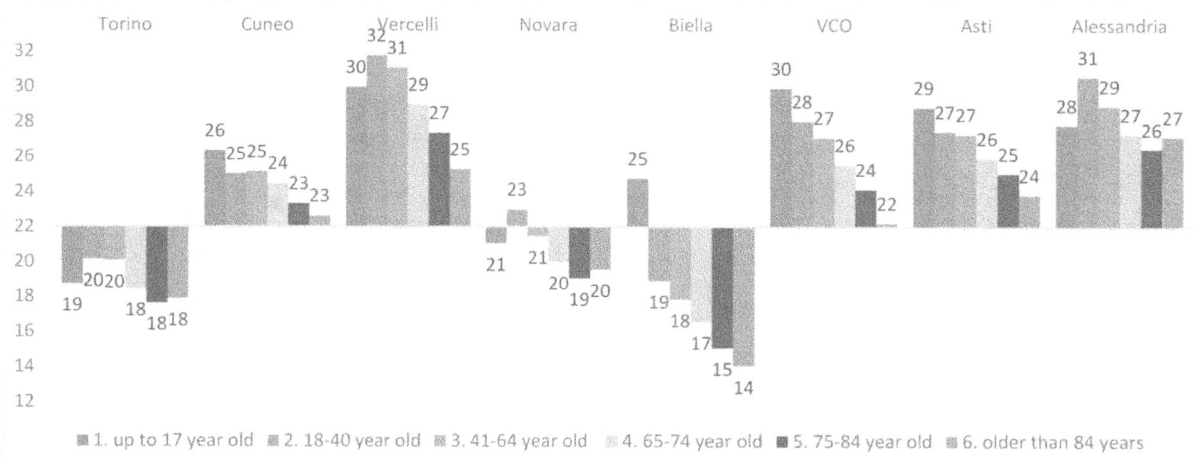

a. Average journey length (kms)

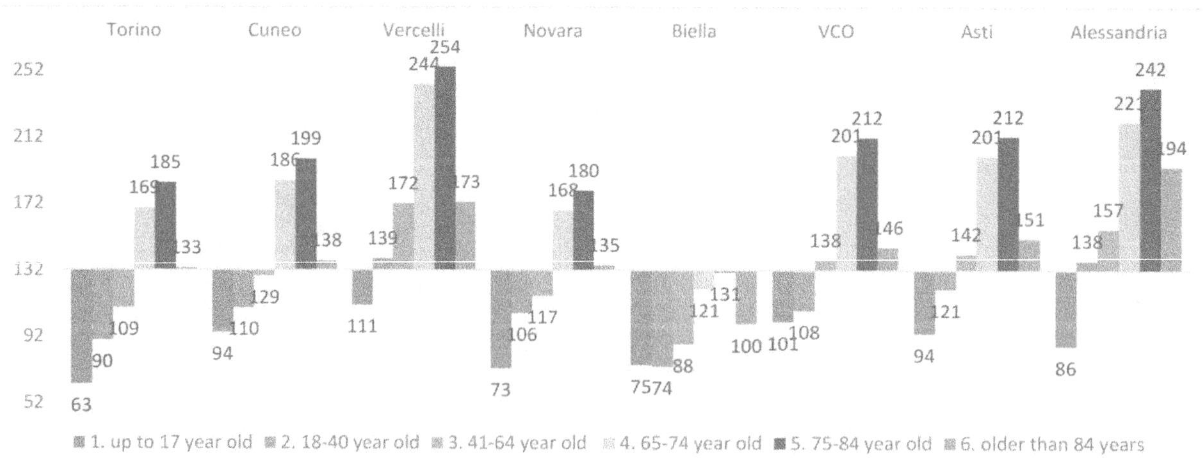

b. Distance yearly covered by a patient (kms)

Insights from the Tested Hypotheses and Discussion

The hypotheses explored in this study can be operationalised in many ways. Here the following choices have been made:

HP1: TM is used for reaching those services which are located at a distance from the residence zone which is greater than:

1. the regional average distance (22 kms) (*HP1.1 regional average threshold*);
2. the journey lengths defined by age brackets and provinces, shown in Fig. 8a (*HP1.2 prov&age threshold*).

HP2: TM is used for reaching the regional healthcare facilities entitled to deliver TM services:

1. no matter how far they are from the patients' home (*HP2.1 facility threshold*);
2. and which are located in places at a distance from home greater than the journey lengths defined by age brackets and provinces (Fig. 8a) (*HP2.2 facility&prov&age threshold*).

They include the (119) healthcare facilities which provided TM services by 2020. They are situated in 62 municipalities.

HP3: TM is used by those patients

1. who have to travel frequently, that is who make, on the average, more than 2.8 journeys a year to get medical care outside their residence zone (see Fig.7) (*HP3.1 journey frequency threshold*);
2. who have to travel frequently to healthcare facilities in places at a distance from home greater than the journey lengths defined by age brackets and provinces Fig.8a) (*HP3.2 journey freq&prov&age threshold*).

Table 2 summarizes the main effects resulting from the six hypotheses; they account for:

a. the percentage reductions observed for patients, journeys and travelled distances. Throughout the discussion, these reductions are interpreted as proxy indicators for the likely benefits to be accrued by the utilization of TM services. The higher these values are, therefore, the greater the positive impact of the hypothesis is expected to be;
b. the sub regional areas (provinces) most and least impacted by health mobility changes associated with the hypotheses;
c. the share of journey and distance reductions of senior patients (those aged 65 and more). The larger the share, the more seniors would take advantage from the changes produced by the hypotheses;
d. some features of health mobility (average journey length, patient average distance, number of journeys per patient) for those patients living in zones not affected by the hypotheses and thus not likely to use TM services. The lower the values, the more positively impactful the tested hypotheses would be.

A graphical comparison of the mobility reductions also distinguished by type of zones (urban and rural) is offered in Fig. 9. Not unexpectedly, given the spatial distribution of the regional health flows (see Figs. 5 and 6) all the hypotheses yield a reduction in health mobility which, except for the *HP 2.1 facility threshold,* is relatively more noticeable in rural than in urban ones.

Table 2. Overview of the effects produced by the tested hypotheses. a) Overall Results

	% reductions			**impact on the provinces**		**reduction shares of seniors**	
Hypotheses	Patients	Journeys	Distances	Most impacted	Less impacted	Journeys	Distances
HP1.1 Regional distance threshold	40	34	63	Alessandria	Biella	44	43
HP1.2 Prov&age threshold	39	34	61	Asti	Novara	47	44
HP2.1 Facility threshold	79	80	82	Cuneo	Biella	46	44
HP2.2 Facility&Prov&age threshold	33	28	52	Cuneo	Biella	46	43
HP3.1 Journey frequency threshold	44	61	50	Biella	Torino	56	55
HP3.2 Journey freq&prov&age threshold	11	15	24	Asti	Alessandria	60	55
2019 reference data for Piedmont	3011447	8395345	184, 5 (million kms)				

Table 2b. Results for patients not using TM services

	patients not using TM services		
Hypotheses	**Average journey length (kms) (eq. 2b)**	**Journey average distance per patient (kms) (eq. 2d)**	**Number of journeys per patient (eq. 4b)**
HP1.1 Regional distance threshold	12	38	3,1
HP1.2 Prov&age threshold	13	39	3,0
HP2.1 Facility threshold	20	52	2,6
HP2.2 Facility&Prov&age threshold	12	35	2,8
HP3.1 Journey frequency threshold	28	55	1,9
HP3.2 Journey freq&prov&age threshold	14	28	2,0
2019 reference data for Piedmont	22	61	2,8

The *HP 2.1 facility threshold* hypothesis which is associated with the enhanced functional role of healthcare services, is the most influential among the tested hypotheses: the percentage reductions of the number of patients, journeys and distances covered as well reach values as high as 79-82. Although quite extreme, their investigation makes it possible to appreciate the potential extension of the spatial coverage of the surveyed healthcare facilities. As illustrated by the graphs in Fig. 10, in particular, this is almost complete in the Cuneo area and relatively less extended in the Asti and Biella provinces.

The combination of this hypothesis with that introducing a threshold on distance, *H2.2facility&prov&age threshold*, has a much more modest influence on the reduction of the number of patients (33%) and journeys (28%). As in the case of the H2.1 hypothesis, the impact is weakest in Biella and Asti provinces. A significant reduction in health journeys and distances is observed in the Cuneo and to a lesser extent in VCO province, an area situated in North-eastern mountainous part of the region.

Figure 9. Reductions of the number of journeys, travelled distances (kms) and patients resulting from the tested hypotheses in the region and in rural and urban zones

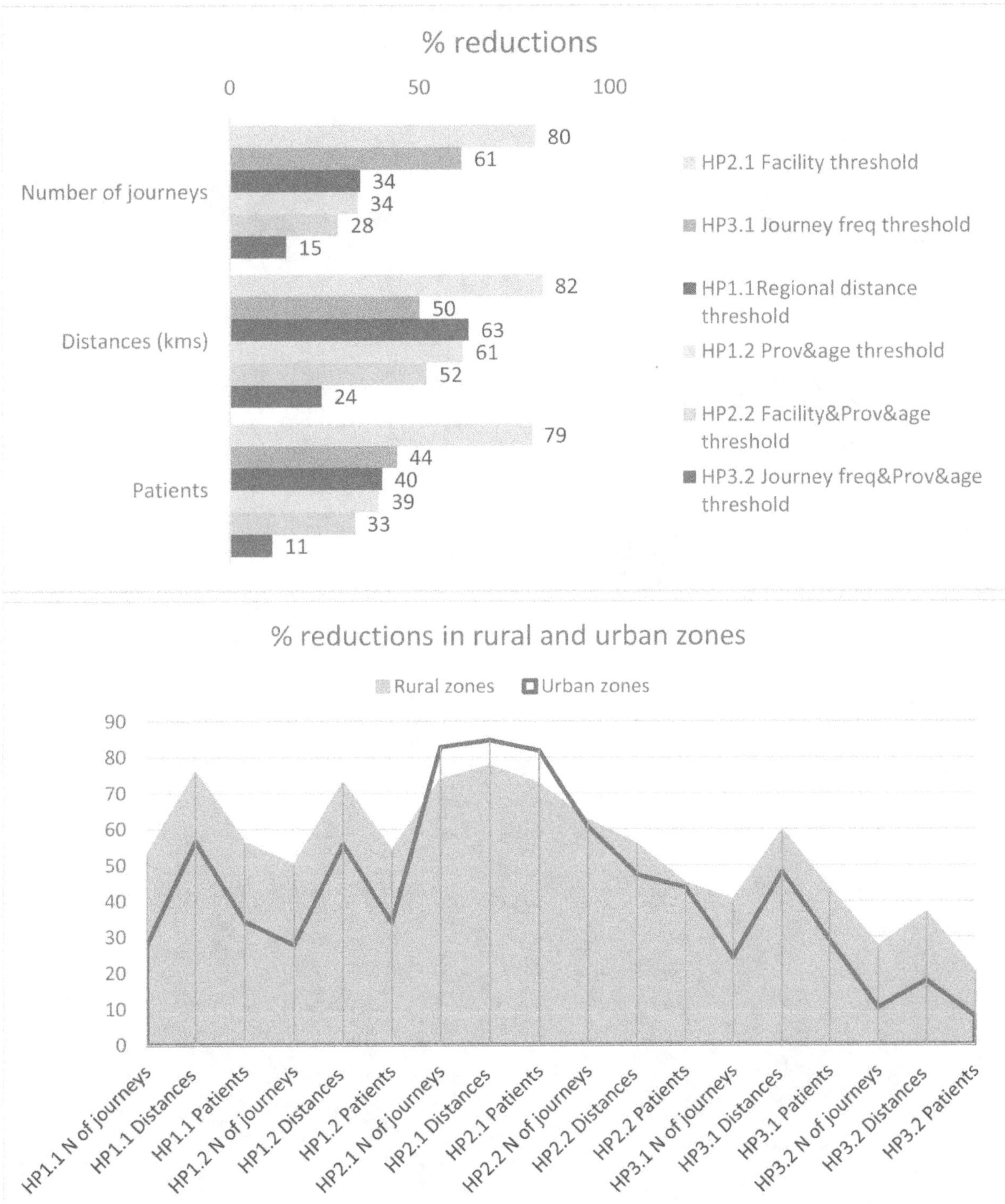

Figure 10. Reductions (%) of the number of journeys and distances (kms) for the tested hypotheses in the Piedmont provinces (*)
(*) Bubble size is proportional to the number of journeys except for the regional value which is the mean province value.

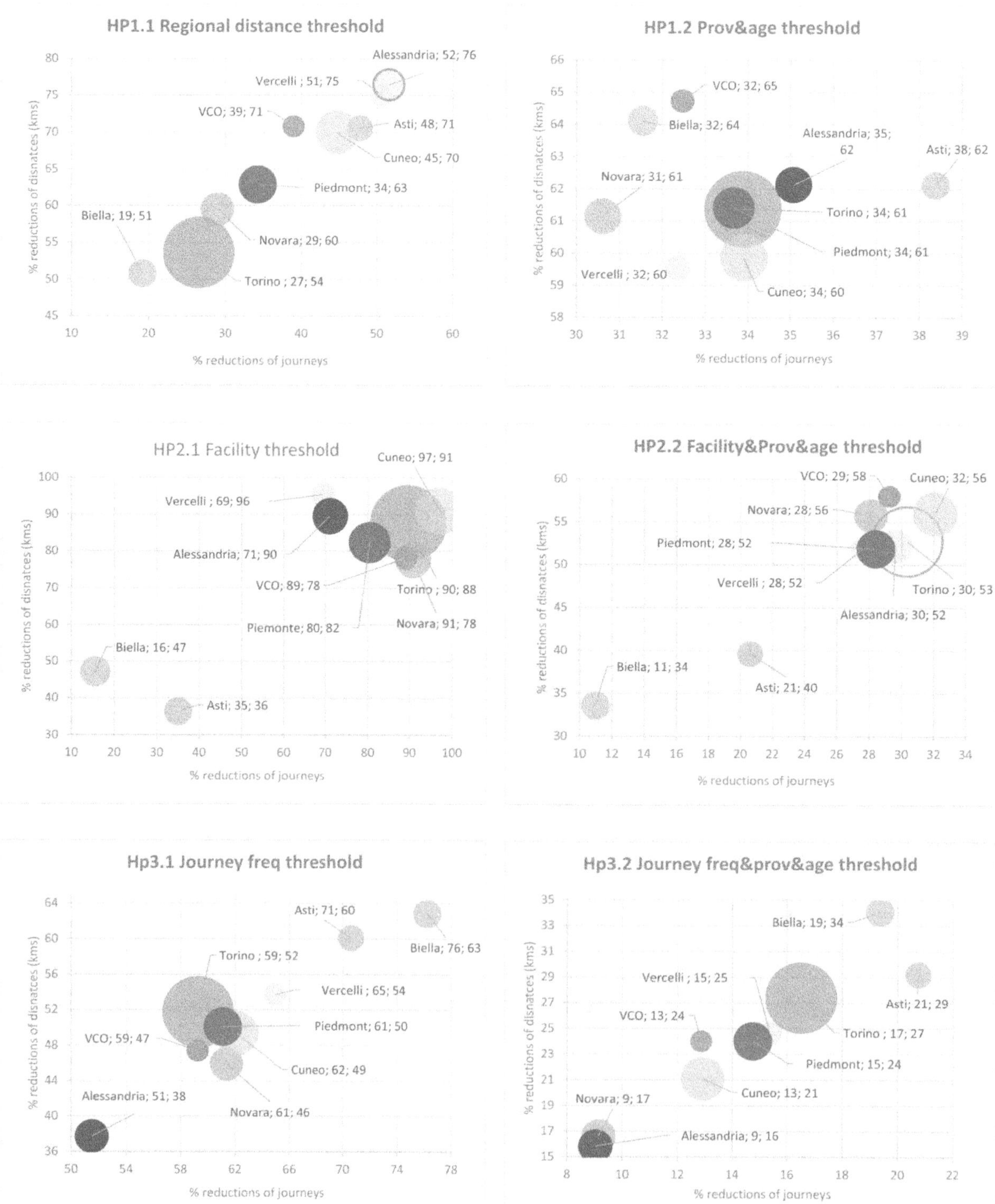

The hypothesis considering the frequency of health journeys (*HP3.1 journey frequency threshold)* produces the second greatest effect on the reduction of both the number of journeys (61%) and patients (44%). Unlike the other hypotheses in which travelled distances decrease more than the number of journeys, in this case the opposite occurs: distances decrease by 50% and journeys by 61%. The average journey length (28 kms), for the patients not likely to use TM services, is also the longest among the tested hypotheses and even longer than the regional average (22 kms, see Tab. 2). Asti and Biella are the provinces which most benefit from the impact of this hypothesis, Fig.10.

The introduction of the threshold on the journey length as an additional constraint, *HP3.2 journey freq&prov&age,* yields the weakest reduction in the number of patients (11%), journeys (15%) and travelled distances (24%). It is worth noting, however, that remoteness from healthcare premises and a high access frequency to care services are zonal attributes that, taken together, identify situations where TM services would likely to be most effective. The fact that the HP3.2 hypothesis has the most relevant influence on seniors' mobility (Tab. 2) reinforces the argument and suggests that a supplement of analysis is needed for the regional sub-areas in these situations.

The two hypotheses accounting for the spatial thresholds on journey lengths (HP1) yield similar results as for the contraction of health mobility and decrease of the average journey length (Tab. 2). At a province level, however, their relative impact is different, Fig. 10.

In the case of the *HP1.1 regional average threshold,* the province situations tend to cluster in two groups, depending on whether their reductions in the number of journeys and distances are above or below the regional averages. As could be expected, the greatest decreases for both the number of journeys and covered distances take place in the provinces of Alessandria and Vercelli, where the journey average length and patient travelled distances are longer, see Tab. 1.

A relatively more diversified pattern across provinces results from the *HP1.2 prov&age threshold,* although the range of variation in the observed changes is smaller. In this case, the greatest decline in the numbers of journeys and covered distances occurs in the province of Asti and VCO, respectively.

Given the relevance of the ageing issue in the Piedmont region, it is worth examining in more depth to which extent seniors' health mobility would be influenced by the hypotheses investigated in this study.

As already made apparent in Tab. 2, the hypotheses regarding the frequency of health journeys (HP3), and particularly the *HP3.2 journey freq&prov&age,* are those most relevant for seniors' health mobility: in these cases, nearly 60% of the reductions in the number of journeys consist of those made by seniors. The graph in Fig. 11 clearly reveals that this occurs in all provinces, but mostly in that of Alessandria, where however the impact of these hypotheses on the overall reduction in the total number of journeys is the lowest (see Fig.10).

A zonal positive effect, e.g. a decrease in the travelled distances, is clearly visible in all cases except for the HP2.1 *Facility threshold.* In spite of the significant reduction produced at regional level in the number of journeys and distances, in this case some zones, in particular in the Cuneo province, do not benefit from it.

Among the tested hypotheses, the one considering thresholds on both the frequency and length of journeys, *HP3.2 journey freq&prov&age,* yields the most appreciable distance gain in almost all zones.

Finally, we give a closer look at the average distances travelled by patients who would not use the TM services, as the profile of their health journeys does not comply with the chosen thresholds. This is offered by Fig. 12, which compares the values in the reference year with those resulting from the mobility changes (reductions) produced by the impact of the different hypotheses.

Figure 11. Share of the reduction in the number of journeys by seniors (patients aged 65 years and older) in the tested hypotheses, in the Piedmont provinces (*)
(*) Reference values are calculated from the 2019 mobility database.

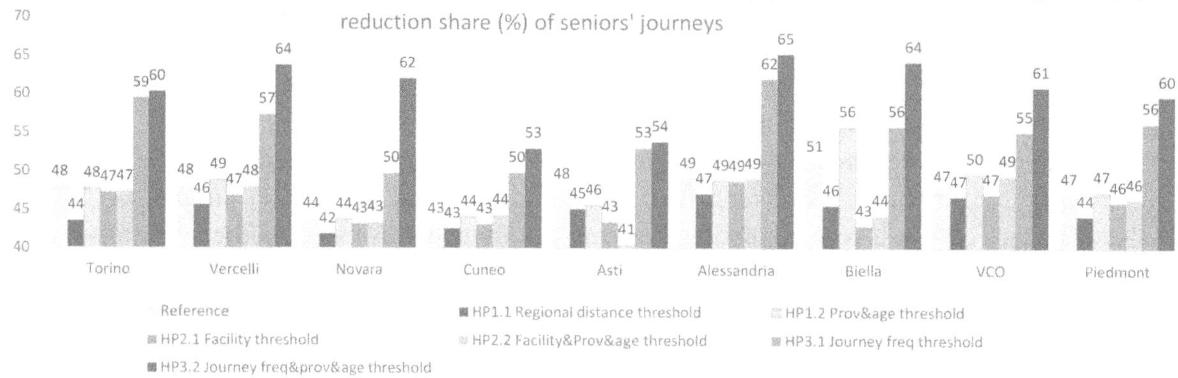

CONCLUDING REMARKS AND SUGGESTIONS FOR FUTURE RESEARCH

In spite of or, to some extent, because of the recent pandemic, ongoing reflections about potentials and risks of innovative healthcare solutions, such as TM services are receiving new impetuses. Attention is especially drawn to the fact that in order to reap the benefits of telemedicine and limit its risks an attentive policy environment is necessary capable of continuous learning and supporting knowledge sharing (Cravo & Hashiguchi, 2020; HIMSS, 20121).

Taking into account issues of healthcare access and sustainability in the analysis of TM is part of this undertaking. This was the case of this enquiry for the Piedmont region, which dealt with the spatial dimension of access in patients' mobility to specialist healthcare.

The study adopted a forward looking perspective and enquired to which extent TM services might overcome the distance gap between patients and services delivering specialist healthcare, given certain hypotheses on the spatial distribution of patients and services in the regional sub-areas.

Although partial, these arguments are likely to gain a prominence in the efforts of regional healthcare providers to restore and/or re-organize the healthcare delivery after the disruptions caused by the Covid-19 pandemic.

For example, they draw attention to the fact that expanding TM services entails to think about the kind of access they afford, primarily enquiring how they overcome remoteness, help delivery integrated and timely treatments, shorten waiting times and enable exploiting health-care services already existing in an area. Digging into the local situations while providing sound evidence about the physical and virtual access profile is a necessary prerequisite for articulating the access notion at conceptual, policy and practical levels (Rouse & Serban, 2014).The experience made with this study can be a part of a knowledge mobilisation effort by the regional health organisations (Haynes et al., 2020; Ekeland, Bowesc & Flottorpd, 2011) to increase capacity for developing TM services which better respond to population health needs across regional sub areas.

The study demonstrated how this could be prompted by applying a set of spatially articulated health performance indicators, based on existing health flow data registered in administrative data pipeline.

Figure 12. Journey average distance travelled by patients in the reference year and as result of the tested hypotheses in the study zones (*)
(*) Zones are ordered by decreasing 2019 values in each macro areas: AM (Turin metropolitan area), CN (Cuneo province), NE (it includes Biella, Vercelli, VCO and Novara provinces), SE (it includes Asti and Alessandria provinces)(see Tab. 1). The lowest values are generally observed in the urban zones, see Fig. 2a.

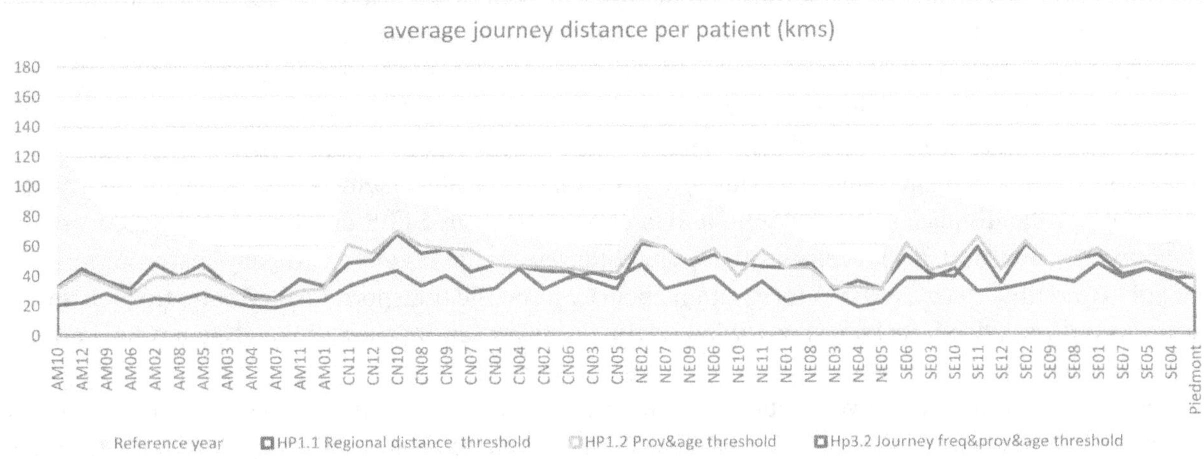

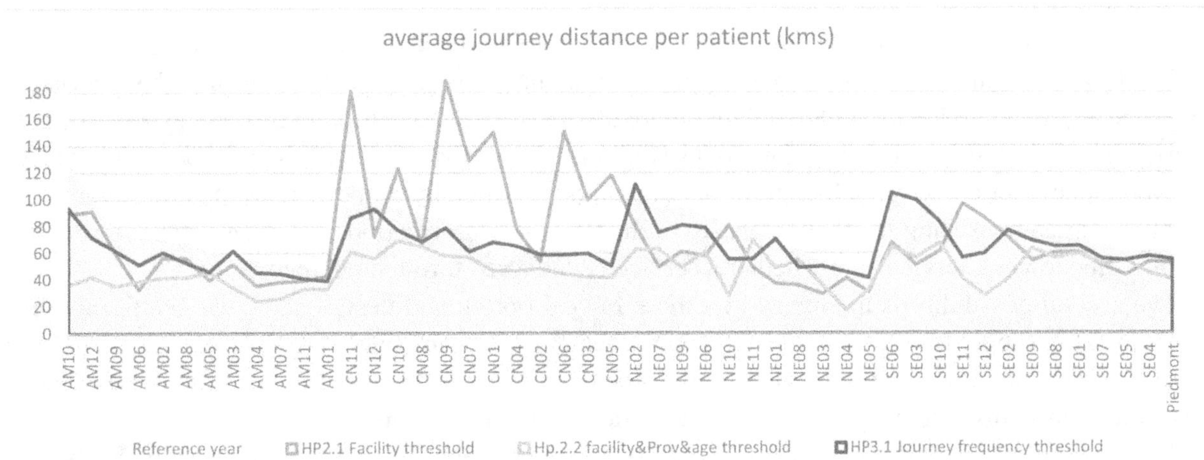

Of course, the approach used in the application can be extended in many ways, and in the following a few suggestions are made to refine hypotheses, indicators and the information basis.

As graphically shown in Fig. 3, a pertinent assemblage of hypotheses about the TM-access relationships ought to take into account a number of factors which, in addition to institutional conditions, enable or constrain the provision of the new services. For example, on the medical ground, attention should be paid at the type of treatments and medical specialties which in the study area are most likely to be delivered digitally. As for patients, their digital literacy and familiarity with digital tools cannot be neglected as these socio-demographic characteristics affect the use of the new services. On the supply side, as well, the organization of healthcare providers has a prominent role to ensure the quality of the delivered services and has to be taken into account. Finally, future hypotheses will have to properly consider the condition of broadband connectivity across local areas and of virtual interactions between patients and healthcare providers.

Extending spatial information indicators (SPIs) draws attention to the twofold questions of aligning the objectives of these indicators with those of the health functions regional governments are responsible for (Duran et al., 2012) and of supporting communities for long-term TM implementation and use (Otto & Whitehouse, 2021). In fact, SPIs belong to the set of measures which are being developed to gauge how useful, beneficial, and efficient are new methods for delivering existing models of healthcare (Dubai Health Authority, 2021; Hollander & Neinstein, 2020; World Health Organization, 2016a). To this end, SPI could be articulated by medical disciplines and type of healthcare structures providing specialty care.

As for the measurement of journey length, in particular, several refinements are possible, such as the calculation of the distance covered within a municipality through a GIS analysis, and the inclusion of travel time by transportation means, as these are collected by the regional company managing private and public mobility (http://www.5t.torino.it/en/about). For public transport, furthermore, frequencies of services could be added, making it possible to define a richer set of criteria to analyse physical access to healthcare premises.

Results of the empirical investigation also indicate that investigations of TM-access relationships would benefit from a more robust design of the information platform and in this regard two possibilities can be mentioned.

The first aims at sharpening the profiles of origins (patients) and destinations (health providers) of health flows. This means to have a more appropriate description of individuals' health characteristics and clinical pathways, for example applying a population/patient case-mix adjustment system for categorizing patients' illnesses such as The Johns Hopkins ACG® System (2014). On the destination side, information about the delivered health services could be enriched by properly merging data about their operations already existing in other databases of the Regional Information System, e.g. type of service, waiting time to get a service, and whether a service is delivered through telemedicine.

The second possibility is to engage in a more radical option and design anew the health mobility platform. This should be a part of a more comprehensive effort in healthcare management to deliver more accurate and timely health service statistics (Jannati et al., 2018; Kalra et al., 2020), and back up surveillance systems more responsive to risk situations, such as pandemics.

In this direction, a promising strategy would focus on patients and understand health mobility as a comprehensive perspective for accounting healthcare events in patients' life courses. In this direction, attention should be paid at the overall design of the health mobility platform, appropriately selecting the type of data regularly made available in the Regional Information System, according to the levels of the healthcare system (people, processes, organization and ecosystem) and their specific representation (medical, administrative, and research and development) (Ainsworth & Buchan, 2015; Rouse & Serban, 2014). The resulting information platform would enable to perform in a coherent way a number of much wanted knowledge-oriented functions, such as analysing patient decisions and assessing cost-effectiveness of the healthcare processes (OECD, 2019a), understanding disparities in healthcare coverage (OECD, 2019b) and designing innovative and better integrated health services at local levels (World Health Organization, 2016b) to be delivered through every possible channels (Dahl, Milne & Peltier, 2021; Damurski, 2021).

Finally, a broader research theme to be addressed in future research concerns how and to what extent the adoption of technology supported models of care might affect the location of new healthcare structures and/or the spatial re-organization of existing services. Challenging questions, specifically involving the territorial dimension include, for example, territorial medicine, linkages among primary, secondary and tertiary levels of care, and community engagement in supporting healthcare provision in local areas.

ACKNOWLEDGMENT

We gratefully acknowledge the help of Alessandra Casano from **Servizio sovrazonale di epidemiologia ASL TO3** (Regione Piemonte) in preparing the figures and maps for this publication.

REFERENCES

Aday, L. A., & Andersen, R. (1974). A Framework for the Study of Access to Medical Care. *Health Services Research*, *9*(3), 208–220. PMID:4436074

Ainsworth, J., & Buchan, I. (2015). Combining Health Data Uses to Ignite Health System Learning. *Methods of Information in Medicine*, *54*(6), 479–487. doi:10.3414/ME15-01-0064 PMID:26395036

Bashshur, R., Doarn, C. R., Frenk, J. M., Kvedar, J. C., & Woolliscroft, J. O. (2020). Telemedicine and the COVID-19 pandemic, lessons for the future. *Telemedicine Journal and e-Health*, *26*(5), 571–573. doi:10.1089/tmj.2020.29040.rb PMID:32275485

Clarke, G. P., & Wilson, A. G. (1994). A new geography of performance indicators for urban planning. In C.S. Bertuglia, G.P. Clarke G.P., & A.G. Wilson (Eds.), Modelling the city. Performance, policy and planning (pp. 55-81). Routledge.

Costa, G., Stroscia, M., Zengarini, N., & Demaria, M. (2017). *40 anni di salute a Torino. Spunti per leggere i bisogni e le politiche a Torino*. Inferenze.

Cravo, T., & Hashiguchi, O. (2020). *Bringing health care to the patient: An overview of the use of telemedicine in OECD countries*, OECD Health Working Papers, N. 116. doi:10.1787/18152015

Dahl, A. J., Milne, G. R., & Peltier, J. W. (2021). Digital health information seeking in an omni-channel environment: A shared decision-making and service-dominant logic perspective. *Journal of Business Research*, *125*, 840–850. doi:10.1016/j.jbusres.2019.02.025

Damurski, L. (2021). How to Include Omnichannel Services in Land-Use Policy? E-Planning Holds the Key. *International Journal of E-Planning Research*, *10*(3), 70–85. doi:10.4018/IJEPR.20210701.oa5

Dubai Health Authority. (2021). *Guidelines for Reporting Telehealth Key Performance Indicators*. https://www.dha.gov.ae/en/HealthRegulation/Documents/Guidelines%20for%20Reporting%20Telehealth%20KPIs.pdf

Durán, A., Kutzin, J., Martin-Moreno, J.-M., & Travis, P. (2012). Understanding health systems: scope, functions and objectives. In J. Figueras & M. McKee (Eds.), *Health Systems, Health, Wealth and Societal Well-being. Assessing the case for investing in health systems* (pp. 19–36). Mc Maidenhead Graw Hill Open University Press.

Ekeland, A. G., Bowesc, A., & Flottorpd, S. (2011). Methodologies for assessing telemedicine: A systematic review of reviews. *International Journal of Medical Informatics*, *81*(1), 1–11. doi:10.1016/j.ijmedinf.2011.10.009 PMID:22104370

Esper, G. J., Sweeney, R. L., Winchell, E., Duffell, J. M., Kier, S. C., Lukens, H. W., & Krupinski, E. A. (2020). Rapid System Wide Implementation of Outpatient Telehealth in Response to the COVID-19 Pandemic. *Journal of Healthcare Management*, *65*(6), 443–452. PMID:33074968

European Commission. (2018). *Market study on telemedicine. Luxembourg: Publications Office of the European Union.* https://ec.europa.eu/health/sites/default/files/ehealth/docs/2018_ provision_market-study_telemedicine_en.pdf

European Commission. (2020). A European strategy for data. COM(2020) 66 final.

Evans, D. B., Hsua, J., & Boermaa, T. (2013). Universal health coverage and universal access. *Bulletin of the World Health Organization*, *91*(8), 546–546a. doi:10.2471/BLT.13.125450 PMID:23940398

Ferrara, F. M. (2021). *Osservatorio sulla Telemedicina Operativa.* https://www.dati-sanita.it/wp/wp-content/uploads/2021/08/101-osservatorio-telemedicina.pdf

Feuillet, T., Charreire, H., Hercberg, S., & Oppert, J.-M. (2019). Pour une géographie régionale des spécificités loc ales. L'exemple des liens entre environnement urbain et comportements individuels de santé. In V. Chasles & L. De Biaggi (Eds.), *Santé et territoire, une réflexion géographique* (Vol. 3, pp. 13–13). Université Jean Moulin.

Gatrell, A. C., & Wood, D. J. (2012). Variation in geographic access to specialist inpatient hospices in England. *Health & Place*, *18*(4), 812–840. doi:10.1016/j.healthplace.2012.03.009 PMID:22522100

Haynes, A., Rychetnik, L., Finegood, D., Irving, M., Freebairn, L., & Hawe, P. (2020). Applying systems thinking to knowledge mobilisation in public health. *Health Research Policy and Systems*, *18*(1), 134. doi:10.118612961-020-00600-1 PMID:33203438

HIMSS. (2021) *Future Healthcare Report.* https://www.himss.org/sites/hde/files/handout-FutureOf-HealthcareReport-EMC-0821-FINAL_156.pdf

Hollander, J., & Neinstein, A. (2020). *Maturation from Adoption-Based to Quality-Based Telehealth Metrics.* https://catalyst.nejm.org/doi/pdf/10.1056/CAT.20.0408

Jannati, A., Sadeghi, V., Imani, A., & Saadati, M. (2018). Effective coverage as a new approach to health system performance assessment: A scoping review. *BMC Health Services Research*, *18*(1), 886. doi:10.118612913-018-3692-7 PMID:30470214

Joint Action to support the eHealth Network. (2017). *Report on EU state of play on telemedicine services and uptake recommendations.* https://ec.europa.eu/health/sites/default/files/ ehealth/docs/ev_20171128_co09_en.pdf

Kalra, D., Kolitsi, Z., Zobell, O., Bujok, S., & Stroetmann, V. (2020). *Better Utilisation of Data Infrastructures to Support Secondary Uses of Health Data.* https://digitalhealtheurope.eu/wp-content/uploads/2020/10/Better-Utilisation-of-Data-Infrastructures.pdf

Kidholm, K., Ekeland, A. G., Jensen, L. K., Rasmussen, J., Pedersen, C. D., Bowes, A., Flottorp, S. A., & Bech, M. (2012). A model for assessment of telemedicine applications: Mast. *International Journal of Technology Assessment in Health Care*, *28*(1), 44–51. doi:10.1017/S0266462311000638 PMID:22617736

Landini, S., Occelli, S., Dalmasso, M., Durando, M., & Montano, G. (2019). *Elementi di analisi per una mobilità sostenibile*. http://ires.Piemonte.it/relazione2019/ Capitolo_5_allegato01.pdf

Levesque, J.-F., Harris, M. F., & Russell, G. (2013). Patient-centred access to health care: Conceptualising access at the interface of health systems and populations. *International Journal for Equity in Health*, *12*(18), 1–9. doi:10.1186/1475-9276-12-18 PMID:23496984

Ministero della Salute. (2018). *Mappatura delle esperienze di Telemedicina sul territorio nell'anno 2018*. https://www.salute.gov.it/imgs/C_17_pagineAree_2515_2_file.pdf

National Academies of Sciences, Engineering, and Medicine. (2021). *Population Health in Challenging Times: Insights from Key Domains. Proceedings of a Workshop*. The National Academies Press.

Occelli, S., & Dalmasso, M. (2019). Probing patient mobility in a regional setting: the case of Piedmont. In V. Chasles & L. De Biaggi (Eds.), Santé et territoire, une réflexion géographique (pp. 123-138). Université Jean Moulin, Lyon3.

Occelli, S., & Dalmasso, M. (2021). *Health mobility in the Piedmont region: an outlook in the middle of the tens* [Paper presentation]. MoHeaP 2021: Virtual Workshop on Mobility, Health, and Place.

Occelli, S., & Scelfo, B. (2020). Can We Enable Digital Transition in Healthcare Delivery?: Insights From a Survey of Telemedicine Services in the Piedmont Region. *International Journal of E-Planning Research*, *9*(2), 43–65. doi:10.4018/IJEPR.2020040103

OECD (2019a). *Health in the 21st Century: Putting Data to Work for Stronger Health Systems*. OECD Health Policy Studies, OECD Publishing.

OECD. (2019b). Health for Everyone?: Social Inequalities in Health and Health Systems. OECD Health Policy Studies, OECD Publishing.

OECD. (2021). *State of Health in the EU. Italy Country Health Profile 2021*. https://ec.europa.eu/health/system/files/2021-12/2021_chp_it_english.pdf

Oliveira, T. C. O., Barlow, J., Gonçalves, L., & Bayer, S. (2013). Teleconsultations reduce greenhouse gas emissions. *Journal of Health Services Research & Policy*, *18*(4), 209–214. doi:10.1177/1355819613492717 PMID:23945677

Osservatorio GIMBE. (2019). *La mobilità sanitaria interregionale nel 2017*. Fondazione GIMBE. https://www.gimbe.org/osservatorio/Report_Osservatorio_GIMBE_2019.06_Mobilita_sanitaria_2017.pdf

Osservatorio demografico territoriale del Piemonte. (2022). https://www.demos.piemonte.it/previsioni-demografiche/indicatori-demografici

Otto, L., & Whitehouse, D. (2021). Empowering Communities to Support Telemedicine and its Business Model. *Health Management*, *21*(4), 219–222.

Otto, L., Whitehouse, D., & Schlieter, H. (2019). On the Road to Telemedicine Maturity: A Systematic Review and Classification of Telemedicine Maturity Models. In A. Pucihar, M.K. Borstnar, R.W.H. Bons, J. Seitz, H. Cripps, & D. Vidmar (Eds.), 32nd Bled eConference: Humanizing Technology for a Sustainable Society (pp. 485-506). University of Maribor Press. doi:10.18690/978-961-286-280-0.26

Penchansky, R., & Thomas, J. W. (1981). The concept of access: Definition and relationship to consumer satisfaction. *Medical Care*, *19*(2), 127–140. doi:10.1097/00005650-198102000-00001 PMID:7206846

Perchoux, C., Kestens, Y., Thomas, F., Van Hulst, A., Thierry, B., & Chaix, B. (2014). Assessing patterns of spatial behavior in health studies: Their socio-demographic determinants and associations with transportation modes (the RECORD Cohort Study). *Social Science & Medicine*, *119*, 64–73. doi:10.1016/j.socscimed.2014.07.026 PMID:25150652

Peters, D. H. (2014). The application of systems thinking in health: Why use systems thinking? *Health Research Policy and Systems*, *12*(51), 1–6. doi:10.1186/1478-4505-12-51 PMID:25160707

Purohit, A., Smith, J., & Hibble, A. (2021). Does telemedicine reduce the carbon footprint of healthcare? A systematic review. *Future Healthcare Journal*, *8*(1), 85–91. doi:10.7861/fhj.2020-0080 PMID:33791483

Rouse, W. B., & Serban, N. (2014). *Understanding and Managing the Complexity of Healthcare*. The MIT Press.

The Johns Hopkins ACG® System. (2014). *Excerpt from Version 11.0 Technical Reference Guide*. https://www2.gov.bc.ca/assets/gov/health/conducting-health-research/data-access/johns-hopkins-acg-system-technical-reference-guide.pdf

van Dyk, L. A. (2014). Review of telehealth service implementation frameworks. *International Journal of Environmental Research and Public Health*, *11*(2), 1279–1298. doi:10.3390/ijerph110201279 PMID:24464237

World Health Organization. (2016a). *Defining evaluation indicators for telemedicine as a tool for reducing health inequities*. Pan American Health Organization. https://iris.paho.org/bitstream/handle/10665.2/28562/PAHOKBR16006eng.pdf?sequence=8

World Health Organization. (2016b). *Integrated care models: an overview.* https://www.euro.who.int/en/health-topics/Health-systems/health-services- delivery/publications/2016/integrated-care-models-an-overview-2016

ENDNOTES

1 Telemedicine is part of the healthcare strategy in the 2021 Recovery and Resilience Plan, laid out by the Italian government to recover the economy after the Covid pandemic and support a green and digital development.

2 In Italy, inter-regional health mobility is regularly investigated to assess the delivery capability and financial balance of regional health systems, as these have a main responsibility in the healthcare provision (Osservatorio Gimbe, 2019).

Chapter 9
COVID-19 and Citizen Engagement in Sustainable Mobility Planning:
A Case Study From the City of Banska Bystrica

Marian Gogola
https://orcid.org/0000-0001-6206-4665
University of Zilina, Slovakia

Stanislav Kubaľák
University of Zilina, Slovakia

Mikulas Cerny
University of Zilina, Slovakia

ABSTRACT

The situation over the COVID-19 pandemic has affected and continues to affect the mobility of the population in several countries. Traditional approaches in transport planning have had to be replaced by new methods using digital technologies and e-planning methods. The COVID-19 pandemic brought the opportunity to use such a situation to enhance feedback from the citizen's perspective. This chapter describes the approach of population involvement through e-planning in the field of sustainable mobility. The case study itself is made based on data from the town of Banská Bystrica, where the process of preparing the strategic document Sustainable Urban Mobility Plan began last year. It can be stated that perhaps also thanks to the COVID-19 pandemic, the population was involved to a much greater extent than in other periods.

DOI: 10.4018/978-1-7998-9090-4.ch009

INTRODUCTION

Unlike in the past, when transport planning was carried out mainly from the table without public participation, the current trend involves the public in all planning phases. The traditional planning approach represents so-called centralised planning, where the responsible institution or authority decides to act often without discussion with the citizens. If this occurs, the citizens demand changes and hear their voices. The evidence provided various examples (Polletta, 2016) when the citizens' participation took place even in the natural form (public protests, official letters, etc.) as well as the official promotion for acting at the local level, e. g Local Agenda 21 (Coenen, 2009). Citizens engagement is a highly recommended method in participatory budgeting on various other issues. For instance, participation in policy planning has a long tradition (Reed, J. et al., 2006). But in some way, it was just a formal process without further improvement. Although the public could formally participate in the past, not all residents had the opportunity to participate. This was mainly due to a different form of commenting when the notice of the construction or planned measures was hung only on the notice board of the competent authority. Although several cities have to inform their citizens about planned activities, the approaches vary from city to city. Sometimes only active citizens are searching for the information and take the responsibility to comment on the city actions in various areas as planning, financing, public budget spending, etc.

However, the Sustainable Urban Mobility Plan (SUMP) has a different approach. The SUMP consists of various steps[1] (Eltis). As part of its preparation, individual activities are divided into several events, where public participation is required. Of course, even in this case, this can only be done informally, with information being communicated and residents expected to get involved. Alternatively, proactive communication can be implemented, meaning that knowledge and information are actively disseminated to the public to collect feedback. Increasingly, citizen engagement is used more frequently due to the digital era and information and communication technologies (ICT). Citizens could instantly see what is going on, watch the streaming, read the legislation or see the planning activities. Some citizens prefer to engage in every potential problem solving; others prefer to stay outside the planning process or not participate. We can say that no one foresaw and set up involving the population in sustainable mobility planning. The recommended process considers the common condition without any limits. There are various excellent and inspiring examples of citizens engagement during the development of SUMP (e.g. Brussel Mobility; Bremen transport plan). However, the COVID pandemic reversed most of these processes. Traditional approaches could not be used, and therefore the methods available in terms of minimal risk to the population were used. Thus, the planning process moved to a digital environment that allowed certain people to be involved in the mobility planning process.

The study presented in this chapter aims to address the following research questions:

RQ1 - Is the pandemic a reason for a higher engagement of citizens in urban planning?
RQ2 - Does the digital platform make the process of citizens engagement easier in comparison to traditional participation tools?

The chapter is organised as follows. Sections 2 and 3 present the context of mobility planning, emphasising citizen engagement and engagement tools. Section 4 elaborates the topic of the Covid 19 pandemic and its effect on the ongoing planning process. The methodology followed in this study is described in section 5. The case study is presented in section 6.

MOBILITY PLANNING AND CITIZENS ENGAGEMENT AND PARTICIPATION

Nowadays, many cities are challenged to plan sustainable transport systems, and therefore they are conducting or following the goals in sustainable urban mobility plans. The citizen's participation and involvement of the population are essential in preparing the SUMP plan. Many actors, including the experts, stakeholders and citizens, plan the future vision of mobility. Sometimes, attracting citizens to engage in this mobility issue is very hard; therefore, some studies or projects present how to enhance citizens' participation in transport planning (e.g. ELAN report). In general, the SUMP planning process across various cities shows the different levels of citizens engagement or participation and considers the citizens' engagement as a precondition of Sustainable Urban mobility (Lindenau et al.,2014). There is also evidence to invite citizens to create various types of policies (e.g. in the city of Berlin - Schneidemesser.). The study Myrovali et al. (2018) presents the MOTIVATE interactive environment based on the motivation and reward of citizens participating in the planning process. The traditional SUMP process collects data and invites citizens to participate through personal events like workshops, discussions, and roundtables. But the adoption of digital platform tools was already described in various studies related to urban planning and smart city (Pham et al. (2016). There are different digital platforms such as official municipality or government websites, social networks or smartphone apps where citizens can participate. Bonsón et al. (2019), in their study, researched the communication of municipalities and citizen involvement via Twitter through the analysis of tweets from the largest Andalusian municipalities. The main reason to participate in the content can be spreading in the various digital platforms. Some of them provide just the basic information; others allow active engagement. This can lead to e-government within the smart city concept. But for engaging citizens in the smart city concept, it is necessary to develop the entire network. Stratigea (2012) describes the critical factors that help go smart: deployment of broadband infrastructure, education and training, innovation capacity, marketing of smart communities, and promotion of digital democracy. But we should not forget that there are still citizens with no accessibility to e-government or digital services or are willing not to participate. Digital technologies are, therefore, a perfect tool for citizens using them, but there are also groups, sometimes called marginalised, where ICT penetration is low (Liao et al.,2000); Ortiz et al., 2019).

The case studies from various cities confirmed the active role of citizens during the strategic plan preparation, for instance, in Barcelona – (Satorras et al., 2020), Netherlands (Lin and Benneker,2021) or Cork (Cork Smart Gateway). The role of citizens' engagement in transport and mobility planning is gaining greater attention. Cascetta & Pagliara (2013) provide insights related to Italy's citizens' engagement. Al-Nasrawi (2019) presented a validated Model for Citizen Engagement and the Smartness of Cities.

The smooth process of e-governance with the active participation of citizens can lead to a win-win model for citizens and government (Alam et al., 2008). It is already an inseparable part of the Smart City in some cities. There are also critical voices in the citizens' engagement methods in a hierarchical political structure that is not open to citizen engagement (Reddelet al., 2004). On the other side, there are trials of combining the traditional planning processes with innovative ones, for instance, digital platform usage in the SUMP planning process in some Greek cities (Bakogiannis et al., 2019). There are also critical voices in the citizens' engagement methods in a hierarchical political structure that is not open to citizen engagement (Reddelet al., 2004). Some authors don't consider the number of participants essential in planning. For these authors, the main goal of citizen engagement is more qualitative than quantitative (e. g. Bickerstaff, Tolley & Walker, 2002). Since there is currently no SUMP in the digital format, we have compared the results of previously developed plans from particular cities. For instance, in project Dynamo, the cities Gdynia and Aachen tested the new digital approach along the traditional communication channels as events, workshops, etc. The same did the city of Brno with citizens' engagement in the future city vision strategy. Table 1 provides the overview of the number of participants and the number of proposals.

Table 1. The comparison of several participants and proposals in chosen cities

City	Nr. of participants	Nr. of proposals	Population (in thousands)
Gdynia*	2700	500	250
Aachen*	270	215	250
Brno[2]	500	2500	380

*Project Dynamo

Despite the population size, there were not so many participants (excluding the city of Gdynia), but it was not essential to have a vast number of participants in that context. More likely, the cities have focused on the quality of the proposals.

PUBLIC ENGAGEMENT TOOLS

The citizen's participation tools are designed to provide directions for engaging citizens in the decision-making process to solve a problem, e. g., in the field of transportation or mobility and numerous other aspects of urban planning (Kleinhans et al., 2015). These tools provide several options to engage the public or citizens. There are some options on using them, first, as a personal tool, which can be used on their own—the second as a series of exercises for larger groups or some projects. Falco & Kleinhans (2018) study explains that despite the vast number of digital platforms, actual citizens' engagement depends on contextual factors such as internet accessibility, illiteracy, technological factors, and organisational factors.

Therefore, the success of the whole decision-making process depends on selecting the most effective public engagement technique. The choice and the use of inappropriate techniques can give poor results. In addition, it can create unnecessary barriers to the project if the decision-makers are selective in who or how they engage in some circumstances.

Different techniques may be used to engage people in the process. No one right technique will suit every issue. Very rarely are perfect models adhered to. Using more than one technique may increase the possibility of gaining a more representative response. In each situation must be made an appropriate choice and the method need to be used for the particular purpose of the engagement and who is being engaged. Waeterloos et al. (2021) examine how local online and offline social integration contributes to civic participation, above and beyond typical predictors such as news consumption and civic talk. In recent years, digital platforms have become an integral part of government and smart cities. Some authors, as Slaev et al. (2019), also warn on the potential problems with a low level of citizens participation and engagement.

The successfully developed ICT services are valuable for citizens from their perspectives (Yeh, 2017). And citizens can react to urban changes with the help of digital media. Together with a standard regular planning process, the non-governmental sector, or third sector, used to comment and participate in public issues to enhance citizens' current status and general good. There are already many tools that municipalities or NGOs use. The most common mechanism is participation via website or municipality channels, apps, etc. (Evans-Cowley et al., 2012). Cegarra-Navarro et al. (2016) present the role of micro involvement with social media, mainly Twitter. The relevance of social media is also confirmed in the Marino et al. (2018) study. It is essential to distinguish the comments on social media (complaints, proposals, negative comments without added value). Pereira et al. (2017) present ways to increase participation and collaboration. Many cities are constrained in ICT due to limited budgets and human resources.

These tools are divided into some groups depending on the approach to engage citizens, for example, printed material (letter, newspaper, and poster); media and telephone (telephone, TV shows or radio), Internet (social media, internet sites, forum/chat); surveys (questionnaire, direct surveys) and others (public meetings, focus group, stakeholders conference). There are various citizens engagement platforms for digital democracy (DemocracyOS[3], VoteIT[4], Polis[5], CitizenLab[6], Decidim[7], ArcGis Hub[8], Consider it[9], CivicPlus[10]). On the other side, media are showing the environmental data intelligence - Ortelium[11], suitable, for instance, in the Smart city framework. These platforms allow automatic data gathering and provision in the city dashboard. But there are limits to citizens engagement. An important feature is the cost of such a platform or technology. In the case of the cities with enough budget, it is possible to purchase a professional digital platform. On the other, the lack of sufficient funding in the NGOs and other institutions makes it necessary to look at opensource or low-cost solutions that are also at disposal. The direct comparison of various platforms in table 2 displays the attributes of the media such as open-source, enabling discussion, voting or having intelligent features (IoT or AI).

Table 2. The platforms for citizens engagement [prepared by the author]

Platform	Free/OpenSource	Discussion	Voting	Assign Point	Smart
DemocracyOS	YES	YES	YES	N/A	N/A
VoteIT	YES	YES	YES	YES	N/A
Polis	YES	YES	NO	N/A	YES
CitizenLab	YES	YES	YES	YES	YES
Decidim	YES	YES	YES	YES	N/A
ArcGIS Hub	NO	YES	YES	N/A	N/A
Considerit	YES	YES	YES	N/A	N/A

Those platforms provide an effective means for greater citizen engagement. The citizens can express their opinion, discuss problems and make their proposals. For example, the CitizenLab platform enables users to customise the environment based on their ideas and requirements. The citizens can visualise ideas on a particular map and share it with everyone or see at a glance what is happening in each district and discuss it, like urban or mobility plans. The platform was used in various countries (e., g. Belgium and the Netherlands), as referred to in Van Aeken (2016).

DemocracyOS is the first platform that connects the two formal kinds of code known to the public. The digital software (the net) and the legal contractual operate most governmental processes. This platform provides the following benefits to citizens: a new tool to empower themselves as valid participants in public debates; an effective tool to control the changes their representatives make to legal bills while they go through the different commissions. In addition, citizens can also participate in control processes, e. g. budget control and deciding which projects should be used. For instance, this platform was used in the city of Buenos Aires (Cerrato-Pargmanet al., 2016). Some tools allow the visualisation based on the GIS data as the ARCGISHUB (Lafia et al., 2018), so citizens and government can see the ideas on the map.

However, one of the crucial problems that the implementation of such platforms faces is the low level of computer and digital literacy, especially in particular groups of citizens. Another is the low level of credibility in digital communication.

THE ROLE OF CITIZENS ENGAGEMENT AND COVID 19

Many studies have been related to transport or urban planning and citizens engagement or participation in the last decades. When the COVID pandemic occurred, many citizens initiatives, formal or informal, became more active, primarily concerned with sharing information about the COVID, policy measures or lockdown. Although ICTs are increasingly being used in the planning process, during the COVID-19 pandemic, this mainly was the only way to involve citizens in public affairs.

The limited mobility and reduction of personal contacts worldwide was also a negative outcome of the pandemic. A positive consequence was the increase in digital technologies by citizens worldwide to communicate with each other. For this reason, the pandemic enabled a greater activation and involvement on the part of the population, as personal contacts were limited or wholly suspended. Due to the limiting of physical encounters and mobility, countries worldwide began to use digital media to inform the population about measures to reduce the spread of the virus (Chen et al., 2020).

The role of citizens engagement and participation is perceived as the critical tool to prepare a better future for citizens (Alford, J., 2009). This applies to different age groups, such as youngsters dealing with climate change (Widholm & Appelgren, 2020) or older adults (Falanga, R. et al.,2021). Falanga (2020) presents an overview of citizens engagement in several European cities. The citizen's attention can be perceived positively and negatively (Jost et al., 2018). Citizens have been engaged in various ways. Some of them are just passive (Loukis et al., 2015), listeners, commenting on the object of interest. More valuable is participation in which citizens act actively (Robinson, 2015), thus enhancing the entire urban governance process.

The Covid-19 pandemic led to higher use of digital media mainly due to isolation or quarantine. Although the COVID-19 pandemic significantly impacted our activities, it did not stop life. Many planned activities had to be postponed or implemented in another way. This was also the case with SUMP plan processing. Of course, some pandemic measures arose negative feedback due to lockdown actions from many governments with negative citizen feedback (Anttiroiko, 2021)). Other studies suggest that citizens' engagement through the digital platforms helped reduce the spread of COVID-19, as in the case of Spain and Italy (Santoveña-Casal et al.,2021 & Graffigna et al., 2020). The use of digital platforms increased due to the pandemic (Cellini et al., 2020; Paschke et al., (2021). The World bank[12] also refers increase in public engagement and consultations through digital tools. These results only confirm that the citizens tend to use these communication technologies that are currently available to them. This will also be the case in other areas that need the population's involvement.

METHODOLOGY

We researched with citizens' digital platforms, thematic stakeholders groups, and consultation with the city. It should be emphasised that due to the Covid pandemic, all actions were carried out in the digital format. We chose a simple form to mark the location when considering the digital environment. Despite the benefits that ICT tools provide, there are also potential threats that pose differences in the operating system of individual mobile phones. Therefore, we decided to use only a digital platform that will be user-friendly and straightforward, allowing comments and suggestions from citizens. In addition to quantitative evaluation, they can also express their own opinions, describing the details, determining the location and what they see as the problem and what should be improved. The primary condition to participate in the data collection was internet access (Figure 1).

Figure 1. The difference between traditional process and only digital

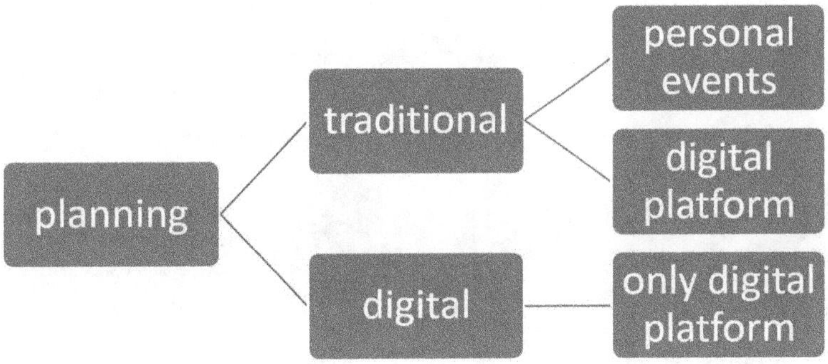

The process started with the consultation with city representatives responsible for urban and transport planning. These consultations took place regularly. But the SUMP process also requires the contribution of expert's groups (nonmotorised persons, transportation, public vehicles, car traffic and parking, freight transport and city logistic). These experts' groups included members of various companies such as transport operators, NGOs, regional transport authorities and various other experts. The role of these groups is to control and steer the process of mobility planning. The first phase of the mobility plan development also required the problem statement and citizens participation. Therefore, we had to use the website platform to review the content of the citizens' ideas, proposals, or critics.

The promotion was done through digital media, social networks and municipal newspapers. Moreover, the campaign was also promoted on Radio and local TV. To identify the problems related to mobility, the citizens were asked to participate in a digital survey with the opportunity to mark and mention the location. The usage of digital platforms such as websites, social networks, and smartphone apps is one of the most used information sources, mainly for the young and middle-aged generation. The basic model of citizens engagement relies on commenting and proposing the change, and they are expecting that their voices will be heard. The process of using data from citizens was divided into several steps. First, citizens could propose the problems and location they want to improve. These data help us understand how individual citizens also perceive other modes of transport and improve them. The second step consists in reviewing and voting for the best ideas that fit or meet the criteria of citizens. Moreover, the essential characteristics of individual inhabitants are also known. For instance, they indicate their age and location and the kind of transport they use.

For example, we can see the needs of pedestrians compared to car users. How do drivers' requirements differ towards modes of transport that they do not use, for example, public transport or bicycle transport?

The evaluation of the current situation was focused on individual modes of transport. At the same time, we asked about the overall assessment using a scale from 1 to 5 (1 is the worst and five is the best). Residents could also mention the location (e.g., street, crossroads, dangerous sections) with which they are dissatisfied and other comments on the problem.

An essential step in this process is how to evaluate the feedback from citizens. Citizens are invited to assess the most critical issues and compromise with experts about the future transport systems in Banská Bystrica. The principle of engagement in mobility issues is depicted in Figure 2.

Figure 2. The personal engagement in the first phase of the SUMP process [prepared by the author]

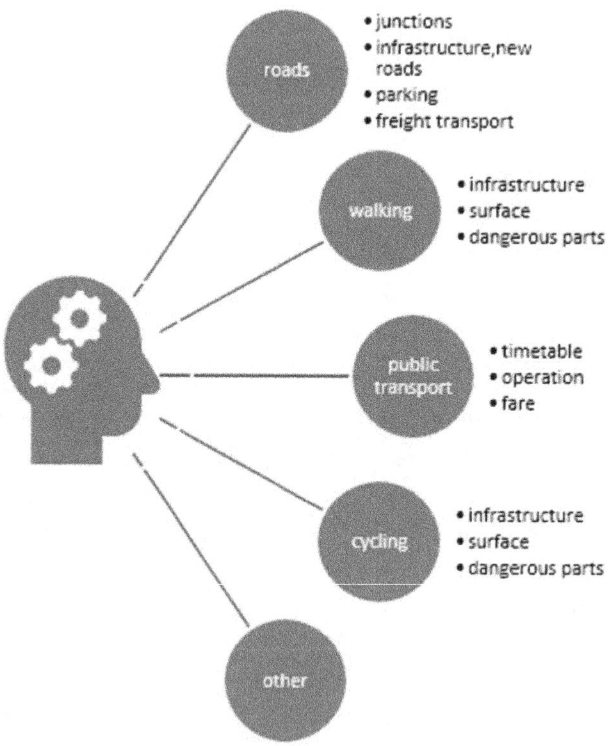

Moreover, citizens could also mention other related problems related to transport but not grouped in these main areas (Figure 2). This can be, for example, the problem of vegetation along the streets or noise, etc.

THE CASE STUDY SUMP BANSKA BYSTRICA

The town of Banská Bystrica, like other smaller towns, wants to activate their inhabitants with extensive planned measures of spatial planning and land use. The city has some experience with citizen participation and engagement in the participatory budget (Participative budget of Banská Bystrica 2021), collecting feedback about various issues related to the city government. Most recently, the city was interested in the views of its citizens regarding SNP Square and the consideration of this area in the city's zoning plan. They collected feedback in ideas, comments or new thoughts focused on the central city zone and the square. The population was active, and more than 3,000 participated (City of Banska Bystrica 1). In a similar spirit, the preparation of the Plan for Sustainable Mobility of the Functional Area of the City was defined. The plan was rolled out in 2020, with initial work beginning in early 2021. But the COVID pandemic limited the potential engagement tools due to lockdown. Therefore, the city and the University developed a digital questionnaire platform to engage citizens. One of the first forms of involvement was proposing and voting for three essential characteristics concerning the city. These characteristics represented what kind of city the inhabitants want in the future.

Out of more than a dozen ideas, residents said they wanted a green, safe and ecological city.

One of the most important aims was to collect suggestions from the residents. As face-to-face contact and mass events were impossible, residents could send their problems and comments through a digital application and web interface.

In the first month, several suggestions were collected, which determined more than 2,000 specific localities submitted by more than 600 citizens. From the statistical point of view, 51% women and 48% of men, 11% in age 15-30, 50% in the period between 31-45 years, 29% in age 46-62 and 9% older than 62.

The commented areas were divided into pedestrian traffic, cycling, car traffic, parking, public transport, and other transport and mobility issues.

In addition, they contained sub-questions regarding the quality of roads, quiet zones, etc.

The locations proposed by the citizens to enhance are shown in Figure 3 (a - f). We can see these aggregated locations submitted by citizens as the initial problems to be solved.

Of course, the submitted locations and problems need to be verified with the municipality's help and from its feasibility. Let's look at the issues and areas that citizens have marked—for instance, the sites submitted as problematic from the parking perspective, as indicated in Figure 4.

Residents of the city of Banska Bystrica stated that they would welcome more parking spaces in the city districts of Sásová (56), Fončorda (36) and Radvaň (19), as well as on Tatranská Street (15). However, residents said in the questionnaire that they were not willing to pay for parking. They added that they would welcome the construction of new parking garages, but not at the expense of greenery. Respondents expressed interest in free parking in the city centre for a short period (e.g., 30 minutes).

Respondents in the survey stated that the worst location in terms of road infrastructure quality is Strieborné námestie (40 respondents). According to many respondents, the reconstruction of Silver Square is necessary. Other areas with the worst road infrastructure include Partizánska cesta (21) and Bernolákova ulica (22). In general, residents of Banska Bystrica are not satisfied with the quality of the road network and demand to increase the quality and repair of most roads in and around the city (Figure 6).

Figure 3. The location proposed by the citizens to enhance [prepared by the author]

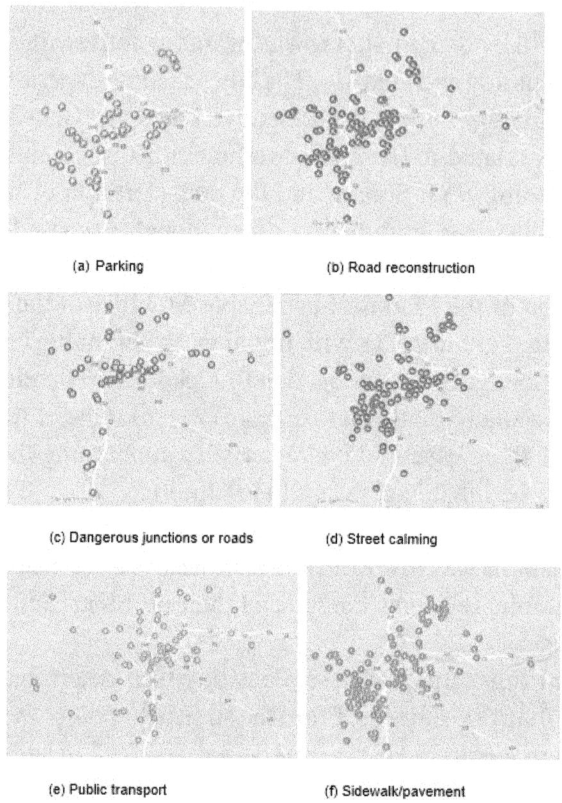

Figure 4. Parking [prepared by the author]

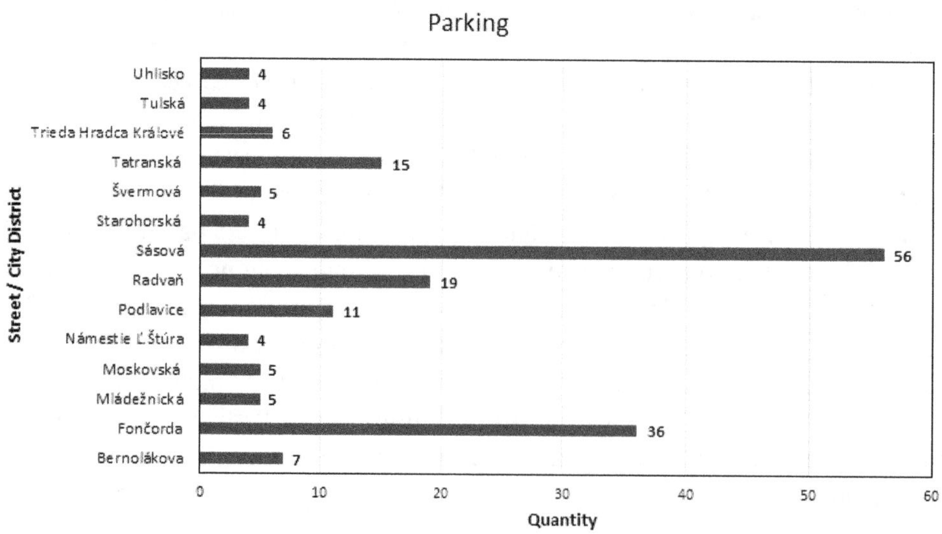

Figure 5. Locations proposed for road reconstruction [prepared by the author]

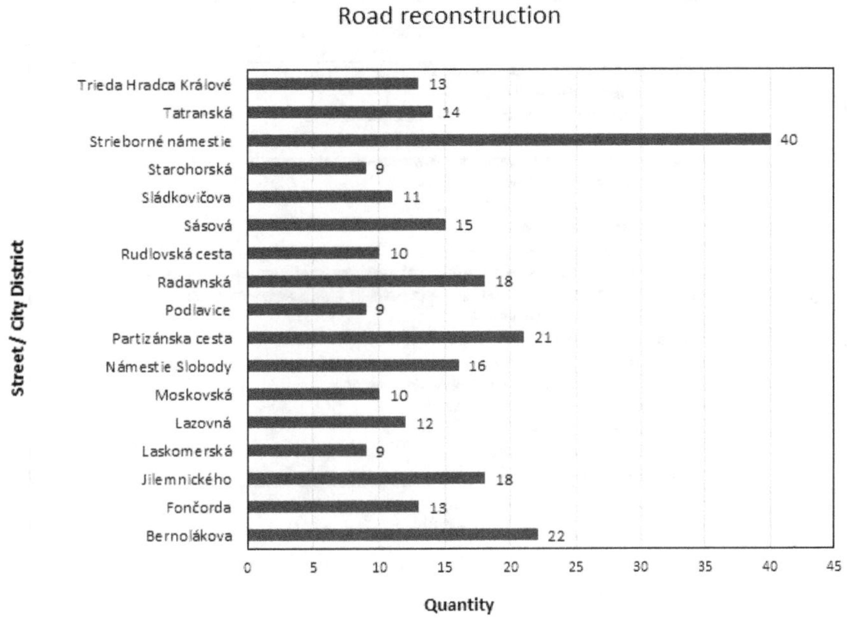

Figure 6. Dangerous junctions or streets perceived by citizens [prepared by the author]

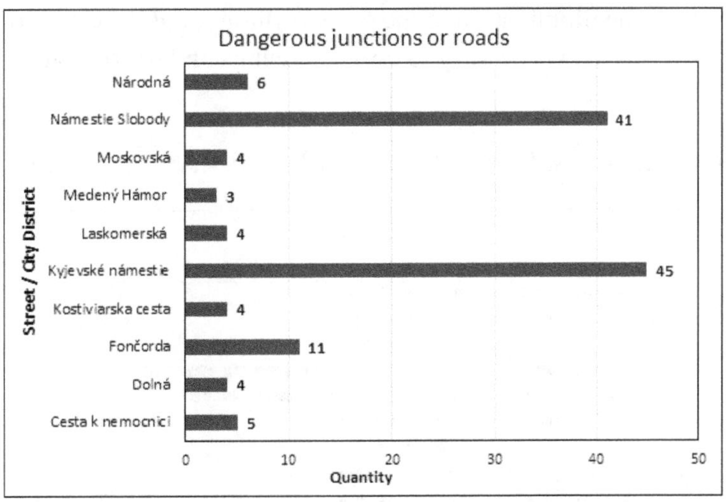

The risky and dangerous locations from the point of view of pedestrians and drivers were described in Námestie Slobody (41), Kyiv Square, and Finčorda (11). The respondents' answers show that Námestie Slobody is considered a locality where frequent accidents occur. Residents also point to an inappropriate solution to the traffic situation around the Europa and Point shopping centres. In addition, they would welcome more excellent pedestrian safety, notably by building level crossings for pedestrians (e.g., underpass or overpass), as indicated in Figure 7.

Figure 7. Locations of sidewalk/pavement to be enhanced [prepared by the author]

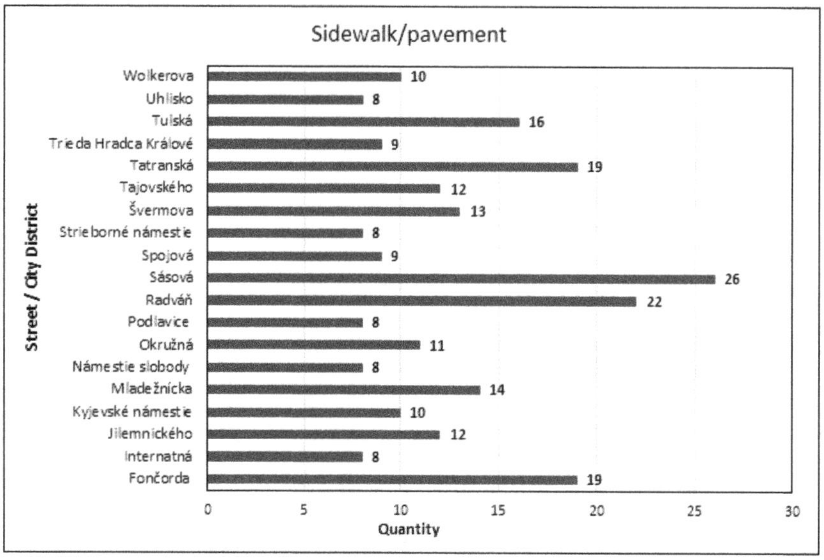

The respondents would welcome most of the sidewalks suitable for reconstruction in the urban areas of Sásová (26), Radváň (22), Fončorda (19), and Tatranská Street (19). Respondents demand the construction of new sidewalks, barrier-free crossings between the sidewalk and the road and pedestrian crossing. Furthermore, they welcome the illumination of the corners throughout the city and reinforce the already existing illuminated hubs. It is also necessary to build crossings and traffic lights on multi-lane roads.

Figure 8. Street calming locations submitted by citizens [prepared by the author]

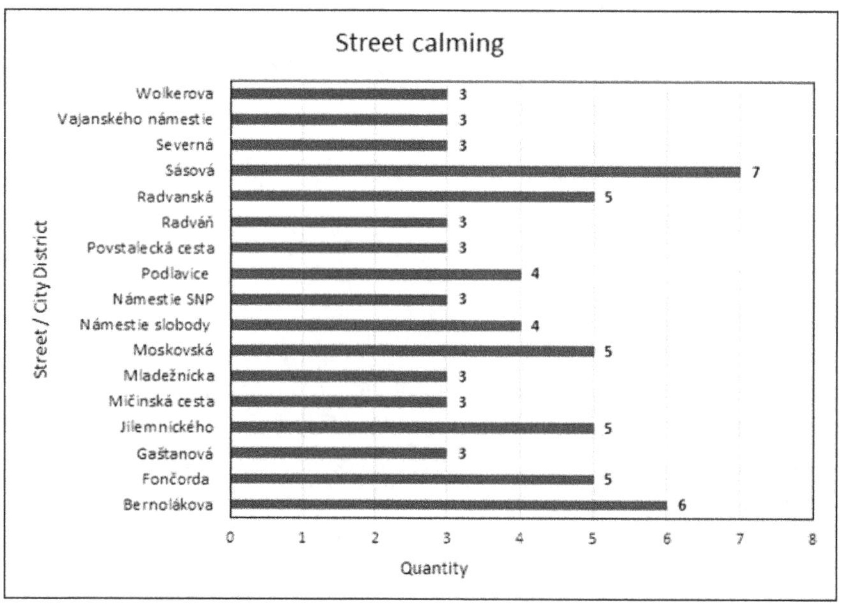

Most respondents would welcome the establishment of calming zones, residential zones, and zones 30 - on the streets Trieda Hradca Králové (10), Tatranská (7), Tulská (6) and Bernolákova (6). They further stated that they would like such zones in all residential areas with more significant movement of children and pensioners (Figure 8). They also point out that the prescribed speed is not observed in zones where the speed is already reduced, which they consider dangerous.

Figure 9. Demand for public transport enhancement [prepared by the author]

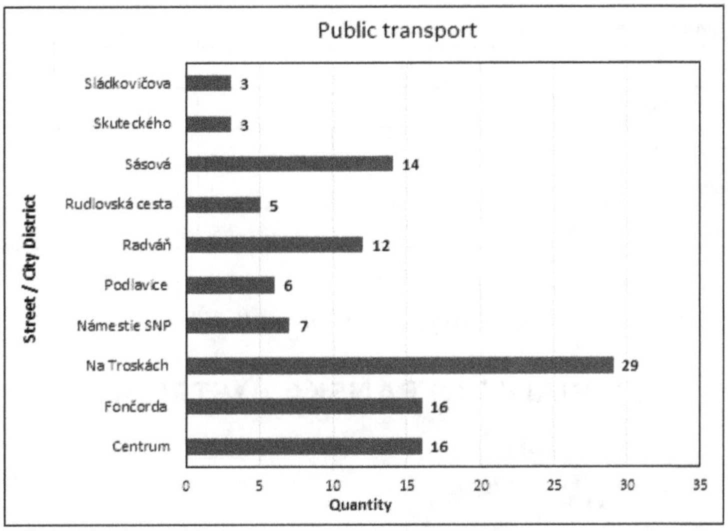

Respondents stated that they would welcome more connections and stops in the area of Na Troskách Street (29), in the Sásová (14) and Fončorda (16) districts, as well as in the centre of Banská Bystrica. Residents have evaluated the quality of public transport vehicles (Figure 9). Most of them don't like delays of PT service and the drivers' driving style. They welcome better connections and other lines, as well as railway lines. They would also include a better ticketing system (simple fare system, payment by ATM card, etc.) and a lower price for transport.

Statistical Analysis

We performed a statistical analysis of the mean, mode, median, and standard deviation. The results of the statistical analysis are stated in Table 3.

The satisfaction of respondents with sidewalks in Banská Bystrica is shown in Figure 10.

Up to 33% of respondents are dissatisfied with Banská Bystrica's sidewalks. 31% are instead on average connected, 24% are depressed, and only more than 1% of respondents are delighted with the sidewalks in the city.

Figure 11 shows the overall satisfaction with the maintenance of the sidewalks.

40% of respondents consider sidewalk maintenance sufficient, representing 120 answers. Only 1% of respondents rate the maintenance of sidewalks as the best. The following Figure12 shows the satisfaction with cycling.

Table 3. The results of the statistical analysis [prepared by the author]

Descriptive Statistics					
question	mean	median	mode	standard deviation	variance
satisfaction of sidewalks	2,33	2	2	0,99	0,99
sidewalks maintenance	2,69	3	3	0,97	0,95
cycling conditions	1,95	2	1	1,06	1,12
transport infrastructure for road transport	2,72	3	3	0,97	0,94
satisfaction with road maintenance	2,77	3	3	1,05	1,10
parking in the city centre	2,46	3	3	1,09	1,18
parking in residential areas, housing estates	2,13	2	1	1,08	1,16
satisfaction with public transport	3,18	3	4	1,12	1,26
satisfaction with suburban bus transport	3,03	3	3	0,85	0,73
satisfaction with rail passenger transport	3,04	3	3	0,98	0,95

Figure 10. Sidewalks in Banská Bystrica [prepared by the author]

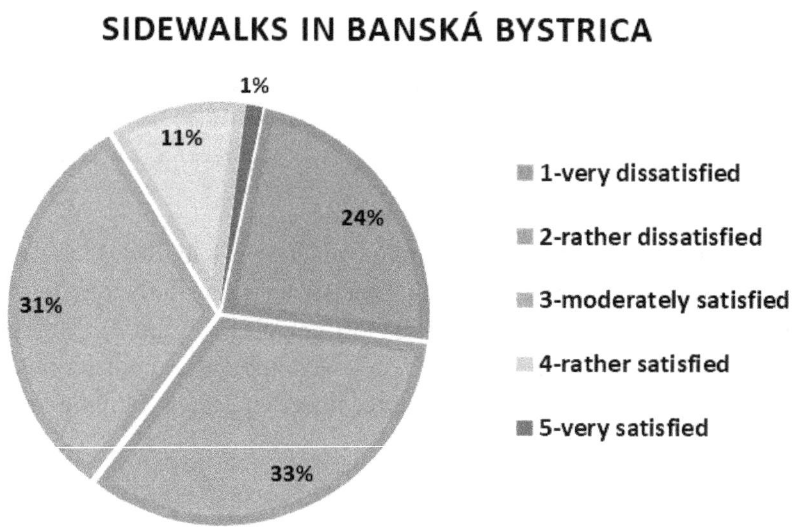

When asked about satisfaction with the conditions for cycling, 47% of respondents were very dissatisfied. 23% are somewhat unhappy, and 22% are on average satisfied. Only 6% of the respondents are reasonably happy with the conditions of bicycle transport in the city of Banská Bystrica. 2%, which represents six respondents, are delighted with the requirements. The satisfaction with car transport infrastructure is depicted in Figure13.

Another issue concerns satisfaction with transport infrastructure. 41% of respondents are satisfied, and 25% are somewhat dissatisfied. Only less than 2% of respondents are delighted with the transport infrastructure in Banská Bystrica.

Figure 11. Maintenance of sidewalks in Banská Bystrica [prepared by the author]

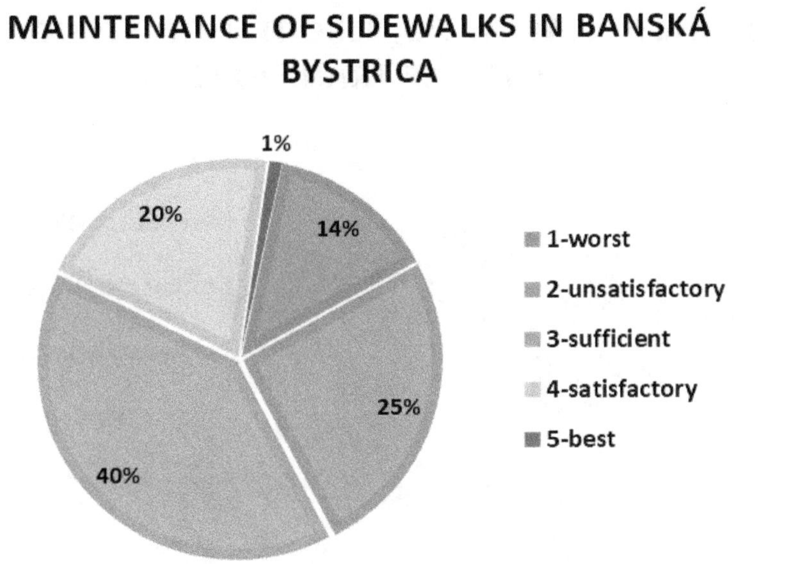

Figure 12. Satisfaction with cycling transport conditions [prepared by the author]

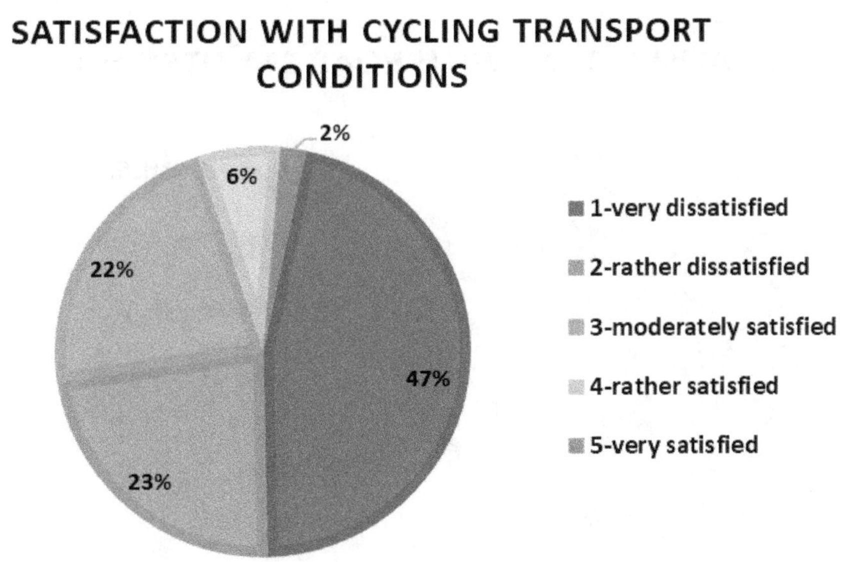

The satisfaction of respondents with road maintenance in Banská Bystrica is shown in Figure 14. Up to 34% of respondents are on average satisfied with road maintenance in Banská Bystrica. 27% are somewhat dissatisfied, 22% choose the answer they are comfortable with. 13% of respondents chose the answer that they were depressed. Only 4% of respondents are delighted with the maintenance of roads in the city.

Figure 13. Satisfaction with transport infrastructures for car transport [prepared by the author]

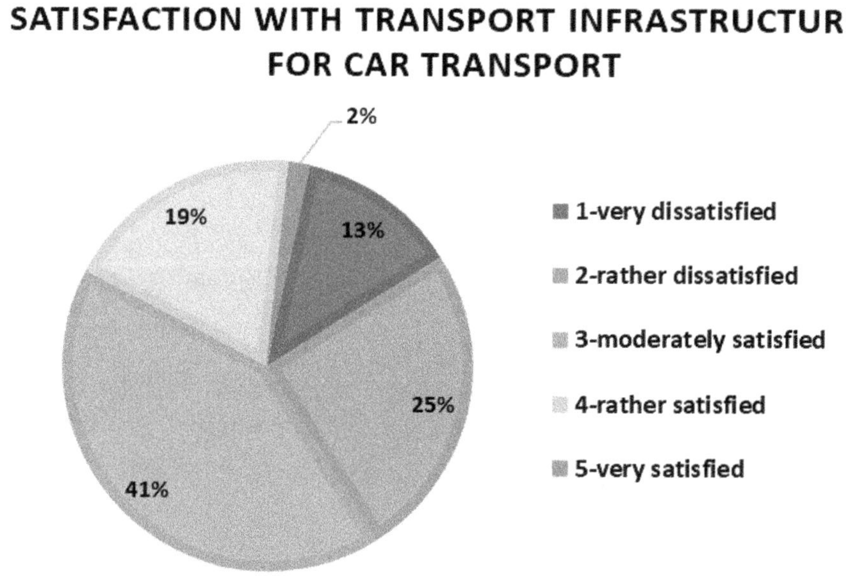

Figure 14. Satisfaction with road maintenance [prepared by the author]

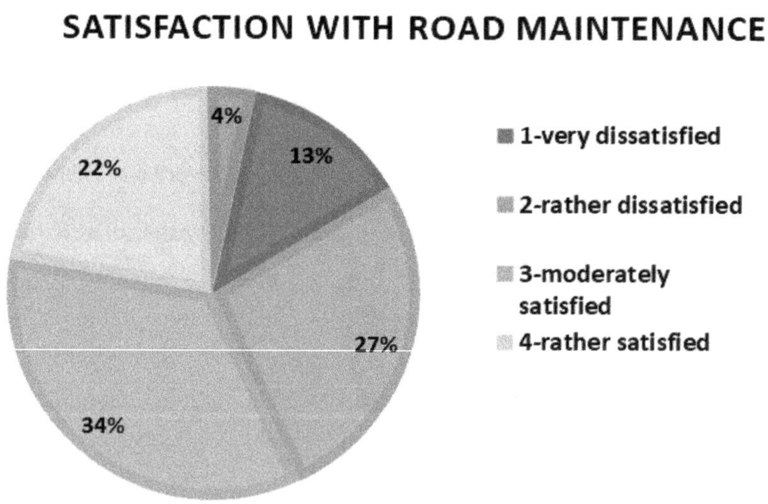

The question relating to satisfaction with the parking solution in the city centre is described in Figure 15. The figure shows that 35% of respondents are satisfied, and 25% are somewhat dissatisfied. Only 3% of respondents are delighted with the parking solution in the city centre.

Figure 16 describes satisfaction with the parking solution in residential areas and housing estates. 35% of respondents are very dissatisfied (106 respondents). 30% are somewhat disappointed, and 25% are on average satisfied. Only 8% of the respondents are relatively happy with parking in residential areas and housing estates in the city of Banská Bystrica. 3%, representing nine respondents, are delighted with parking in the assessed regions.

From Figure 17, we see the overall satisfaction with public transport in Banská Bystrica. 31% are somewhat satisfied with public transport, which represents 93 responses. 30% of respondents are, on average comfortable. Only 8% of respondents rate their satisfaction with public transport as very dissatisfied.

Figure 15. Satisfaction with a parking solution in the city centre [prepared by the author]

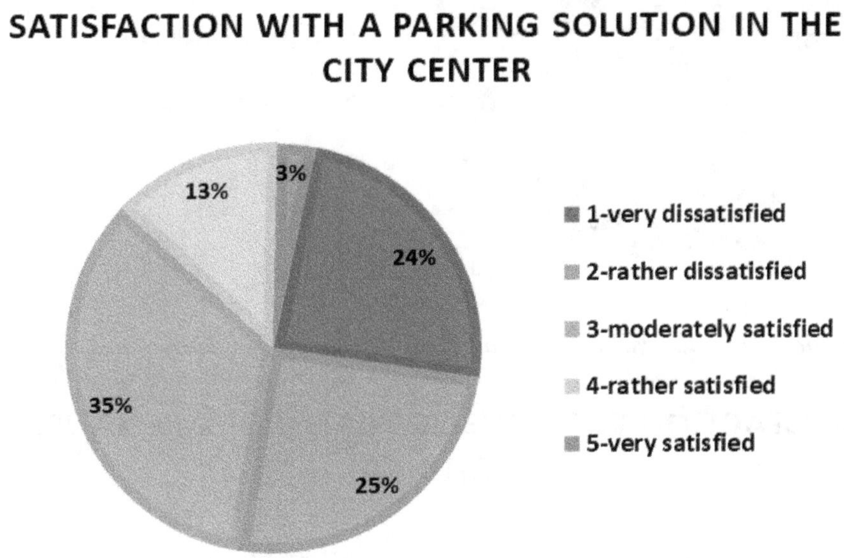

Figure 16. Satisfaction with a parking solution in residential areas and housing estates [prepared by the author]

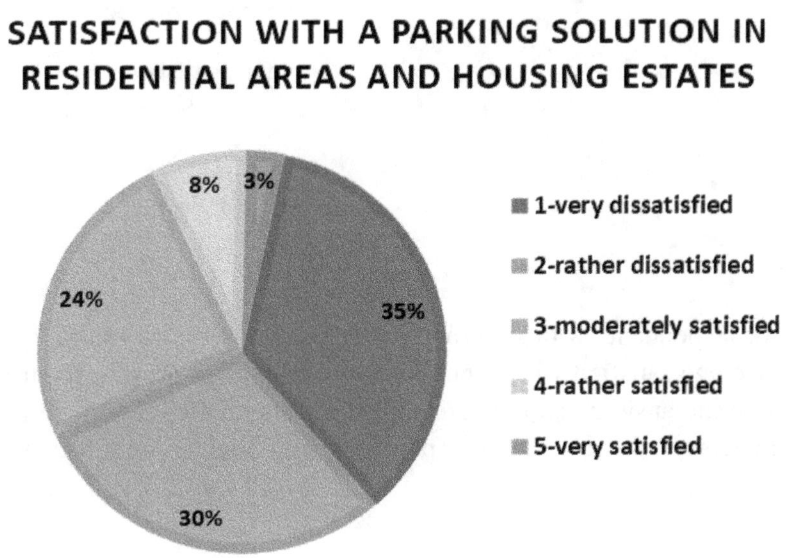

Figure 17. Satisfaction with public transport in Banská Bystrica [prepared by the author]

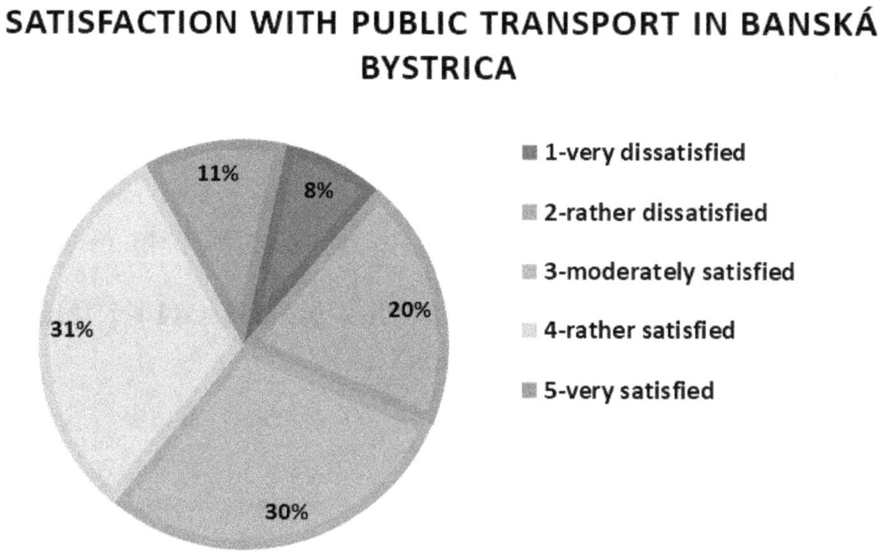

Figure 18. Satisfaction with bus transport in Banská Bystrica [prepared by the author]

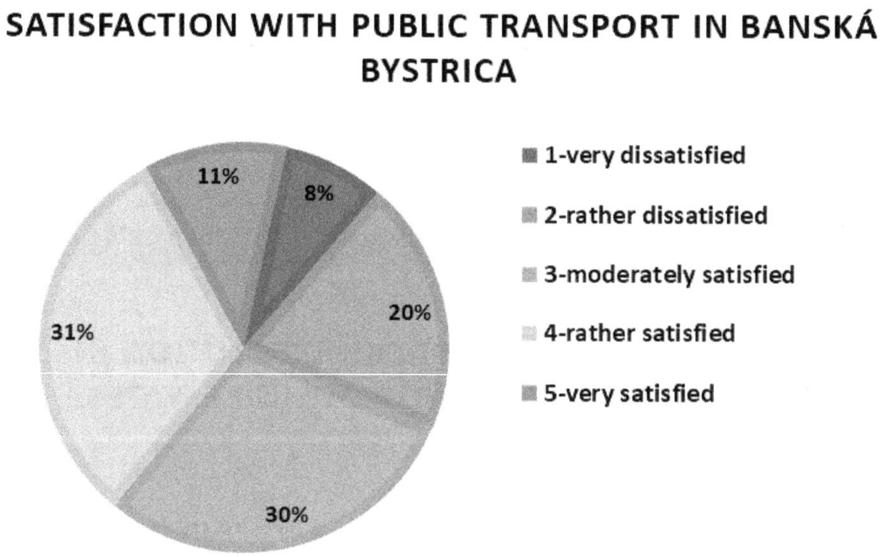

The satisfaction of respondents with suburban bus transport is shown in Figure 18. Up to 53% of respondents are on average satisfied with suburban bus transport in Banská Bystrica. 22% are more connected, 16% choose the answer that they are somewhat dissatisfied. 5% of respondents prefer the solution that they are depressed. Only 4% of respondents are delighted with the suburban bus service in Banská Bystrica.

Figure 19. Satisfaction with rail transport in Banská Bystrica [prepared by the author]

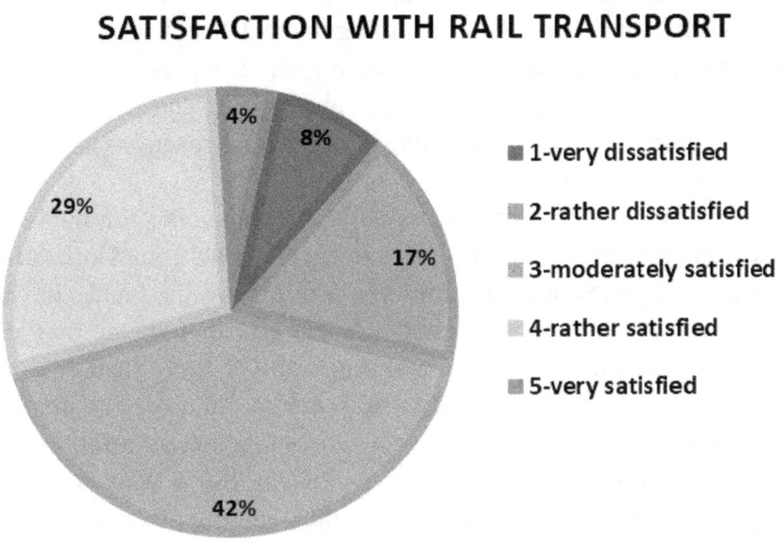

The result of the answers to the last question regarding satisfaction with rail passenger transport is shown in Figure 10. 42% of respondents are, on average satisfied. Eighty-five respondents are somewhat happy, which represents 29%. 4% of respondents are delighted with rail passenger transport (Figure 19).

We can extract interesting insights from this preliminary sample of citizens' submissions. The first insight relates to location and perception of current status and expectation. Secondly, it is possible to see the potential conflicts among various users, for instance, pedestrians vs car drivers, cyclists, etc. Third, subjective perception plays a vital role in how the citizens see the problem and how urgent they want to solve it.

DISCUSSION AND FUTURE WORK

As the SUMP process is spread over several years, the results presented in this chapter are only preliminary and based on data collected through digital means. Local citizens themselves will still submit their suggestions by the end of 2021.

The study also has several limitations. Firstly, the results presented in this chapter are only preliminary. We expect the number of comments to increase. As mentioned above, we have only considered until now electronically submitted ideas. Secondly, we know that the digital platform was used only by those who had access. This means that certain sections of the population have not been able to comment on transport and mobility problems. This was mainly due to the unavailability of the Internet, non-monitoring of the website, social networks, etc.

Compared to different projects with digital participation, this study confirms citizens' high interest and involvement in the planning process. One of the explanations is that the citizens have been at home, and they had time to submit their ideas and locations. On the other side, it can also be the topic of transport and mobility which have attracted the citizens to give feedback and participate. So, it is probably a combination of factors and circumstances that brought citizens to participate.

The research questions have been answered according to the data collected, even if only preliminary. The digital platform for citizens who have access to the internet represents a simple one-stop tool to engage in civic actions. Moreover, citizens probably felt more secure and courageous to record the comment anonymously. Of course, anonymity could also bring adverse reactions and vulgar comments, but this is a part of the risks of digital democracy.

This study confirms that digital platforms are essential for enhancing communication among stakeholders, municipalities, and citizens. Such tools offer an easy way for comment, consultation, and feedback on the various actions or policy measures. In some way, these digital make the process of mobility planning, consultancy and evaluation easier for both citizens and planning authorities.

There are still open questions about the practical usage of ICT tools during the COVID pandemic. For instance, it is not yet clear if the restrictions to movement within the city contributed to more active communication through digital media. It is also not clear whether all groups of citizens have equal access to the city's digital platform and whether they will use it. If not, what communication tools will they have at their disposal if they want to be involved in the planning process? These questions need to be answered from the city's perspective, including citizens and planning authorities.

We aim to examine further whether residents' mobility restrictions led them to use the digital platforms. It will also be essential to determine whether all sections of the population are involved and to what extent and for which areas of the people these digital platforms are inaccessible.

Scenarios will also be subject to citizen involvement, deciding which measures should be implemented and in what time horizon (Figure 20).

Figure 20. The process of using the feedback from citizens

For the visualisation of the inputs from the citizens, this research project is considering the possibility to use additional tools which could offer the option of discussion, voting or a simple overview of the status of the process. For instance, there is a plan to run citizens workshops with gamification features. There are exciting ideas on rewarding citizens that engage in the process (Eleman - https://civitas.eu; ConCensus:Richard et al., 2018), which could eventually be used in this case.

Blockchain technology can be a possible solution for security concerns in these participatory processes since the properties of Blockchain assures that data shared among all stakeholders is transparent, authentic, and verified (Bagloee et al., 2021; (Li, 2018). A blockchain-based digital identity system would ensure that citizens could not use fake identities, nor could their identities be stolen. The persistent storage of data in Blockchain would ensure that all comments, requests, suggestions would be visible to all and thus could not be ignored by local governments or municipalities. It would also be possible to increase the transparency and credibility of the platform. Such implementation could increase the costs of the city. However, it could ultimately increase the involvement of the population.

CONCLUSION

Even though citizen participation should be a vital issue in the planning process, this is not always the case. The COVID-19 pandemic has helped use digital tools for citizen participation and involvement in SUMP planning. This chapter deals with the preliminary participation of the population in planning the SUMP strategic document from several areas. This chapter offers insights on citizens engagement when the planning process goes virtually or online. In this case, the circumstances are quite different from traditional planning processes. The responsible planning authorities have to solve problems related to the technology itself, to citizens access to technology, and the promotion of the participatory process in the various communication channels.

The case study has confirmed that the digital environment can be applied to pedestrians, cyclists, public passenger transport, and car users. But the problems are perceived differently by pedestrians and car users, and it is evident that there are conflicts between these views. Disputes must therefore be resolved sensitively concerning the overall goal of sustainable mobility in the city, in this case, in Banská Bystrica. Citizens could play a more active role if the participatory process is well designed and resourced.

It is expected that citizen engagement will be made more accessible through a more extensive use of digital tools in planning participatory processes.

But one question remains unanswered: how will the city deal with citizens not too attached to digital technologies or tools? And, especially, how will it be done in a time of considerable limitations to physical movement due to pandemics? This fact calls for more innovative cities, for cities to develop an action plan that allows citizens engagement under all circumstances.

REFERENCES

Alam, M., Ahmed, K., & Islam, M. A. (2008, December). E-governance: Challenges and opportunities. In *Proceedings of the 2nd International Conference on Theory and Practice of Electronic Governance* (pp. 264-267). 10.1145/1509096.1509150

Alford, J. (2009). *Engaging public sector clients: From service-delivery to co-production.* Springer. doi:10.1057/9780230235816

Al-Nasrawi, S. (2019, December). A Validated Model for Citizen Engagement and Smartness of Cities. In *2019 International Conference on Smart Applications, Communications and Networking (SmartNets)* (pp. 1-6). IEEE. 10.1109/SmartNets48225.2019.9069794

Anttiroiko, A. V. (2021). Successful government responses to the pandemic: Contextualizing national and urban responses to the COVID-19 outbreak in east and west. *International Journal of E-Planning Research, 10*(2), 1–17. doi:10.4018/IJEPR.20210401.oa1

ArcGIS Hub web. (n.d.). Available online: https://hub.arcgis.com/

Bagloee, S. A., Heshmati, M., Dia, H., Ghaderi, H., Pettit, Ch., & Asadi, M. (2021). Blockchain: The operating system of smart cities. *Cities, 112*. doi:10.1016/j.cities.2021.103104

Bakogiannis, E., Siti, M., Kyriakidis, C., Christodoulopoulou, G., & Vassi, A. (2019). Tools and Technologies for Enhancing Public Engagement in Sustainable Urban Mobility Planning—The Case of Rethymno, Crete. In *Mediterranean Cities and Island Communities* (pp. 237–255). Springer. doi:10.1007/978-3-319-99444-4_10

Bickerstaff, K., Tolley, R., & Walker, G. (2002). Transport planning and participation: the rhetoric and realities of public involvement. *Journal of Transport Geography, 10*(1), 61-73. doi:10.1016/S0966-6923(01)00027-8

Bonsón, E., Perea, D., & Bednárová, M. (2019). Twitter as a tool for citizen engagement: An empirical study of the Andalusian municipalities. *Government Information Quarterly, 36*(3), 480-489. doi:10.1016/j.giq.2019.03.001

Broadband. (n.d.). Available online:https://en.wikipedia.org/wiki/Broadband

Bremen transport plan. (n.d.). Available online: bauumwelt.bremen.de

Brussels Mobility. (n.d.). *Brussels regional public service*. Available online: https://mobilite-mobiliteit.brussels/en/good-move

Boudjelida, A., Mellouli, S., & Lee, J. (2016). Electronic Citizens Participation: Systematic Review. In *Proceedings of the 9th International Conference on Theory and Practice of Electronic Governance (ICEGOV '15-16)*. Association for Computing Machinery. 10.1145/2910019.2910097

Cascetta, E., & Pagliara, F. (2013). Public engagement for planning and designing transportation systems. *Procedia-Social and Behavioral Sciences, 87*, 103-116. doi:10.1016/j.sbspro.2013.10.597

Cegarra-Navarro, J. G., Pachón, J. R. C., & Cegarra, J. L. M. (2012). E-government and citizen's engagement with local affairs through e-websites: The case of Spanish municipalities. *International Journal of Information Management, 32*(5), 469–478. doi:10.1016/j.ijinfomgt.2012.02.008

Cellini, N., Canale, N., Mioni, G., & Costa, S. (2020). Changes in sleep pattern, sense of time and digital media use during COVID-19 lockdown in Italy. *Journal of Sleep Research, 29*(4). doi:10.1111/jsr.13074

Cerratto-Pargman, T., & Joshi, S. (2016). Reconfiguring civic participation: open source software in the political space. In *14th Participatory Design Conference, Aarhus, Denmark, 15-19 August 2016* (pp. 1-6). Retrieved from: http://urn.kb.se/resolve?urn=urn:nbn:se:su:diva-137488

Citizen Engagement and Participation. (n.d.). Available online: https://www.corkcity.ie/en/dbic/smart-city/citizen-engagement-and-participation/

Citizen Engagement and Stakeholder Consultations during COVID-19. (n.d.). Available online: https://pubdocs.worldbank.org/en/875401606251371090/Fact-Sheet-Citizen-Engagement-and-Stakeholder-Consultations-during-COVID-19.pdf

Citizenlab. (n.d.). *An e-democracy platform for stronger communities*. Available online: https://www.citizenlab.co/

Civicplus. (n.d.). *Website and Citizen Engagement Tools To Help You Build A Better Community*. Available online: https://www.civicplus.com/civicengage/local-government-website-design

Civitas Elan Final Evaluation Report. (n.d.). Available online: https://civitas.eu/sites/default/files/d_10_11_final_evaluation_report_civitas_elan.pdf

Coenen, F. (2009). Local Agenda 21: 'meaningful and effective' participation? In *Public participation and better environmental decisions* (pp. 165–182). Springer. doi:10.1007/978-1-4020-9325-8_10

Considerit. (n.d.). *Civil, Organized, and Efficient online Dialofue*. Available online: https://consider.it/

Cork Smart Gateway. (n.d.). Available online: http://www.corksmartgateway.ie/citizenengagement/

Decidim. (n.d.). *Free open-source participatory democracy for cities and organizations*. Available online: https://decidim.org/

Evans-Cowley, J. S., & Griffin, G. (2012). Microparticipation with social media for community engagement in transportation planning. *Transportation Research Record: Journal of the Transportation Research Board*, *2307*(1), 90–98. doi:10.3141/2307-10

Falanga, R. (2020). *Citizen participation during the COVID-19 pandemic: Insights from local practices in European cities*. http://library.fes.de/pdf-files/bueros/lissabon/17148.pdf

Falanga, R., Cebulla, A., Principi, A., & Socci, M. (2021). The participation of senior citizens in policy-making: Patterning initiatives in Europe. *International Journal of Environmental Research and Public Health*, *18*(1), 34. doi:10.3390/ijerph18010034 PMID:33374566

Falco, E., & Reinout Kleinhans, R. (2018). Beyond technology: Identifying local government challenges for using digital platforms for citizen engagement. *International Journal of Information Management*, *40*, 17-20. doi:10.1016/j.ijinfomgt.2018.01.007

Graffigna, G., Barello, S., Savarese, M., Palamenghi, L., Castellini, G., Bonanomi, A., & Lozza, E. (2020). Measuring Italian citizens' engagement in the first wave of the COVID-19 pandemic containment measures: A cross-sectional study. *PLoS One*, *15*(9), e0238613. doi:10.1371/journal.pone.0238613 PMID:32915822

Chen, Q., Min, C., Zhang, W., Wang, G., Ma, X., & Evans, R. (2020). Unpacking the black box: How to promote citizen engagement through government social media during the COVID-19 crisis. *Computers in Human Behavior, 110*. doi:10.1016/j.chb.2020.106380

Jost, J. T., Barberá, P., Bonneau, R., Langer, M., Metzger, M., Nagler, J., Sterling, J., & Tucker, J. A. (2018). How Social Media Facilitates Political Protest: Information, Motivation, and Social Networks. *Political Psychology*, *39*, 85–118. doi:10.1111/pops.12478

Jun, M. S. (2018). Blockchain government - a next form of infrastructure for the twenty-first century. *Journal of Open Innovation*, *4*(1), 7. doi:10.118640852-018-0086-3

Key Factors for Effective Citizens Engagement in Smart City: The Case of Cork City. (n.d.). Available online: https://iot-smartcities.lero.ie/wp-content/uploads/2016/12/CorkCitiEngage-Identifying-key-factors-for-effective-citizens-engagement-in-Smart-City-initiatives.pdf

Kleinhans, R., Van Ham, M., & Evans-Cowley, J. (2015). Using social media and mobile technologies to foster engagement and self-organization in participatory urban planning and neighbourhood governance. *Planning Practice and Research, 30*(3), 237–247. doi:10.1080/02697459.2015.1051320

Lafia, S., Turner, A., & Kuhn, W. (2018). Improving discovery of open civic data. In *10th International Conference on Geographic Information Science (GIScience 2018).* Schloss Dagstuhl--Leibniz-Zentrum fuer Informatik. UC Santa Barbara. Retrieved from https://escholarship.org/uc/item/6rn1b22w doi:10.4230/LIPIcs.GISCIENCE.2018.9

Li, S. (2018). Application of Blockchain Technology in Smart City Infrastructure. *2018 IEEE International Conference on Smart Internet of Things (SmartIoT),* 276-2766. 10.1109/SmartIoT.2018.00056

Lin Y, Benneker K. (2021). *Assessing collaborative planning and the added value of planning support apps in The Netherlands.* doi:10.1177/23998083211009239

Liao, C., Goff, L., Chaparro, A., Chaparro, B., & Stumpfhauser, L. (2000). A comparison of website usage between young adults and the elderly. In *Proceedings of the Human Factors and Ergonomics Society... Annual Meeting* (Vol. 4, p. 101). Sage Publications Ltd. https://www.proquest.com/scholarly-journals/comparison-website-usage-between-young-adults/docview/235449180/se-2?accountid=49401

Lindenau, M., & Böhler-Baedeker, S. (2014). Citizen and stakeholder involvement: A precondition for sustainable urban mobility. *Transportation Research Procedia, 4,* 347–360. doi:10.1016/j.trpro.2014.11.026

Loukis, E., Charalabidis, Y., & Androutsopoulou, A. (2015, August). Evaluating a passive social media citizensourcing innovation. In *International Conference on Electronic Government* (pp. 305-320). Springer. 10.1007/978-3-319-22479-4_23

Marino, V., & Presti, L. L. (2018). From citizens to partners: the role of social media content in fostering citizen engagement. Transforming Government: People. *Process and Policy., 12*(1), 39–60. doi:10.1108/TG-07-2017-0041

Myrovali, G., Tsaples, G., Morfoulaki, M., Aifadopoulou, G., & Papathanasiou, J. (2018, May). An interactive learning environment based on system dynamics methodology for sustainable mobility challenges communication & citizens' engagement. In *International Conference on Decision Support System Technology* (pp. 88-99). Springer. 10.1007/978-3-319-90315-6_8

Open Government Partnership. (n.d.). Available online: https://www.opengovpartnership.org/stories/strengthening-citizen-engagement-during-a-pandemic-a-look-at-sao-paulo-brazil/

Ortelium. (n.d.). *Smart cities & municipalities.* Available online: https://www.ortelium.com/client-solutions/smart-cities?gclid=EAIaIQobChMIy5nMmuWX8QIVEK-yCh1fgwGqEAAYAyAAEgL-W0vD_BwE

Ortiz, J., Young, A., Myers, M. D., Bedeley, R. T., Carbaugh, D., Chughtai, H., ... Wigdor, A. (2019). Giving voice to the voiceless: The use of digital technologies by marginalized groups. *Communications of the Association for Information Systems, 45*(1), 2. doi:10.17705/1CAIS.04502

Paschke, K., Austermann, M. I., Simon-Kutscher, K., & Thomasius, R. (2021). Adolescent gaming and social media usage before and during the COVID-19 pandemic. *Sucht: Zeitschrift für Wissenschaft und Praxis, 67*(1), 13–22. doi:10.1024/0939-5911/a000694

Pham, L., Mai, T. T., & Massey, B. (2016). Key factors for effective citizens engagement in smart city: the case of Cork City. *AIS Pre-ICIS Workshop on "IoT & Smart City Challenges and Applications.*

Polis web. (n.d.). Available online: https://pol.is/home

Project Dynamo. (n.d.). Available online: https://civitas.eu/sites/default/files/participation_2.0_in_the_sump_process_dynamo_web.pdf

Polletta, F. (2016). Participatory enthusiasms: A recent history of citizen engagement initiatives. *Journal of Civil Society*, *12*(3), 231–246. doi:10.1080/17448689.2016.1213505

Reddel, T., & Woolcock, G. (2004). From consultation to participatory governance? A critical review of citizen engagement strategies in Queensland. *Australian Journal of Public Administration*, *63*(3), 75–87. doi:10.1111/j.1467-8500.2004.00392.x

Reed, J., Cook, G., Bolter, V., & Douglas, B. (2006). *Older people 'getting things done 'Involvement in policy and planning initiatives*. Academic Press.

Richard, E., & David, L. F. (2018). The future of citizen engagement in cities—The council of citizen engagement in sustainable urban strategies (ConCensus). *Futures*, *101*, 80–91. doi:10.1016/j.futures.2018.06.012

Robinson, Y., Paraskevopoulou, A., & Hollingworth, S. (2019). Developing 'active citizens': Arts Award, creativity and impact. *British Educational Research Journal*, *45*(6), 1203–1219. doi:10.1002/berj.3558

Salha, R., El-Hallaq, M., & Alastal, A. (2019). Blockchain in Smart Cities: Exploring Possibilities in Terms of Opportunities and Challenges. *Journal of Data Analysis and Information Processing*, *7*(03), 118–139. doi:10.4236/jdaip.2019.73008

Santoveña-Casal, S., Gil-Quintana, J., & Ramos, L. (2021). *Digital Citizens' Feelings in National #Covid19 Campaigns in Spain*. doi:10.20944/preprints202103.0734.v1

Satorras, M., Ruiz-Mallén, I., Monterde, A., & March, H. (2020). Co-production of urban climate planning: Insights from the Barcelona Climate Plan. *Cities (London, England)*, *106*, 102887. doi:10.1016/j.cities.2020.102887

Slaev, A. D., Kovachev, A., Nozharova, B., Daskalova, D., Nikolov, P., & Petrov, P. (2019). Overcoming the failures of citizen participation: The relevance of the liberal approach in planning. *Planning Theory*, *18*(4), 448–469. doi:10.1177/1473095219848472

Standard Deviation and Variance. (n.d.). Available online: https://www.investopedia.com/terms/

Stratigea, A. (2012). The concept of 'smart cities'. Towards community development*? Netcom. Réseaux, communication et territoires, 26*(3/4), 375-388. https://journals.openedition.org/netcom/1105

Sustainable Urban Mobility Plan Bremen 2025. (n.d.). Available online: https://www.bauumwelt.bremen.de/sixcms/media.php/13/SUMP_Bremen2025_web.pdf

The Urban Mobility Observatory. (n.d.). Available online: https://www.eltis.org/mobility-plans/sump-guidelines

Van AekenK. (2016). Digital democracy in Belgium and the Netherlands. A socio-legal analysis of citizenlab. be and consultatie. nl. Koen Van Aeken, Digital Democracy in Belgium and the Netherlands. A Socio-Legal Analysis of Citizenlab. be and Consultatie. nl. In *Digital Democracy in a Globalized World*. https://ssrn.com/abstract=2932962 doi:10.2139/ssrn.2932962

Viale Pereira, G., Cunha, M. A., Lampoltshammer, T. J., Parycek, P., & Testa, M. G. (2017). Increasing collaboration and participation in smart city governance: A cross-case analysis of smart city initiatives. *Information Technology for Development*, *23*(3), 526–553. doi:10.1080/02681102.2017.1353946

VoteIt web. (n.d.). Available online: https://www.voteit.se/in-english/

Waeterloos, C., De Meulenaere, J., Walrave, M., & Ponnet, K. (2021). Tackling COVID-19 from below: civic participation among online neighbourhood network users during the COVID-19 pandemic. *Online Information Review*. https://www.emerald.com/insight/content/doi/10.1108/OIR-08-2020-0379/full/html

Widholm, A., & Appelgren, E. (2020). A softer kind of hard news? Data journalism and the digital renewal of public service news in Sweden. *New Media & Society*.

World Justice Project. (n.d.). Available online: https://worldjusticeproject.org/our-work/programs/democracyos

Yeh, H. (2017). The effects of successful ICT- based smart city services: From citizens' perspectives. *Government Information Quarterly, 34*(3), 556-565. doi:10.1016/j.giq.2017.05.001

ENDNOTES

1 https://www.eltis.org/mobility-plans/sump-guidelines
2 https://www.eltis.org/mobility-plans/activity-14-plan-stakeholder-and-citizen-involvement
3 https://worldjusticeproject.org/our-work/programs/democracyos
4 https://www.voteit.se/in-english
5 https://pol.is/home
6 https://www.citizenlab.co
7 https://decidim.org
8 https://hub.arcgis.com/),
9 https://consider.it
10 https://www.civicplus.com/civicengage/local-government-website-design
11 hhttps://www.ortelium.com/client-solutions
12 https://pubdocs.worldbank.org/en/875401606251371090/Fact-Sheet-Citizen-Engagement-and-Stakeholder-Consultations-during-COVID-19.pdf

Chapter 10
Learning From the Self-Organizing Universe:
Towards Evolutionary E-Planning

Jenni Partanen
Tallinn University of Technology, Estonia

ABSTRACT

Ubiquitous digitalization in our technology-mediated cities has recently pushed the computation from virtual worlds into the cybernetic sensor-guided cites. AI-driven architecture and planning assumedly make cities efficient, sustainable, and "smart" in a linear, mechanical manner. However, cities are complex adaptive, self-organizing systems evolving through transitions. They are hard to control using any planning tool founded on a rational comprehensive model. Computational, generative methods are established tools in the urban design and planning providing novel perspective for smart city planning. Moreover, the paradigm of complex systems and (universal) self-organization enable crossing of ideas in biological systems, cities, and digital systems. This generates intellectual progress in both philosophical and methods development for e-planning tools. Here, the author introduces an e-planning method applying self-organization of information, capable of reflecting the complex urban dynamics. She concludes by illustrating possible futures in self-organizing technology and planners' roles in it.

INTRODUCTION

Background

We have entered a full speed era of technology mediated cities, with ubiquitous digital tools constantly guiding us - not only in the form of computational, virtual networks, but through corporeal cities utilizing myriads of sensors, sondes and probes (Batty 2016, Ridell 2019). The digital applications, tools, and apparatus allegedly make our cities "smarted", that is, more ecologically and economically sustainable, efficient, and more convenient in many senses for us humans. Great expectations lay in architecture and planning praxis which are supposed to create smart buildings and smart urban environment capable of

DOI: 10.4018/978-1-7998-9090-4.ch010

ultimate optimization of the performance of cities, and the lives of the upper, urban middle class. The emphasis is on sheer efficiency and machine-like behavior – often ignoring needs of digitally excluded and vulnerable ones (Carli et al. 2018, Krishnan & Balasubramanian 2016, Tanwar et al 2021).

For architecture, urban planning and design practices computation has been well-established after the *digital twist* in the 1990s, forming today an integral part of these professions (Carpo 2012). Computation affects not only the means to design but the very content, enabling previously impossible solutions both regarding the form and cognitive content (Botazzi 2018). Digital twist in design is not methodological but a qualitative cultural transition. Moreover, many endeavors in computational design strive towards generative approaches that would bridge scales, and more importantly, embrace better understanding of cities and their complexity and embed it into the design worlds (Aguiar & Cardoso 2017, He et al. 2012).

According to current understanding, cities are complex adaptive systems that self-organize from their internal premises and evolve through revolutionary transitions (Portugali 1999, Batty 2007, Mitchell 2009). Consequently, cities resist any attempts to be tamed using even the most sophisticated, digitalized planning tools if they are built on the premises of the rational comprehensive model (Batty & Mashall 2009, Allmendinger 2017). This is often the case of smart city development, which still have not adopted the bottom up perspective and relies largely on neoliberal project-based, top-down thinking (Cardullo & Kitchin 2019).

The urban paradigm embracing complexity enables fast progress, crossbreeding ideas from computation and theories of self-organization in biological systems and cities; urban evolution; and typomorhpological tradition. This enables intellectual progress in methods development, applying new concepts and stances from these disciplines. Here I suggest that first, self-organization forms an underlying common mechanism for almost-universal order formation in all (including non-living and artificial) complex systems, and hence could help us to build understanding and tools, complementing city planning praxis, design and implementation. Secondly, I consider that thorough understanding of nature of self-organization would assist building more considerate planning approaches for complex urbanity.

Outcome of this Study

In this chapter I elaborate these issues on a conceptual level to build a foundation for practical applications and approaches. This conceptualization is illustrated through an introduction of a proposed structure for a digital planning tool capable of reflecting urban self-organization through *liquid thinking* and so-called complementary *rule-based/case-based reasoning*. It is indicated that such a tool could apply self-organization of information in constructing and updating the ontologies on which the rules that guide urban spatio-functional processes are based. I conclude this chapter by discussing the further implications of the "self-organizing digital twist" for the designer's and planner's role.

COMPLEX, DIGITALIZING CITIES

Digitalization of the Planning and Design Tools

Computational tools have been in everyday use for architects, urban planners and designers for decades. Until recently, the field of the digital design devices has however been fragmented regarding the scale and scope (Gil et al 2010). For example, the family of CAD-programs or algorithmic design applications have concentrated on building or site-level scale. The planning and design or analytical methods from urban simulations to procedural or large-scale information models and geodesign have operated on neighborhood, city, regional, or intraregional scale. The operational principles of these large-scale planning tools vary from statistical or parametric to more self-organizing and emergent dynamics, mimicking interactive, computational processes typical for many real-world systems.

There are current endeavors to bridge these gaps of scale and scope, progressing the latter type, more generative thinking, particularly among information models and procedural modeling (Stojanovski et al. 2020, He et al. 2012, Aguiar & Cardoso 2017). Furthermore, the digital tools assisting in planning and design are increasingly applying artificial intelligence and machine learning, enabling integrated process of formation of urban project in concert with environmental factors. Simultaneously, we are starting to gain an appropriate understanding of urban systems characteristics and operation through theories of complex adaptive systems and self-organization. Moreover, this knowledge has forced us to recognize the limitations it sets for traditional, top-down -oriented urban planning practices (deRoo & Hillier 2016, Portugali et al. 2012).

Digital Methods Reflecting Urban Dynamics

The landscape of these digital tools reflects the evolution of planning ideals. Many of the non-generative tools are intellectually founded on theories adopting the mechanistic, rational comprehensive view of cities of Modernity. This stance often appears prevalent in smart city development, digital twin models or many parametric tools, implying ideas of prediction, controllability and linearity of the cities and planning from these (erroneous) premises. Generative models and methods, however - with various philosophical roots in Leibnitzian thinking (Lynn 1998) to certain natural scientific traditions (Casti & Berry 1997) - imply dynamic, evolutionary understanding of urban systems (Almendinger 2017, Partanen 2018). Such applications include microsimulation models, certain procedural models and algorithmic or generative design (Batty 2007, He et al. 2012, Terzidis 2006). Many of these can better embrace the emergent unpredictability of urbanity with different foci. Urban dynamic computer models aim at simulating the behavior of urban (sub)system (traffic, land use, passenger flows) based on rules extracted from the system's prior operation. In procedural models, variance in their operation and resulting output emerges in a trans-scalar manner from so-called shape grammar (Hohmann et al. 2010), implying rules between geometric fundamental entities, iteratively refine a design by creating increasing level of details. In algorithmic, generative design emphasizes formation process resulting from external and internal forces affecting the project (Terzidis 2006).

Urban Complexity

According to many authors, the theories of complex adaptive systems (CAS) would provide a more appropriate frame for smart city planning and discourse (Batty 2007, Portugali 2000, Allmendinger 2017, Cilliers 2005). Complexity theories have potential for replacing or complementing the current, technology- and control-oriented view that prioritizes efficiency, optimization, and top-down control of a single building, block, urban district or even entire cities (Cugurullo 2013), and hence strongly assumes predictability (Partanen 2018, Lom & Pribyl 2021). The emergence of this novel perspective is particularly necessary concerning the climate change mitigation and adaptation, while it highlights the essential non-equilibrium characteristics well established in science of ecosystems in nature. These include tipping points in the nested systems of systems; necessity for redundancy for building evolutionary resilience (Davoudi et al. 2013); and difficulty to optimize complex systems (Walker & Salt 2012, Lenton 2011, Lom & Pribyl 2021, Allen et al. 2014).

Emerging in the 1970s, theories of complex adaptive systems originally cover a set of distinct theories in natural sciences exploring and explaining phenomena priorly challenging to study, such as those in non-linear mathematics, statistical mechanism, and mathematical chaos (Mitchell 2009, Kleick 1997, Casti & Berry 1997). The term complex adaptive system was first introduced by sociologist W.F. Buckley (1968). Resonating with the ideas of town planner Patrich Geddes (1915) about general evolution, his model saw analogies in socio-cultural systems and biological species. CAS imply interactions between the parts that cause unpredictable behavior for the overall system (Portugali 1999, Casti & Berry 1997). Typically, these systems are far from equilibrium and they embrace chaotic dynamics: irreversibility, sensitivity to initial conditions, and a deterministic yet seemingly stochastic behavior (Casti&Berry 1997). Furthermore, CAS evolve through sudden revolutions, in which process the more predictable and less predictable periods alternate (Portugali 1999). In chaotic state, embryonic orders compete, one of which becomes dominant and 'enslaving' the system, forming the new dominant order (from bottom up) for a while - until internal turbulence or external shock pushes the system to another chaotic state and so on. In CAS a transient collective order, that emerges from actors' local relations on the higher scale, is common (Haken 1985).

Current understanding of cities as complex adaptive systems (CAS) has been established already in 1990s (Allen 1997, Portugali 1999, Portugal et al 2012, Batty 2007, deRoo & Silva 2010). Complexity theories form a credible planning paradigm, which can also integrate qualitative and quantitative studies, diverse cultural and social–economic dynamics while considering the spatial and physical urbanity. (Batty 2007, Batty & Marshall 2009, Portugali 1999; 2006, Mitchell 2009).

Classical Urban Theories in the Light of Complexity

Complexity theories have many reference points in how urban systems are described in classical urban theories (see e.g. Portugali 1999). Profound feature of CAS – dynamics embracing (series of) revolutions while the system oscillates on the edge of instabilities where more predictable and chaotic states alternate - resonate with different interpretations of societal ruptures or revolutions in urban theories. These are related to e.g. intertwined technological innovations, economic cycles and societal changes. Castells (2011) recognizes at least two such revolutions bundled with so-called modes of development - technology induced transformation and its social relations. Batty (2018) suggests that certain revolutionary ruptures might be more numerous and getting more frequent reflecting the progress and innovations in technology, and finds a connection point to Kolmogorov waves and Schumpeter's economic cycles distorting the space-time continuum by the changes in communication and transportation technology (Batty 2018, p. 167). Emerging qualitative shifts from internal turbulence generated partly by drastic innovations such as steam engine, combustion engine or internet, are elaborated in detail by David Harvey (1989) and Francois Ascher (2004). Harvey's idea that innovation in communication and transportation technology is one booster of societal change reducing the travel time resonates with Batty and Longley's (1994) findings regarding fractal urban form. They discovered that cities have intrinsic fractal properties that results from the actor's attempt to minimize the transportation and communication costs, and implicit rules responsible for these complex morphologies (Batty and Marshall 2012).

It is also possible to find a factual connection between the circular dynamics of emergent socio-economic or even spatial structures in complex systems, and the structuration theory of Giddens 1989m Portugali 1999). City scale patterns (spatial, functional, cultural, social) form in complex systems from local interactions, to which they feed back, guiding their formation. Giddens argues that while the socially constructed structures influences individual's autonomy they are simultaneously maintained and adapted through the exercise of agency in a dynamic manner, bridging the micro-macro -perspectives (Giddens1989). Structuration, and emergence, takes place in the interface or structure-agency (or pattern-actor) -interaction (Batty and Marshall2009). The two cannot be studied separately: agents compile and consolidate the structures, but also construct and modify them. Hence Giddens' structures, signification, legitimation, and domination, could be considered as a description of emergent properties (Portugali 1999, Giddens 1989).

Complexity Applied in Urban Studies and Planning

Christopher Alexander was one of the first urban designers arguing for complexity of cities in early 1960s. He considered that the natural cities emerge from internal logics of certain ordering principles. He argued that complex, transient composition of overlapping basic components connected as a semilattice, i.e. a considerable number of connections between nodes produce significantly more possibilities than nodes organized as a simple tree diagram (Alexander et al 1977). He also developed an advanced rule-based design system regarding basic urban blocks and their relations respecting such open composition. Another early promoter of urban complexity, and a critic against rational-comprehensive planning practices ignoring it, was Jane Jacobs (1961). Echoing Warren Weaver, she argued that organized complexity emerges in cities spontaneously from numerous variables that interact in a random manner. Jacobs had also recognized the role of social processes in complexity and suggested that spatio-functional features generate and enable human encounters and vivid urban life from bottom up. Hillier (2012) pointed out that the despite the general formation mechanisms in complex cities, the local variation of resulting patterns is locally unique, gradually shaped by a combination of spatial laws and human agency, and cultural norms in a path dependent manner. Local circumstances are hence essential in planning CAS, stressing the need for collaborative planning processes DeRoo&Rauws (2012).

A groundbreaking study illustrating non-linearity and spatial self-organization of urban system was Schelling's (1969) simple segregation microsimulation model, which proved that unintuitively low number of intolerant residents cause spatial segregation pattern that rise from the agents' local interaction. This work has been applied in real cities multiple times (e.g. Hatna and Benenson 2014). Microsimulation models (Batty 2007) have become established tool in dynamically generating geographical patterns emerging from location logics of actors related to land development, previously presented already in early urban economic models of von Thünen, Christaller and Alonso (Capello 2013). Although lacking the dynamics of the recent computer simulation, these simplistic presentations were also based on actors' dissipated gravitation towards preferable facilities or resources forming activity, land use and land price patterns embracing dissipated decision making, similar to those in self-organization processes in complexity thinking.

Overall, two broad approaches can be recognized in recent complexity planning approaches: first an idea of a tolerant rule framework allowing local socio-cultural and economic processes (Moroni 2014), and participatory, collaborative approaches (Innes and Booher 2010.). Moroni (2014) stresses that these can be complementary to each other, and that allocation of public resources (infrastructure, public services and amenities) still takes place through means of conventional spatial planning in a mixed system of development controls and urban codes. (Moroni 2014, Marshall 2012). Hence any complexity planning method, such as the one combining framework and collaborative approaches presented in this paper, is by no means meant to be universal replacing the current planning system, but complementary to conventional planning, providing viewpoints currently absent in planning discourse.

Currently, the intrinsic complexity of cities is only increasing due to digitalization, which introduces a completely new systemic level of digital infrastructures bundled with each other and with other systems (Ridell 2019). Due to the accelerated expansion of the myriads of coupled, redundant, and incrementally build digital technologies, along with their side-product, accumulating sea of data evading the cities and our lives, 'programmable cities' are becoming more and more intertwined, cluttered, and unpredictable (Gabrys 2014, Batty 2016). However, although many phenomena in cities appear stochastic, they often follow an intrinsic mechanism according to which all complex systems form organized structures without external guidance, that can be captured by adopting the concept of self-organization for a new reading stressing surprising impacts of conscious planning actions. (Batty 2007, Portugali 1999). Metaphors are the momentum of new conceptualizations: the transition from one discursive context to another becomes hard or impossible (Pakarinen 2007). Metaphors may provide alternative perspectives to build new, critical readings of contemporary urban planning, along with better understanding cities and the forces behind their formation (Newman 1999, Oswald and Baccini 2003).

Such metaphors from nature are not new in urban design and planning flourishing particularly in the beginning of planning discipline after the mid-1850s. Often these have a notorious reputation, for leading to superficial analogies or naturalistic fallacies. In these analogies, cities were considered as whole organisms, developing and growing to an optimal size. Such metaphors were, however, just another mechanistic, top down and machine-like approaches to planning, seeking for overall control over cities (Batty& Marshall 2009). Recent mental models from network city and urban metabolism to self-organization and evolution differ from these early examples. They reach far beyond the reduction, not only seeking a simplistic resemblance to natural forms, but emphasize the profound ontological or functional similarities between mechanisms and dynamics in natural systems and cities- while recognizing also differences (Batty&Marhshall 2009). The new generation of urban metaphors have become established within the urban discourse, and they accent transitions, recursion, and self-regulation in urban processes. Particularly, contemporary metaphors are concerned with the systems dynamics of interlinked *patterns and processes,* interacting in a circular manner (see also Ascher, Giddens). According to Batty and Marshall (2009), particularly complexity thinking has proved its worth in providing valuable tools for understanding and managing these circular processes.

SELF-ORGANIZATION

Self-Organization in Biological Systems

Self-organization is a core mechanism of how open complex systems form order internally for their own benefit. In self-organization, dynamic order and temporal stability appear within seemingly fuzzy and chaotic systems. It implies that local agents' interactions appear as an unintentional order on a higher scale without guidance from an external force. Self-organization is common in many types of open complex systems that range from molecular or cell scale to variety of organisms and animal behavior (flocks, herds), and is perceived also in urban systems, society and economy (Kauffman & Strohman 1993, Casti & Berry 1997, Krugman, 1996, Mitchell, 2009). Self-organization typically implies the formation of emergent patterns, which then enslave the agents through feedback: agents start to follow one of the emerging orders in fluctuating system, which stabilizes and forms an order parameter, becoming dominant for a certain period of time (Haken 1985, Portugali 1999). Such interactive and selective order formation process share similar principles with evolving organic systems and can be considered as a "pre-evolutionary" principle for non-living entities (Eigen 1971). As the self-organizing pattern enslave the parts, these configurations adapt to the environment (Haken, 1985). Similarly to evolution, the pattern benefits the system by the means of saving energy or material, reducing friction, or maximizing the surface for photosynthesis (Eigen, 1971, Prigogine & Lefever 1973, Bak 1990 Kauffman & Strohman 1993, Casti & Barry 1997, Mitchell 2009). Accepting self-organization as a universal principle in all complex systems (Janch 1980) helps to bridge many gaps in understanding the behavior of assumedly distinct natural and artificial, also digital, systems.

Self-Organization in Technology Inspired by Biological Systems

Advances based on knowledge of self-organization in nature have benefitted the design of digital, computational systems, obscuring the difference between "artificial" and "natural" by highlighting the systemic similarities between these two worlds. Forms of self-organizing (digital) computation can be considered as an important step towards truly generative planning, design, and even implementation applications, and evolving AI outperforming the solution-based linear ones that are still prevalent (Kendall 2018).

An illustrative example would be a well-known, educational toy model the *Game of Life* John H. Conway introduced in 1970s in Scientific American (Gardner 1976). GoL is a simple cellular system that operates in a lattice, the rules of which are based on the neighborhood relations. As the status of a cell can be *on* or *off*, during iterations the system can produce numerous dynamic states and surprising behavior emerging through iterations, based on its initial conditions. While dynamics of cellular automata have later been studied thoroughly (Wolfram 1984) and its applications are well established in urban modelling (Batty 2007), the Game of Life was a prime example that proved the capability of a simple, computational system for both self-replication and universal computing. It appeared that an artificial system could behave in a way previously considered only possible for living organisms.

More serious, yet as remarkable, applications have later drawn from biology. Both heartbeat and neuron activity in the brain is known to emerge through self-organization. On the local, cellular level, interactive neural impulses or cellular contraction give rise to resilient patterns typical for the organ's operation (Haken 1985, Kugler et al. 1980). Based on this brain operation, Teuvo Kohonen (1989) introduced a groundbreaking paradigm for artificial neural networks called *self-organizing maps* (SOM). It implies computing functions similar to those emerging autonomously in the brain, being able to not only adapt, but also form self-organizing transient "clusters" without guidance. SOM still provides a solid paradigm in brain theory, while it has also formed a basis for simplified algorithms used for calculation of real-world operations applying unsupervised learning, (see e.g. May et al. 2010, Olawoyin et al. 2013). Development had recently led to brain-inspired computer architecture applying SOM. They mimic the structural and functional plasticity using the modulation of neurons and synapses for self-organizing hardware structure (Novac et al 2020, Khacef et al. 2018).

On the higher level of brain function, (human) cognitive processes similarly follow self-organizing dynamics. It unfolds in ways the activity areas in brain are activated and alternated, and in emergence of hysteresis as a non-equilibrium phenomenon (Haken et al 1985). Furthermore, higher cognitive processes, such as memory, emerge through self-organization as well. In the so-called *dynamic memory model* (Schank 1983) the cognitive processes of understanding, learning, and memorizing are intertwined. Individual events are organized as self-organizing clusters, *remindings*, which can later be recalled for problem solving in a new situation. In some cases, recalling exactly similar past events or experiences may be impossible, and remindings from former, typologically different cases may help to resolve these anomalies in novel situations (De Mántaras et al. 2005, Schank 1983, Riesbeck and Schank 2013). In computing, this cognitive process has inspired so-called *case-based systems* (de Mántaras et al 2005), in which the prior solutions systematically build a library that can be adjusted, combined, and adapted to solve new problems (Golding & Rosenbloom 1996).

Self-Organization in Cities and Nature are Similar

For the science of cities and for urban planning, self-organization provides a strategic lens for perceiving the human activity and some of its unintentional consequences from a new viewpoint. Instead of intentionality and control, such a new reading stresses emergent, seemingly autonomous nature and output of many urban processes resulting from the actors' inability to acquire complete knowledge of the whole. Adopting right strategic decisions for the study objectives, such perspective may help to reveal inequalities: emergence of grassroot activities, behavior of vulnerable people, or clustering of non-privileged groups. This enables adopting of appropriate necessary policies to guide the dynamics to a preferable direction. Overall, self-organization helps to perceive and concentrate on the collective urban dynamics emerging from interrelated individual, conscious decisions, to better guide it.

In everyday life, self-organization may appear, for example, in human actors' attempts to find the best locations for their activities; sudden emergence of mobility patterns resulting from individual choices regarding the transport mode or route; prevalent cultural conventions; or housing trends that feed back to the citizens' individual choices. The systemic enslaving might emerge in collective life as a group view or social pressure, or in the companies' attempt to benefit from collaboration with or competition between their neighbors. The order parameter that guides the system may appear as quantifiable entities, for example in the latter example, as land prices (Haken&Portugali 2003, 2019). Self-organization – or complexity frame in general – can also be applied in scrutinizing variety in social issues, from (informal) settlement patterns in developing countries to education, along with inequity issues related to health, race, or crises management (see e.g. Dovey 2012, Moore 2011, Agyepong et al. 2012, Sarriot & Kouletio 2015, Clark-Ginsberg 2020, Buckley 1998, Bobo 2018, Auerbach 2009).

Overall, self-organization can be considered as an essential, resilient, and dynamically stabilizing mechanism in complex (urban) systems. It might help the system to renew at the time of crises (Walker & Salt 2012, Portugali 1999). Planning should acknowledge and allow preferable types of self-organization, hinder the harmful ones (segregation, monofunctional areas) and leave the rest to operate – similarly to the ecosystems in the nature (Allen 1997, Partanen 2018).

URBAN EVOLUTION

Regarding the ways the relations between entities produce ordered rule-based forms, urban (self-organizing) evolution[1] can be considered as a type of computation (Salingaros 2012; 2000). In such computation, urban tissue emerges through dynamic repetition of continuous and interactive adaptations of built elements. Receiving feedback from their environment, the system produces adapted urban spatio-functional configuration and form, resulting from variety of conditions and limitations human necessities and desires produce. This urban morphogenesis takes place as a sequence of extended iterative events. In this sequential computation, each building action adapts to local circumstances, that may consider mobility flows, built structure, uses and activities, topography, or micro-climates. Furthermore, individual or cultural habits and needs, privacy or necessary social connection guide these dynamics. Such a process is the very manner the traditional settlements were historically constructed, and still to an extent until the WW2 in West. It is also the mode for constructing informal settlements globally, which makes it a common tradition in building in developing countries even today (Salingaros 2012, 2000, Shane 2011).

A typical procedure of the emergence of a traditional settlement is recognized also within certain scholars in the urban typomorphological school of thought (Caniggia and Maffei 2001). The settlement formation can be considered as self-organizing, non-centralized activity involving intuition and tacit knowledge of local building tradition. Cognitively, in such processes human minds have a collective capacity for calculation by applying remindings also regarding urban environment (Haken & Portugali 2003, de Mántaras et al. 2005, Salingaros 2012). In addition, these maneuvers were executed following so-called spontaneous consciousness, which in vernacular architecture implies mainly applying the established building types. In this urban progress, the way to evaluate the operation of the local 'plan' occurs through the robustness of the implemented urban tissue, which can be perceived even in the case of informal settlements. Configurations of their everyday activities appear to self-organize relatively successfully via generative computation.

Salingaros (2012) considers that while all human settlements represent a certain mode or level of urban computation, their degrees range from non-computational or random to fixed rules little use for vital cities, to interactive computing producing and generating ordered complexity more likely to produce viable urban environment (Salingaros 2012). Alfasi and Portugali (2007), who call for sea change in planning, promote a return to the evolutionary planning mode for ordered complexity abandoned by modernistic and rational planning paradigms. A comprehensive vision of a city is not possible nor appropriate, and setting the spatio-functional configuration cannot be a task of a single planner – either state, a municipal planner, or an investor. Instead, echoing Alexander (1977), they argue that it is necessary to reflect and play with the qualitative relationships of built elements to each other and to the city structure. As a collective action of dealing with constituent elements of urban space this would also contribute with equity and justice, and add a collective perspective to the political process bringing substance to communicative planning (Alfasi&Portugali 2007, Feinstein 2000, Cambell and Marshall 2002).

For such generative planning, this view sets new requirements regarding the urban environment's capability to evolve, learn and adapt in ways resembling the human cognitive processes or neural networks. Non-adaptive or computerized planning methods, for example simple parametric systems, produce formal output or intentional stochasticity detached from the actual process from which the variation should emerge.

Politics and Power

It is noteworthy that this spontaneous calculation need guidance since self-organization is a value-free mechanism: inequality or other socially intolerable results are possible. Planning as a steering mechanism is necessary in the society for ethical reasons and for equity's sake.

Planning is intrinsically political activity (Allmendinger 2017). Planning is however not democratic: it has no elected representation, it embraces specific non-general rules, and stresses end-result instead of universal procedure (Alfasi 2021). Power in this system can take two forms, that of institutionalized power of a planner or a politician, or informal tactics of stakeholders - often planners, but also (economically) powerful actor bending the explicit planning rules with no transparency (Fox-Rogers and Murphy 2014). This implies that we cannot rely solely on rationality since power and knowledge relationship is not symmetrical (Flyvbjerg 2002). Moreover, the power of planner lies in socially constructed discourse framing how phenomena are defined and talked about within a certain cultural frame (Foucault 1969). Intentional knowledge distortion of those who benefit (such as stakeholders distorting rules) becomes an issue.

While complexity provides a new viewpoint towards uncertainty and how to deal with autonomous urban socio-economic and cultural processes with the rule-framework thinking in planning, the questions of power likely retain in traditional planning context (along with infra/social planning) in the mixed systems proposed by Moroni (2014). All in all, e-planning tools (even those built on complexity thinking) are not panacea for planning; instead, they may provide a complementary approach for considering complex features of urbanity regarding humans' benefit. At best, e-planning tools developed consider to promote maximal transparency, fact-checking, and reflection of motives and emerging topics as a feedback.

Furthermore, although self-organization mechanism resonates well with agglomeration economics (Fujita 1996), the role of political economy cannot be ignored in creating the framework or operation for economic actors (laws, regulation, governmental policies, and political agendas). Political economy promoted the current complex operating environment in cities. Starting from the late 1980s', the global economy and information/knowledge-based society was deliberately politically constituted through policies of deregulation privatization and liberalization of trade and investment enabled by international institutions (IMF, WTO) and enacted by local government in the pursue of their own (economic or political, state's or personal) interest. Neoliberalism was a political project (Harvey 2007). However, the resulting very complex networked system is irreducible and cannot easily be undone; furthermore, as this complex system behaves nonlinearly, it is nearly impossible to predict and is hence beyond (political) control. Individual states' decoupling from this network would imply closing the access to growth (Harvey 2007, Castells 2011).

Self-Organization in Generative, Algorithmic Design

Generative or algorithmic design methods aim at simulating the formation process of buildings. They consist of a family of approaches capable of reflecting the dynamic, evolutionary features and relationships of the complex urbanity. These approaches rose from the 1990s philosophical design movement and they intent to formulate a new paradigm for building design posing a critical view against, on the one hand, modernistic "empty-site"-design ignorant of local circumstances, and on the other, pure form, Euclidian paradigm, and the designer's supremacy in architecture. Instead, generative design suggests that architecture should be embedded in and capable of reflecting the evolutionary environment in constant state of flux, not separate itself from it (Lynn 1998, Terzidis 2006, Jencks & Kropf 2006).

For that, generative design introduces forces, multiple environmental factors affecting the formation of the design, resembling morphological formation processes in species evolution in the nature. The forces that are considered to impact the design project may range from micro- and macroclimate, energy production and/or consumption, building physics and materials, to laws and regulations, accessibility, uses on site and surroundings, project program, political will, local stakeholders' and other participants' preferences, and so on. These forces mold the architectural form: the rules set border condition within which the geometrical attached entities adjust, stretch, and form "autonomously" organized structure - the design output (Lynn 1998, Terzidis 2006, Hensel 2013).

The capacity of generative approach for adaptation and coevolution enables truly sustainable and resilient design, while allowing (often more visible) non-conservative form-giving. For example, forms and structures that are self-sufficient with water, heat or energy become possible. As in urban simulation models, time is a parameter, and hence these models provide a platform for experiments with form. Furthermore, they offer educational potential to learn about the interaction between the form and forces (Terzidis 2006).

Philosophically, algorithmic or generative architecture does not originally emerge from theories of complex adaptive system. Instead, it has its roots in Leibnizian philosophy and the paradigm of mathematical topological space (Lynn 1998). Due to recognized resemblance in profound concepts and processes between the two are however visible in the ways the conceptual ideas from theories of CAS are borrowed into generative design discourse (Weinstock 2004). Similarities include the characteristics in interactions between components and with the environment, dynamic processes, and the potential for evolutionary rules (Terzidis 2006, Lynn 1998).

Urban Rules

Overall, all these evolutionary and self-organizing - biological and artificial - processes and mechanisms return to one key variable – the rule. Rules in urban systems are of two types: first, emergent, descriptive rules, and second, normative, prescriptive rules (Alfasi & Portugali 2007, Partanen 2020).

Emergent rules form, for example, in urban evolutionary processes in traditional and informal settlements as each building is iteratively adapted to the existing ones and the environment, and economic geographical patterns. Gradually, certain local rules emerge that start to steer the system, and new buildings will follow these rules. They might regard relation to adjacent buildings, routes, slopes, views, or other factors. The rules guiding the relationships and characteristics of urban elements, building types, and their uses, have often emerged autonomously during the building of cities over time, in a process taking decades, centuries, or even millennia. While these traditional processes have been disrupted and no more prevalent in late modern societies in the West, resilient economic geographical patterns and spatial trends in e.g. retail, housing and clustering of industries can be recognized, often yielding master plans (Alfasi et al 2012) (Kuusela & Partanen 2021).

Such urban evolution can be considered as a self-organizing process in many senses. Similarly to the self-organizing process of clustering or other activity-based location patterns, each maneuver - occupying a certain location in a city or design and adjusting a house in prevalent type of the urban tissue - results from careful planning and design actions. However, none of the actors have a complete picture of the whole, resulting surprising patterns on a higher scale of observation guiding the future building maneuvers. Higher level could refer to a neighborhood, a city, or urban region.

As was said, such an autonomous process is valueless *per se*, and in society and city it requires guiding for ruling out undesirable effects. Normative, statutory rules are necessary, but they must not hinder the preferable forms of self-organization producing, at best, economically, socially and culturally viable cities, and interesting and diverse built environment (Alfasi&Portugali 2007, Partanen 2018). Hence the normative rules should provide such a framework that allows the self-organization of variety of urban activities, restricting the harmful forms of action allowing others, while updated infrequently in case of changing urban dynamics (Partanen 2018). These rules or building codes guide the urban system by regulating the interaction of urban components (Alfasi & Portugali 2007) and introduce the permissible options. Typically these rules are laws, regulation and plans. Essential is that they should be able to reflect the emergent rules, not to hinder the desirable forms of self-organization. Since rules cannot be intrinsically adapting – overly flexible rule would not be a rule - they must be underdefined for case-by-case interpretations and only updated occasionally (Schauer 1991).

To conclude essential elements that need to be considered in constructing, updating and using e-planning and -design tools are:

- Self-organization
  - Self-organization of the urban complex system
  - Self-organization of the construction of the tool to reflect the dynamic reality
- Rules
  - Emergent rules (autonomously in the process)
  - Prescriptive rules (normative framework capable of reflecting the emergent rules)

PROPOSED SELF-ORGANIZING E-PLANNING TOOL

Overall Structure

Overall, the system consists of first, a loose digital, statutory rule-based system of conditional plans (*if-then*), that is built to allow the perceived self-organizing behavior. These prescriptive rules are constructed and updated infrequently through self-organization of information in a digital *Liquid Process* by local actors, stakeholders, municipal planners and specialists, applying artificial intelligence on a varying degree (from pattern recognition to more self-organizing, dynamic learning in the future). Cases (i.e. projects) are collected and indexed in the case library, and emerging patterns in them are evaluated through machine learning for recognizing embryonic patterns and changes in trends. This case-based activity may give rise to the need for update in plans potentially applying generative design methods. The conceptual structure of this tool is presented in the Figure 1.

(1) Liquid process is responsible for producing order through self-organization (introducing classes or ontologies) out of entropy of various, proposed entities (problem statements, challenges in urban system at issue). Emergent rules emerge within this process, and they are transformed into prescriptive, statutory frames for building maneuvers in the actual city, (2) against which each case (i.e. project) is reflected to estimate its fit. In case the project does not follow the rule, it is either rejected and/or requested changes, where algorithmic process can be applied for adaptation. Then the case is stored into

the case-base labelled as exception from the rule (3B); those projects following the rule get implemented and will participate in the pattern formation monitored by AI. As certain attributes start to pile up in the case base indicating potential emergence of a trend, or new embryonic patterns are recognized on site, a new round of liquid evaluation (1) of prescriptive rules might take turn infrequently, on demand, and initiated also for specific societal needs. This last procedure is equivalent to updating the traditional plan and it enables emergence of novelties and corrections, instead of just replicating the existing patterns.

Figure 1. Structure of the conceptual system consisting of Liquid and Rule Based/Case based systems and feedback. (details in the Figures 2 and 3)

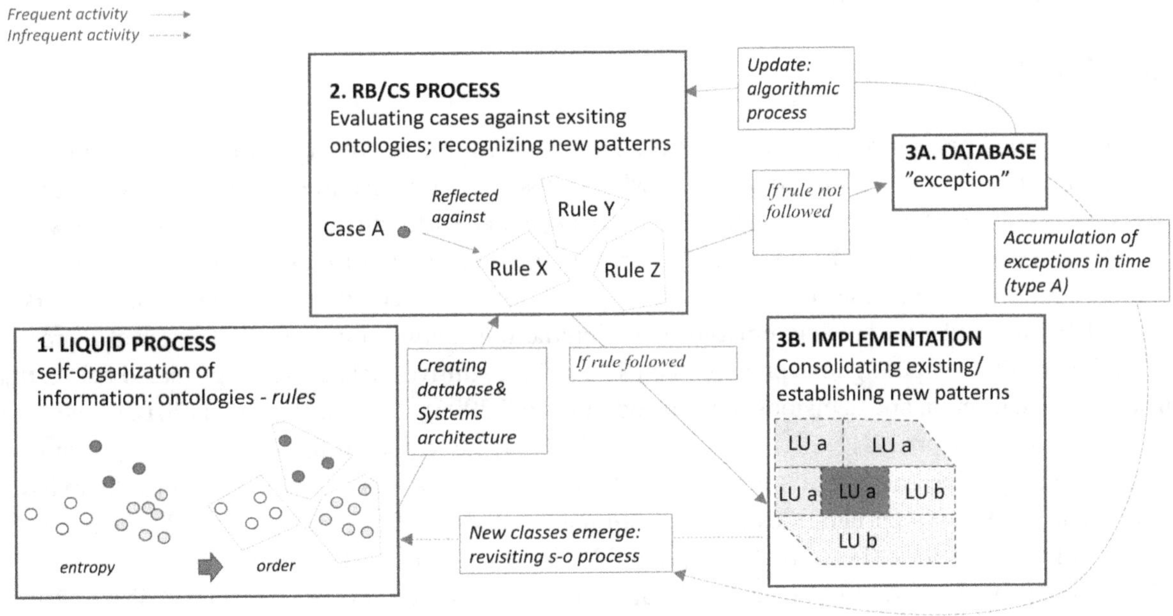

Requirement Elicitation: Self-Organizing Ontologies

Prior to this procedure, requirements for the system need to be recognized. In complex digitalized cities, every event leaves a digital trace which accumulates as big data. Big data is a byproduct resulting from our digital activities virtually but also corporeally, and it intrinsically lacks the system structure (Batty 2016). Therefore, it requires processing, manipulation and combining with other datasets for applicability in planning. For big data, the context is everything, along with relation to other data. (Kitchin & McArdle 2014, Batty 2016).

To tame the big urban data, for any digital platform collecting or storing data for urban planning it is necessary to organize the information - gather and form ontologies for the data in a manner that reflects the complex reality. In the philosophy of computing, ontological approach of l*evel of abstraction* presents a hierarchy of increasing abstraction in a procedural manner, from (client's) intention to *specification* and algorithm, and to computer programming levels. Here particularly the specification implying *requirements elicitation* would become important in case of extracting the knowledge of urban planners, policymakers, specialists, and local stakeholders in a multidisciplinary manner to construct the indicators. This procedure would also explicate the implicit assumptions probably affecting the use of the tool. With other words, the entropy of data needs to be ordered.

Currently applied methods for requirement elicitation vary, embracing interviews, ethnography, surveys and Delphi, some resembling participatory live or e-planning approaches (Healey 2003, Iqbal et al. 2019, Okesola et al. 2019). However, for the sake of reflecting complexities in cities, the requirement elicitation method should be based on information self-organization, to reflect the real-world phenomena of also hidden patterns and trends. Here I suggest that it is possible to learn from self-organizing planning methods, e.g. self-organizing City Games (Tan & Portugali, 2012) or Liquid Planning (Partanen 2016) to enrich the methodology. Such methods also serve the need for system updates/revising the ontological structure in the case of new (potentially prevalent) types emerge.

Liquid Planning

Self-organization or grouping of information is an essential cognitive procedure through which we perceive the city. This occurs not only in terms of what is observed, but also in terms of potential patterns, giving them simultaneously a contextual meaning (Haken and Portugali 2003). Overall, to perceive the city information needs to be self-organized. According to Haken and Portugali (2006; Portugali 2012), this process takes place in a multi-level process called SIRN (synergetic inter-representation networks). Within SIRN, an "order parameter" emerges from interactions between internal and external representations. One of the competing interpretations enslaves the cognitive system creating a steady state that appears as human action and decisions (Haken and Portugali 2006). Tan and Portugali (2012) have demonstrated this in a live process called City Game. In the City Game, the group solves planning problems gradually: each participant places a mock-up building in the model city, adjusting it into the existing environment. As a result, certain rules emerge within the process regarding relations and limitations for the new building locations. These rules start to guide the system (Tan &Portugali 2012).

The liquid process draws from so-called liquid democracy approach (see e.g. Blum & Zuber 2016). Conceptually, the liquid process follows the basic principles of the SIRN: the self-organization of problems statements (i.e. variety of types of unorganized data) first emerge among participating stakeholders, and develop within an intra- and interpersonal process, and as soon as coherent entities (ontologies) start to emerge, in the reference of this "reservoir". Data processing (combining, dividing classes, proposing reformulation) takes place among the stakeholders either live or online (Partanen 2016). The digital process is assisted by computed pattern recognition tools, which enables combining tacit knowledge and (big) digital data.

Partanen (2016) suggested a structure for a liquid spatial planning process which adopts a relatively high level of human participation. This basic structure is presented in the Figure 2 and also applied here.

In more detail, this collaborative process starts by dynamically formulating an appropriate problem statement(s) among participants for the case at issue, e.g. regarding current urban challenges (concerning, e.g. congestion, urban poverty, declining neighborhoods or natural resources, need to support grass root economic activity). Framework for the task could arise for example from city strategy, or within the process. The participants (stakeholders, specialist, decision makers, landowners, residents and other users) each on their turn claim their main interests freely on a shared space (live or on the online forum), keeping their statements concise. Each person may propose the one most essential issue first with no restrictions for the subject.

Figure 2. Conceptual structure of the liquid process. The process would consist of five stages: first, claiming the interests ("entropy creation"; secondly, grouping of the interests (self-organization); thirdly, proposing solutions; fourthly, using these to construct ontologies; and, fifthly, potentially proxy voting on the popularity of each of the future views (Figure 1)
(Adapted from Partanen 2016)

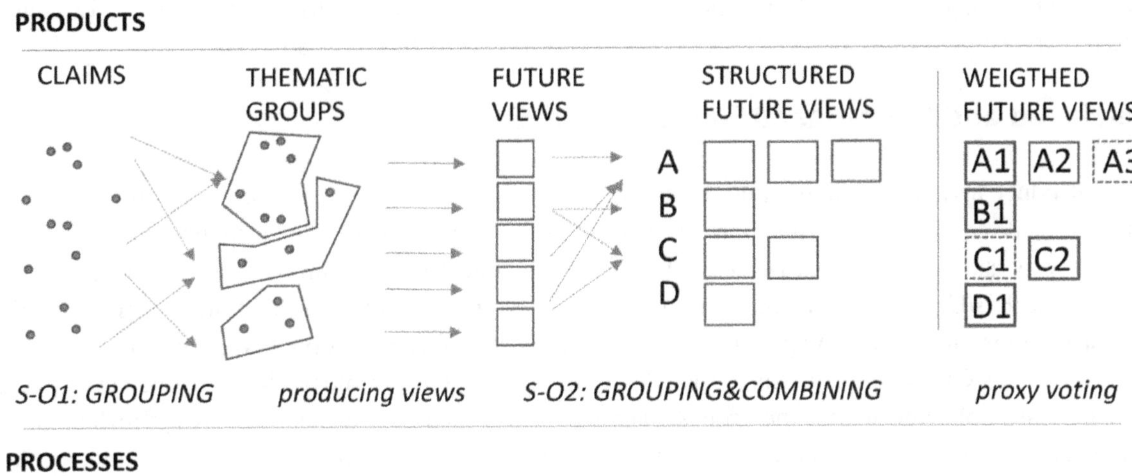

Next, the people themselves group these claims according to their shared similarities. The network architecture should allow people to a) create and read the claims b) to group and re-group the claims using hashtags or similar metadata tags (Crooks et al. 2013), and c) review the overall structure (Partanen 2016). Interest groups form simultaneously though saturation within this dynamic process. The group members formulate a problem statement regarding their thematic interests, and a loose work plan for the next steps. The thematic groups may study their topic using existing online data assisted by computer scientists for any online freeware according to their needs (e.g. GoogleMaps, MyStreet). The groups are encouraged to communicate and share material with each other, and anyone can join any number of groups.

Next, short statements are produced based on the emerging ideas that concern the spatio-functional city on the online platform. They are combined together by the participants according to similar qualities (e.g. "green corridors" and "small business"; "pedestrian accessibility" and "sports facilities"). This process resembles the project Wikipedia in which any user may combine proposed views to form an "article". These can be divided, re-combined, or operations cancelled. The process should produce structured ontologies (urban history, pedestrian/cycling space, green urban space, economic and business spaces and alike). This procedure enables the utilization of existing data analyses and saturation of similarities in views, allowing pinpointing of differences between solutions. In the end modified simulations could be used; for example, learning collaborative model building (Zelner et al. 2012), to modify existing open-source urban models. The phase produces ontologies, in the form of "wiki-articles", possibly interlinked to each other and additional material. At this point the system is frozen and a proxy voting (implying possibility to delegate the vote) procedure is possible on proposed future views to estimate its popularity. In the end the process produces a selection of various potential ontologies, with weights promoting its preference.

The resulting, emergent ontologies can be considered to represent the timely, essential phenomena in the city or a certain target district. The statutory rules are formulated from these in a described computerized process. For example, for a green corridor patterns and dependencies are discovered (e.g. minimum distances from buildings for species diversity for the green corridor; density and types of different uses in an economic hub for viability; urban form for preferable micro-climate or visibility analyses; street corner density to increase accessibility and hence vitality, etc.). These statutory rules form the foundation for the next phase: *the rule-based/case-based system* guiding the building processes.

Rule-Based and Case-Based System (RB/CB)

Next, I introduce a system structure that, first, allows evaluation of fitness of cases, i.e. building projects of different scale and type, to the prevalent, typical spatio-functional profile that is manifested as prescriptive, (relatively) static rules. These rules result from prior liquid process of information self-organization. Secondly, the RB/CB system collects data of changes of the activity profiles that are manifested as prescriptive rules, discovers new patterns, and may suggest (infrequent) updates to the rule.

Generally, RB/CB system consists of two elements: statutory inflexible but loose framework of rules, and a case-base collected within the process, against which each new project (a case) is reflected, and either accepted as such or after adjusting, or rejected. The case-base collects data of implemented (and rejected) project categorizing their characteristics (industries, sizes, volumes; users, employees; physical features such as height, proximity to street/other buildings, granulation, accessibility, and so on), patterns and dynamics of these in time. Next, I will explain this system and its justification in more detail. The applied approach is based on proposals by Golding and Rosenbloom (1996) and Partanen (2020).

The type of dual system represented here that consisting of rigid framework and interpretations, is actually prevalent in many fields that had encountered the limitations of solely rule-based reasoning. The reason is that by definition, rules cannot be adaptive, but they must always be under- or over-defined (Schauer 1991) – an adaptive, non-general rule would be an oxymoron (as a real-world example, there cannot be varying speed limits for more or less skillful drivers, or those concerning only some citizens). Hence, the flexibility that is needed to fit the real-world phenomena into the rule framework needs to be sought elsewhere. In medicine, the diagnostic criteria (i.e. rules) form treatment guidelines within which majority of cases fall; from time to time, case descriptions of exceptions are used for reflecting a new patient (Lau 2006). In law, the statute rule sets an under-defined framework, which is then interpreted to reflect the previous cases, and adapting to the rulings of the Supreme Court (a certain case-base) (Schauer 1991). Furthermore, as stated above, this is common also in human cognition and certain neural networks: they adaptively learn by applying existing knowledge in the new situation (Schank et al 1993). In planning, traditional statuary rules (plans) are interpreted in favor of previous implementations to ensure the equal treatment of the stakeholders. Regarding their relationship to the mechanism of drawing solutions, these examples resemble each other. The rule forms a frame for the cases, which most of the time fit into it. In case of potential failures, another mode of reasoning come into effect, namely the case-based reasoning.

In CRB, the role of experiences is stressed in finding new solutions, which usually adapt to the existing body of knowledge, applying different mechanisms depending on the discipline. These concern e.g. explanatory, interpretational, critical, or methods of drawing equitable solutions from prior cases (Kolodrel 1992, De Mántaras et al. 2005). As was previously explained, in CBR, as a cognitive theory of dynamic memory, self-organizing remindings are retrieved to be applied in a new situation.

Computational system often applies the *K-nearest neighbors* algorithm to explore the case base and recognize similar cases (Craw et al. 2006, De Mántaras et al. 2005). The case base consists of selected or exhaustively covering cases, presenting prototypical, missing, or redundant cases, exceptional ones (Craw et al. 2006). The stages of the procedure include *retrieving* (i.e. analyzing similarities in cases); *reusing*, (i.e. attempting to use it as such); *revising* (gathering suitability of the solution, or go back to start); and r*etaining (*including storing and classifying its attributes) the cases (Aamodt and Plaza 1994, De Mántaras et al. 2005, Craw 2006). The cases may be used as such (for simple classification tasks, e.g. medicine), or refined in a process resembling collective, adaptive problem-solving mode in more complex situations (e.g. planning) (Burkhard 1998, Craw 2006, Partanen 2020).

For planning, prescriptive rules must be 1) stationary, 2) under-defined, and 3) operate in concert with emergent rules (Partanen 2020). The rule-based system is not able to respond to all situations for all cases even if they follow the justification of the rule. However, applying solely CBR in large scale urban planning would lead to an overly complicated, unpredictably operating and heavy system (Partanen 2020). Instead, I suggest her that CBR could provide a complementary method for rule-based system improving its operation. Only approximately correct rules would provide tentative solutions. In exceptions or unresolved problems, analogies could be drawn from the cases allowed by coupled, complementary approach. The rules capture trends and CBR fills the gaps in recursive and dynamic manner (Golding and Rosenbloom 1996, Partanen 2020). Simplified structure for a coupled rule-base/case-based system is presented in the Figure 3.

Figure 3. Simplified structure for a coupled rule-base/case-based system
(Adapted from Partanen 2020)

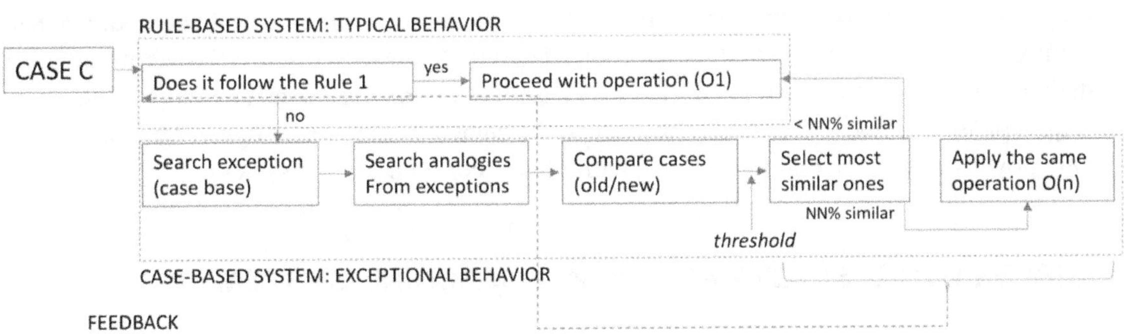

First for a default solution, the rules are applied in a new case. Second, the case base is explored for testing analogies. Thirdly, either the rule or the case analogy is followed. The rules must be accurate, and the case base properly indexed. The case base is analogical to complex reality often departing the statistics. Exploration of case analogy is a form of validation of the result. A contrasting case is in case base could mean that a more representative case regarding relevant attributes may exist, that could provide a new case an appropriate operation for application. The new case stored case will then conform with or deviate from the rule. This procedure is iterative for each separate rule (Partanen 2020, Golding & Rosenbloom 1996).

The system applies knowledge from first, so-called *weak theory* (Golding & Rosenbloom 1996). It consists of statistical or generalized knowledge of the systems' operation and defines the basis for the rule, its operators, and legal actions. Secondly, counterexamples are searched from the case library where the stored cases include the problem, its solution, and the operators used (Golding & Rosenbloom 1996). A feature vector for a building project could consist of uses, and operators that lead to the final result (Partanen 2020). Calling on the case-base resembles triangulation, reflecting statistics against the empirical data. For retrieving the cases, similarity metrics would estimate how similar two problems are regarding how a certain operator was applied following the set thresholds. Indexing of cases is essential and the cases are stored, as conforming or contrasting cases, by the indexing module. The data of the actual solution for analogies is included, and cases are used to test the for or against the rules (Golding & Rosenbloom 1996).

Similarity metrics propose analogies between the case and negative examples consisting of target problem, source problem, and the operator (to be adapted from source to target). The metrics returns a numerical similarity score and the analogical rule. This so-called a-rule (Golding & Rosenbloom 1996) implies a set of features that actually made the cases similar – such as the volume, working places, revenue, industries, customers (Partanen 2020). The a-rule defines how compelling the analogy is. Moreover, empirical verification would test the accuracy (or stochasticity) of the a-rule by revising how well it works for other cases in the library.

Basically, the value of CBS is that it seeks exceptions to the rule; if none is found, the initial rule simply applies – which would be the case for most of the time, if the rule is appropriately constructed in the liquid process. The case-base provides a backup method for embryonic of hidden self-organizing probably important for systems dynamics not visible using e.g. statistical metrics.

Occasionally, urban complex systems go through transitions concerning (intertwined) economic, social, ecological, or health subsystems, or a combination of these. Such ruptures probably require changes in the statutory rule set. A complementary case-based system helps to recognize these emerging needs, and to delineate novel rules from drawing from the case base. However, in case of major shifts, revisiting the liquid phase becomes necessary to recognize the changed behavior of the system, and to draw a completely novel set of rules to guide it along with updates of the RB/CB -system.

PLAUSIBLE FUTURES: TOWARDS HUMAN-MACHINE SYSTEMS?

In this chapter I have discussed the underlying principle of self-organization as an almost-universal autonomous order-producing computing mechanism in complex systems from biological to "artificial" ones, along with its implications in progress of urban technology. Illustrating this stance with an example of a self-organizing planning approach applying liquid thinking, and rule-base and case-based reasoning, I argue that the concept of self-organization would play a key role in both understanding the cities' dynamics and constructing more autonomous programs and structures for their planning and implementation. Liquid RB/CB system introduces a conceptual, yet concrete possibility to apply self-organization that would shed light on a smooth transition from human-led approach towards more AI-led urban e-planning. However, the future possibilities for applications of self-organization will entail solutions that might perhaps still appearing imaginary.

Currently we live in increasingly "smart" or digitalizing cities, in which humans and machines, and corporeal and virtual urbanity bundle in forms of AI, smart systems, data tools like digital twins and different dashboards, and planning and designing tools that are created, developed, and used (to a varying degree) by humans. This threshold resembles a phase transition typical in complex systems, one during which we might need critical consciousness for new urban "types", in very wide sense, for spatio-functional built structure in various scales, and hence new normative rules as well. Basically, and in traditional context, that would require creative, innovative human mind (of a designer); today, in the light of the progress in artificial intelligence, learning algorithms and such, and regarding deepening understanding of universality self-organization and its applicability as a profound mechanism for efficient order-production, the question of capability of future AI to apply also procedures for genuinely creative solutions, i.e. art, becomes tangible.

In biological systems, evolutionary and self-organizing, autonomous processes are capable of producing astonishingly rich and endlessly varying fractal forms hovering between chaos and order, but most of all, viable and economic structures and ecosystems through computation without a central designer. With a reference to capacity of simple CA for reproduction and universal computation (Gardner 1976), this raises a question of what would make, in principle, a future computational system of incompetent for producing, to an extent, results resembling the variance and purposefulness of the autonomous artwork in nature in a framework of ethically set normative ground-rules.

Moreover, in the crossing point of traditional urban evolution and generative, algorithmic design both relying in computation, there is a short mental leap to actual self-building structures and cities already experimented (see for example https://www.youtube.com/watch?v=FbUU8SsrdU8). Such systems would not just obey our simple normative rules as simple robotic structures, but they have capacity to build their own self-organizing emergent rules and realities, communicate and respond to each other and the environment in a holistic manner like new lifeforms, obscuring the difference between humans, machines, and animals. The concept of 'artificial intelligent' would eventually become a contradiction in terms, with relevance only in the 'intelligence'.

One aspect regarding the plausibility of such vision inhabited by (semi-)autonomous bundle of corporeal and virtual universes is about its practical implications: our ability to program self-programming with capacity to self-organization that could operate efficiently, dynamically optimized, and benefiting the entire entwined biosphere (also humans). This is about technical expertise, and likely to become eventually possible: essential obstacles holding evolutionary computation back appear banal, often related to the fact that the problems research focuses on are not necessarily recognized by the industry, or vise versa (Kendall 2018).

Another issue lays deep down in the human cognitive characteristics of perceiving and recognizing familiar objects. Basically, humans seem to have a tendency to feel drastically more eerie about artificial (non-living) objects and entities the more "natural" (i.e. non-artificial, living) they appear; the most uncanny object for them would be live-like completely robots, or zombies (Mori 1970). So-called uncanny valley unravels this relationship. This human relation towards tangible objects could be extended to cover machine intelligence regarding its level of autonomous behavior, from Facebook bots discussing with their own language humans cannot understand (Clark 2017), to (imaginary, potentially malicious) singularity like Skynet from the movie Terminator. In visual fields, so-called "dreams" of Google's AI - artwork produced by neural networks - easily fall into this class as well, producing tortuous chimeras, oddly almost resembling natural lifeforms, sceneries, or artefacts. Regarding architecture or urban design as art form, machine-produced imagery might face similar uncanniness difficult to accept. While humans adapt eventually into stylistic changes and fashions, it remains to be seen whether the valley of uncanny, i.e. gap filled with objects that look almost "natural" yet somewhat alien, can be mentally bridged. Noteworthy is that we are, with increasing speed, giving away the power to the myriads of (privately owned, anonymous and potentially vicious) programs and application without hesitation.

CHANGE IN THE ROLE AND ATTITUDE OF THE DESIGNER

For this approaching future of planning by (and for) human-machines, I suggest that a more holistic paradigm for understanding the world and all the life in it should be adopted for two reasons. First, a serious thread lies in the wider – techno-utopistic - adoption of the autonomous systems and methods. To avoid the (continuation) of overexploitation of the biosphere resources through more and more efficient technology in the name of progress, a sea change is necessary regarding our subjective position in and relation to surrounding (eco)systems. Building more ecological habits may not be sufficient, but a more profound deep ecological view would be required, considering all life forms equally important (Naess xxxx, Thompson& Barton 1994). Whether this would in the future apply not only for cybernetic, networked human-machines, or even machine-machines, or (A)I, is to be seen.

Secondly, I consider that in future, commonly used computerized self-building structures and self-programming systems, with superior capacity to dynamic optimization (Hoffmann & Payton 2018), are not only possible but before long probable. This is not due to the technological determinism alone. Contrarily, adopting the above world view stresses that they are on the logical continuum of self-organizing, complex systems, and as part of the Universe, i.e. not outside the "nature", they would be able to follow the same principles than the majority of the existing complex systems.

For a designer, the autonomous, generative systems would mean the change of their role, from collaborative specialist towards a method developer, with increasingly more requirement in skills concerning e.g. programming or urban research for exploring dependences, patterns and mechanism. Moreover, a designer would also need to maintain their visionary role, however revised. Instead of a traditional mastermind behind the project, they would need sensitivity to explore and discover what can be done with the new methodology. Echoing Bottazzi (2018), a design tool is not just a medium between the designer and the design, instead, it affects the cognitive structure of design, creating true novelties. We do not just design with different method, but also the substance is dramatically different. While we need to leap beyond the Anthropocene into more holistic, *ecocentric* (or *techno-ecocentric*) understanding of the universe to see the profoundly coupled nature of all systems, there is more to come. For the bundled effect of the novel tools and the digital phase transition, the reality is likely to be beyond our linear predictions, still requiring the creativity of an architect to recognize it.

REFERENCES

Aguiar, R., & Cardoso, C. (2017). Algorithmic design and analysis fusing disciplines. *Proceedings of the 37th Annual Conference of the ACADIA*, 28-37.

Agyepong, I. A., Kodua, A., Adjei, S., & Adam, T. (2012). When 'solutions of yesterday become problems of today': Crisis-ridden decision making in a complex adaptive system (CAS)—the Additional Duty Hours Allowance in Ghana. *Health Policy and Planning*, *27*(suppl_4), iv20–iv31. doi:10.1093/heapol/czs083 PMID:23014150

Alexander, C., Ishikawa, S., Silverstein, M., Jacobson, M., Fiksdahl-King, I., & Angel, S. (1977). *A city is not a Tree.* A Pattern Language.

Alfasi, N., & Portugali, J. (2007). Planning rules for a self-planned city. *Planning Theory*, *6*(2), 164–182. doi:10.1177/1473095207077587

Alfasi, N. (2021). Why public participation isn'ta tool for democratizing planning. A comment. *Planning Theory*, *20*(2), 175–178. doi:10.1177/1473095221991487

Alfasi, N., Almagor, J., & Benenson, I. (2012). The actual impact of comprehensive land-use plans: Insights from high resolution observations. *Land Use Policy*, *29*(4), 862–877. doi:10.1016/j.landusepol.2012.01.003

Allen, C. R., Angeler, D. G., Garmestani, A. S., Gunderson, L. H., & Holling, C. S. (2014). Panarchy: Theory and application. *Ecosystems (New York, N.Y.)*, *17*(4), 578–589. doi:10.100710021-013-9744-2

Allen, P. M. (1997). *Cities and regions as self-organizing systems: Models of complexity.* Gordon and Breach Science Publishers.

Allmendinger, P. (2017). *Planning theory.* Macmillan International Higher Education. doi:10.1057/978-0-230-38004-2

Ascher, F. (2004). Metapolis. A Third Modern Urban Revolution. Change in Urban Scale and Shape in France. In Mitten am Rand. Auf dem Weg von der Vorstadt über die Zwischenstadt zur Regionalen Stadtlandschaft (Zwischenstadt Band 1). Wuppertal: Verlag Müller + Busmann KG.

Auerbach, R. (2009). *The seed pod: facilitating self organizing robust settlement patterns in developing countries* (Master's thesis). University of Cape Town.

Bak, P. (1990). Self-Organized Criticality. *Physica A*, *163*(1), 403–409. doi:10.1016/0378-4371(90)90348-V

Batty, M. (2007). *Cities and Complexity.* MIT Press.

Batty, M. (2016). Big data and the city. *Built Environment*, *42*(3), 321–337. doi:10.2148/benv.42.3.321

Batty, M. (2018). *Inventing future cities.* MIT Press.

Marshall, S. (2012). Planning, design and the complexity of cities. In *Complexity theories of cities have come of age: An overview with implications to urban planning and design* (pp. 191–205). Springer., doi:10.7551/mitpress/11923.001.0001.

Batty, M., & Longley, P. A. (1994). *Fractal cities: a geometry of form and function.* Academic press.

Batty, M., & Marshall, S. (2009). Centenary paper: The evolution of cities: Geddes, Abercrombie and the new physicalism. *The Town Planning Review*, *80*(6), 551–575. doi:10.3828/tpr.2009.12

Blum, C., & Zuber, C. I. (2016). Liquid democracy: Potentials, problems, and perspectives. *Journal of Political Philosophy*, *24*(2), 162–182. doi:10.1111/jopp.12065

Bobo, L. D. (2018). Race as a complex adaptive system. *Du Bois Review*, *15*(2), 211–215. doi:10.1017/S1742058X19000043

Bottazzi, R. (2018). *Digital Architecture Beyond Computers: Fragments of a Cultural History of Computational Design*. Bloomsbury Publishing. doi:10.5040/9781474258159

Bryan, C. (2017). *Facebook's AI accidentally created its own language*. TNW News.

Buckley, W. (1968). *Modern systems research for the behavioral scientist*. Aldine.

Buckley, W. F. (1998). *Society—a complex adaptive system: essays in social theory* (Vol. 9). Taylor & Francis.

Burkhard, H. D. (1998). Extending some concepts of CBR-Foundations of case retrieval nets. In *Case-Based Reasoning Technology* (pp. 17–50). Springer. doi:10.1007/3-540-69351-3_2

Campbell, H., & Marshall, R. (2002). ` Utilitarianism's Bad Breath? A Re-evaluation of the Public Interest Justification for Planning '. *Planning Theory*, *1*(2), 163–187. doi:10.1177/147309520200100205

Caniggia, G., & Maffei, G. L. (2001). *Interpreting basic building: Architectural composition and building typology*. Alinea.

Capello, R. (2013). *Classical Contributions: Von Thünen, Weber, Christaller, Lösch*. Academic Press.

Cardullo, P., & Kitchin, R. (2019). Smart urbanism and smart citizenship: The neoliberal logic of 'citizen-focused' smart cities in Europe. Environment and Planning C. *Politics and Space*, *37*(5), 813–830.

Carli, R., Dotoli, M., & Cianci, E. (2017). An optimization tool for energy efficiency of street lighting systems in smart cities. *IFAC-PapersOnLine*, *50*(1), 14460–14464. doi:10.1016/j.ifacol.2017.08.2292

Carpo, M. (Ed.). (2013). *The digital turn in architecture 1992-2012*. John Wiley & Sons. doi:10.1002/9781118795811

Castells, M. (2011). *The rise of the network society* (Vol. 12). John Wiley & Sons.

Casti, J. L., & Berry, M. (1997). *Would-be worlds: How simulation is changing the frontiers of science*. J. Wiley.

Cilliers, P. (2005). Knowing complex systems. In K. Richardson (Ed.), *Managing Organizational Complexity: philosophy, theory, and application* (pp. 7–20). Information Age Publishers.

Clark-Ginsberg, A. (2020). A Complexity Approach for Reducing Disaster Risks for Marginalized Urban Populations: Comparing DRR Interventions Across Four Cities. In *External Interventions for Disaster Risk Reduction* (pp. 171–192). Springer. doi:10.1007/978-981-15-4948-9_10

Craw, S., Wiratunga, N., & Rowe, R. C. (2006). Learning adaptation knowledge to improve case-based reasoning. *Artificial Intelligence*, *170*(16-17), 1175–1192. doi:10.1016/j.artint.2006.09.001

Cugurullo, F. (2013). How to build a sandcastle: An analysis of the genesis and development of Masdar City. *Journal of Urban Technology*, *20*(1), 23–37. doi:10.1080/10630732.2012.735105

Davoudi, S., Brooks, E., & Mehmood, A. (2013). Evolutionary resilience and strategies for climate adaptation. *Planning Practice and Research*, *28*(3), 307–322. doi:10.1080/02697459.2013.787695

De Mántaras, R. L., McSherry, D., Bridge, D., Leake, D., Smyth, B., Craw, S., & Keane, M. (2005). Retrieval, reuse, revision and retention in case-based reasoning. *The Knowledge Engineering Review*, *20*(3), 215–240. doi:10.1017/S0269888906000646

de Roo, G., & Rauws, W. (2012). Positioning planning in the world of order, chaos and complexity: On perspectives, behaviour and interventions in a non-linear environment. In Complexity theories of cities have come of age: An overview with implications to urban planning and design (pp. 207-220). Springer.

De Roo, G., & Hillier, J. (2016). *Complexity and planning: Systems, assemblages and simulations*. Routledge. doi:10.4324/9781315573199

Deleuze, G., & Guattari, F. (1987). A thousand plateaus: Capitalism and schizophrenia. University of Minnesota Press.

Dovey, K. (2012). Informal settlement and complex adaptive assemblage. *International Development Planning Review*, *34*(3), 371–390.

Eigen, M. (1971). Molecular self-organization and the early stages of evolution. *Quarterly Reviews of Biophysics*, *4*(2-3), 149–212. doi:10.1017/S0033583500000627 PMID:5134461

Fainstein, S. S. (2000). ` New Directions in Planning Theory '. *Urban Affairs Review*, *35*(4), 451–478. doi:10.1177/107808740003500401

Flyvbjerg, B. (2002). Bringing power to planning research: One researcher's praxis story. *Journal of Planning Education and Research*, *21*(4), 353–366. doi:10.1177/0739456X0202100401

Foucault, M. (1969). *Madness and Civilization– The Archaeology of knowledge and the Discourse on Language*. Academic Press.

Fox-Rogers, L., & Murphy, E. (2014). Informal strategies of power in the local planning system. *Planning Theory*, *13*(3), 244–268. doi:10.1177/1473095213492512

Fujita, M., & Thisse, J. F. (1996). Economics of agglomeration. *Planning Theory*, *14*(3), 248–267.

Gabrys, J. (2014). Programming environments: Environmentality and citizen sensing in the smart city. *Environment and Planning. D, Society & Space*, *32*(1), 30–48. doi:10.1068/d16812

Gardner, M. (1976). Mathematical games. *Scientific American*, *234*(6), 120–125. doi:10.1038cientific american0676-120

Geddes, P. (1915). *Cities in evolution: An introduction to the town planning movement and to the study of civics*. Williams.

Giddens, A. (1987). *Social theory today*. Stanford University Press.

Gil, J., Montenegro, N., & Duarte, J. (2010). Assessing computational tools for urban design – towards a 'city information model'. *Proceedings of the 28th Conference (eCAADe)*, 361–361.

Gleick, J. (2008). *Chaos: Making a new science*. Penguin Books.

Golding, A. R., & Rosenbloom, P. S. (1996). Improving accuracy by combining rule-based and case-based reasoning. *Artificial Intelligence*, *87*(1-2), 215–254. doi:10.1016/0004-3702(95)00120-4

Haken, H. (1985). Synergetics—An interdisciplinary approach to phenomena of self-organization. *Geoforum*, *16*(2), 205–211. doi:10.1016/0016-7185(85)90029-6

Haken, H., & Portugali, J. (2003). The face of the city is its information. *Journal of Environmental Psychology*, *23*(4), 385–408. doi:10.1016/S0272-4944(03)00003-3

Haken, H., & Portugali, J. (2019). A synergetic perspective on urban scaling, urban regulatory focus and their interrelations. *Royal Society Open Science*, *6*(8), 191087. doi:10.1098/rsos.191087 PMID:31598267

Harvey, D. (1989). *The condition of postmodernity an enquiry into the origins of cultural change*. Academic Press.

Harvey, D. (2007). *A brief history of neoliberalism*. Oxford University Press.

Hatna, E., & Benenson, I. (2014). *Combining segregation and integration: Schelling model dynamics for heterogeneous population*. arXiv preprint arXiv:1406.5215.

He, S., Besuievsky, G., Tourre, V., Patow, G., & Moreau, G. (2012). *All range and heterogeneous multi-scale 3D city models*. Usage, Usability, and Utility of 3D City Models–European COST Action TU0801, 02006.

Hensel, M., Menges, A., & Weinstock, M. (2013). *Emergent technologies and design: towards a biological paradigm for architecture*. Routledge. doi:10.4324/9781315881294

Hillier, B. (2012). The genetic code for cities: Is it simpler than we think? In Complexity theories of cities have come of age: An overview with implications to urban planning and design (pp. 129-152). Springer.

Hoffmann, H., & Payton, D. W. (2018). Optimization by Self-Organized Criticality. *Scientific Reports*, *8*(1), 2358. doi:10.103841598-018-20275-7 PMID:29402956

Hohmann, B., Havemann, S., Krispel, U., & Fellner, D. (2010). A GML shape grammar for semantically enriched 3D building models. *Computers & Graphics*, *34*(4), 322–334. doi:10.1016/j.cag.2010.05.007

Innes, J. E., & Booher, D. E. (2010). *Planning with complexity: An introduction to collaborative rationality for public policy*. Routledge. doi:10.4324/9780203864302

Iqbal, A., Khan, I. A., & Jan, S. (2019). *A Review and Comparison of the Traditional Collaborative and Online Collaborative Techniques for Software Requirement Elicitation. In 2019 ICACS*. IEEE.

Jacobs, J. (2016). *The death and life of great American cities*. Vintage.

Jantsch, E. (1980). *The self-organizing universe: Scientific and human implications of the emerging paradigm of evolution*. Pergamon Press.

Jencks & Kropf. (Eds.). (2006). Theories and Manifestoes of Contemporary Architecture. Wiley-Academy.

Kauffman, S. A., & Strohman, R. C. (1994). *The Origins of Order: self-organization and selection in evolution*. Oxford University Press.

Kendall, G. (2018). Is Evolutionary Computation evolving fast enough? *IEEE Computational Intelligence Magazine*, *13*(2), 42–51. doi:10.1109/MCI.2018.2807019

Khacef, L., Girau, B., Rougier, N., Upegui, A., & Miramond, B. (2018). *Neuromorphic hardware as a self-organizing computing system*. https://arxiv.org/abs/1810.12640

Kohonen, T. (1989). Self-organizing feature maps. In *Self-organization and associative memory* (pp. 119–157). Springer. doi:10.1007/978-3-642-88163-3_5

Kolodner, J. L. (1992). An introduction to case-based reasoning. *Artificial Intelligence Review*, *6*(1), 3–34. doi:10.1007/BF00155578

Krishnan, S., & Balasubramanian, T. (2016). Traffic flow optimization and vehicle safety in smart cities. *Traffic (Copenhagen, Denmark)*, *5*(5), 7814–7820.

Kropf, K. S. (2001). Conceptions of change in the built environment. *Urban Morphology*, *5*(1), 29–46.

Krugman, P. (1996). *The self-organizing economy*. Blackwel Publishers.

Kugler, P. N., Kelso, J. S., & Turvey, M. T. (1980). On the concept of coordinative structures as dissipative structures. Advances in Psychology, (1), 3-47.

Kuusela & Partanen. (2022). *Plenty of Planning, Scanty Guidance: The case of Tampere*. In Progress.

Latour, B. (2007). *Reassembling the social – An introduction to Actor-Network Theory*. Oxford University Press.

Lau, C. (2006). *Problem-based Medical Case Management*. Hong Kong University Press.

Lenton, T. M. (2011). Early warning of climate tipping points. *Nature Climate Change*, *1*(4), 201–209. doi:10.1038/nclimate1143

Lom, M., & Pribyl, O. (2021). Smart city model based on systems theory. *International Journal of Information Management*, *56*, 102092. doi:10.1016/j.ijinfomgt.2020.102092

Lynn, G. (1998). *Folds, Bodies and Blobs–Collected Essays*. La Lettre Volee.

May, R. J., Maier, H. R., & Dandy, G. C. (2010). Data splitting for artificial neural networks using SOM-based stratified sampling. *Neural Networks*, *23*(2), 283–294. doi:10.1016/j.neunet.2009.11.009 PMID:19959327

Mitchell, M. (2009). *Complexity: A guided tour*. Oxford University Press.

Moore, J. (2011). *Exploring the moral underground: self-organizing teachers in the complex adaptive system of a school* (Doctoral dissertation). Pepperdine University.

Mori, M. (1970). The uncanny valley. *Energy*, *7*(4), 33–35.

Moroni, S. (2014). Complexity and the inherent limits of explanation and prediction: Urban codes for self-Newman. *Landscape and Urban Planning, 44*(4), 219-226.

Novac, P. E., Upegui, A., Barrientos, D., Sousa, C., Rodriguez, L., & Miramond, B. (2020). Dynamic Structural and Computational Resource Allocation for Self-Organizing Architectures. In *2020 27th IEEE International Conference on Electronics, Circuits and Systems (ICECS)* (pp. 1-4) 10.1109/ICECS49266.2020.9294983

Okesola, O. J., Okokpujie, K. O., Goddy-Worlu, R., Ogunbanwo, A., & Olamma, I. (2019). Qualitative comparisons of elicitation techniques in requirement engineering. *Qualitative Comparisons of Elicitation Techniques in Requirement Engineering*, *14*(2), 565–570.

Olawoyin, R., Nieto, A., Grayson, R. L., Hardisty, F., & Oyewole, S. (2013). Application of artificial neural network (ANN)–self-organizing map (SOM) for the categorization of water, soil and sediment quality in petrochemical regions. *Expert Systems with Applications*, *40*(9), 3634–3648. doi:10.1016/j.eswa.2012.12.069

Partanen, J. (2016). Liquid planning, wiki-design—Learning from the Case Pispala. *Environment and Planning. B, Planning & Design*, *43*(6), 997–1018. doi:10.1177/0265813516647965

Partanen, J. (2018). *'Don't Fix It if It Ain't Broke': Encounters with Planning for Complex Self-Organizing Cities*. Tampere University of Technology.

Partanen, J. (2020). Guiding urban self-organization: Combining rule-based and case-based planning. *Environment and Planning. B, Urban Analytics and City Science*, *47*(2), 304–320. doi:10.1177/2399808319893687

Portugali, J. (1999). *Self-organization and the city*. Springer-Verlag.

Portugali, J. (2006). Complexity theory as a link between space and place. *Environment & Planning A*, *38*(4), 647–664. doi:10.1068/a37260

Portugali, J., Meyer, H., Stolk, E., & Tan, E. (Eds.). (2012). *Complexity theories of cities have come of age: an overview with implications to urban planning and design*. Springer. doi:10.1007/978-3-642-24544-2

Prigogine, I., & Lefever, R. (1974). *Stability and self-organization in open systems*. Wiley.

Ridell, S. (2019). Mediated bodily routines as infrastructure in the algorhythmic city. *Media Theory*, *3*(2), 27–62.

Riesbeck, C. K., & Schank, R. C. (2013). *Inside case-based reasoning*. Psychology Press. doi:10.4324/9780203781821

Salingaros, N. A. (2000). Complexity and urban coherence. *Journal of Urban Design*, *5*(3), 291–31. doi:10.1080/713683969

Salingaros, N. A. (2012). Urbanism as Computation. In *Complexity Theories of Cities Have Come of Age* (pp. 245–268). Springer. doi:10.1007/978-3-642-24544-2_14

Sarriot, E., & Kouletio, M. (2015). Community health systems as complex adaptive systems: Ontology and praxis lessons from an urban health experience with demonstrated sustainability. *Systemic Practice and Action Research*, *28*(3), 255–272. doi:10.100711213-014-9329-9

Schank, R., Riesbeck, C., & Kass, A. (1994). *Inside Case-Based Explanation*. Lawrence Erlbaum Associates.

Schank, R. C. (1983). *Dynamic memory: A theory of reminding and learning in computers and people*. Cambridge university press.

Schauer, F. (1991). *Playing by the rules: A philosophical examination of rule-based decision-making in law and in life*. Clarendon Press.

Schelling, T. C. (1969). Models of segregation. *The American Economic Review, 59*(2), 488–493.

Shane, D. G. (2011). *Urban Design since 1945. A Global Perspective*. Wiley.

Steinitz, C. (2012). *A Framework for Geodesign: Changing Geography By Design*. ESRI Press.

Stojanovski, T., Partanen, J., Samuels, I., Sanders, P., & Peters, C. (2020). City Information Modelling (CIM) and Digitizing Urban Design Practices. *Built Environment, 46*(4), 637–646. doi:10.2148/benv.46.4.637

Tan, E., & Portugali, J. (2012). The responsive city design game. In *Complexity Theories of Cities have come of Age* (pp. 369–390). Springer. doi:10.1007/978-3-642-24544-2_20

Tanwar, S., Popat, A., Bhattacharya, P., Gupta, R., & Kumar, N. (2021). A taxonomy of energy optimization techniques for smart cities: Architecture and future directions. *Expert Systems: International Journal of Knowledge Engineering and Neural Networks*, 12703. doi:10.1111/exsy.12703

Terzidis, K. (2006). *Algorithmic architecture*. Elsevier. doi:10.4324/9780080461298

Thompson, S. C. G., & Barton, M. A. (1994). Ecocentric and anthropocentric attitudes toward the environment. *Journal of Environmental Psychology, 14*(2), 149–157. doi:10.1016/S0272-4944(05)80168-9

Walker, B., & Salt, D. (2012). *Resilience practice: building capacity to absorb disturbance and maintain function*. Island Press. doi:10.5822/978-1-61091-231-0

Weinstock, M. (2004). Morphogenesis and the Mathematics of Emergence. *Architectural Design*, (3), 10-17.

Wolfram, S. (1984). Cellular automata as models of complexity. *Nature, 311*(5985), 419–424. doi:10.1038/311419a0

Zellner, M., Lyons, L., Hoch, C. J., Weizeorick, J., Kunda, C., & Milz, D. C. (2012). Modeling, learning, and Planning together: An Application of Participatory Agent-based Modeling to Environmental Planning. *URISA Journal, 1*, 77–92.

ENDNOTE

1 Evolution is established concept particularly in typomogphology, describing incremental continuous and adaptive change in cities, regarding the of urban form and functions - implying Bundelend pattern-processes. Differs from the Darwinian use, implying that evolution is not equal to progress; making distinction between individual and aggregate level processes; and identifying circular chains and other features in evolving system (see more: Kropf 2001).

Compilation of References

Aalders, H. J., & Moellering, H. (2001). Spatial data infrastructure. *Proceedings of the 20th International Cartographic Conference*, 4.

Aday, L. A., & Andersen, R. (1974). A Framework for the Study of Access to Medical Care. *Health Services Research*, *9*(3), 208–220. PMID:4436074

ADEME. (2017). *Etudes des liens entre données individuelles de la qualité de l'air, changements de comportement et mises en oeuvre de pratiques favorables à l'air*. Academic Press.

Aguiar, R., & Cardoso, C. (2017). Algorithmic design and analysis fusing disciplines. *Proceedings of the 37th Annual Conference of the ACADIA*, 28-37.

Agyepong, I. A., Kodua, A., Adjei, S., & Adam, T. (2012). When 'solutions of yesterday become problems of today': Crisis-ridden decision making in a complex adaptive system (CAS)—the Additional Duty Hours Allowance in Ghana. *Health Policy and Planning*, *27*(suppl_4), iv20–iv31. doi:10.1093/heapol/czs083 PMID:23014150

Ahonen, P. (2018). Learning from opening data in the context of e-governance: Finland, with special reference to government location data. In L. Muñoz & M. Bolivar (Eds.), *International e-government development* (pp. 271–290). Palgrave MacMillan. doi:10.1007/978-3-319-63284-1_11

Ainsworth, J., & Buchan, I. (2015). Combining Health Data Uses to Ignite Health System Learning. *Methods of Information in Medicine*, *54*(6), 479–487. doi:10.3414/ME15-01-0064 PMID:26395036

Alam, M., Ahmed, K., & Islam, M. A. (2008, December). E-governance: Challenges and opportunities. In *Proceedings of the 2nd International Conference on Theory and Practice of Electronic Governance* (pp. 264-267). 10.1145/1509096.1509150

AlAwadhi, S., & Scholl, H. J. (2013). Aspirations and realizations: The smart city of Seattle. *2013 46th Hawaii International Conference on System Sciences*, 1695–1703. 10.1109/HICSS.2013.102

Albino, V., Berardi, U., & Dangelico, R. M. (2015). Smart Cities: Definitions, Dimensions, Performance, and Initiatives. *Journal of Urban Technology*, *22*(1), 3–21. doi:10.1080/10630732.2014.942092

Alexander, C., Ishikawa, S., Silverstein, M., Jacobson, M., Fiksdahl-King, I., & Angel, S. (1977). *A city is not a Tree*. A Pattern Language.

Alexopoulos, C., Lachana, Z., Androutsopoulou, A., Diamantopoulou, V., Charalabidis, Y., & Loutsaris, M. (2019). How machine learning is changing e-government. In *Proceedings of the 12th international conference on theory and practice of electronic governance* (pp. 354-363). Berlin: Springer. 10.1145/3326365.3326412

Alfasi, N. (2021). Why public participation isn'ta tool for democratizing planning. A comment. *Planning Theory*, *20*(2), 175–178. doi:10.1177/1473095221991487

Alfasi, N., Almagor, J., & Benenson, I. (2012). The actual impact of comprehensive land-use plans: Insights from high resolution observations. *Land Use Policy*, *29*(4), 862–877. doi:10.1016/j.landusepol.2012.01.003

Alfasi, N., & Portugali, J. (2007). Planning rules for a self-planned city. *Planning Theory*, *6*(2), 164–182. doi:10.1177/1473095207077587

Alford, J. (2009). *Engaging public sector clients: From service-delivery to co-production*. Springer. doi:10.1057/9780230235816

Allen, C. R., Angeler, D. G., Garmestani, A. S., Gunderson, L. H., & Holling, C. S. (2014). Panarchy: Theory and application. *Ecosystems (New York, N.Y.)*, *17*(4), 578–589. doi:10.100710021-013-9744-2

Allen, P. M. (1997). *Cities and regions as self-organizing systems: Models of complexity*. Gordon and Breach Science Publishers.

Allmendinger, P. (2017). *Planning theory*. Macmillan International Higher Education. doi:10.1057/978-0-230-38004-2

Almarabeh, T., Majdalawi, Y., & Mohammad, H. (2016). Cloud computing of e-government. *Communications and Network*, *8*(01), 1–8. doi:10.4236/cn.2016.81001

Al-Nasrawi, S. (2019, December). A Validated Model for Citizen Engagement and Smartness of Cities. In *2019 International Conference on Smart Applications, Communications and Networking (SmartNets)* (pp. 1-6). IEEE. 10.1109/SmartNets48225.2019.9069794

Andersen, T. (2009). E-Government as an anti-corruption strategy. *Information Economics and Policy*, *21*(3), 201–210. doi:10.1016/j.infoecopol.2008.11.003

Anthes, G. (2015). Estonia: A model for e-government. *Communications of the ACM*, *58*(6), 18–20. doi:10.1145/2754951

Anttiroiko, A. V. (2021). Successful government responses to the pandemic: Contextualizing national and urban responses to the COVID-19 outbreak in east and west. *International Journal of E-Planning Research*, *10*(2), 1–17. doi:10.4018/IJEPR.20210401.oa1

Araujo, R., Moura, A., & Nogueira, T. (2018). Creating Collaborative Environments for the Development of Slum Upgrading and Illegal Settlement Regularization Plans in Brazil: The MariaTereza Neighborhood Case in Belo Horizonte. *International Journal of E-Planning Research*, *7*(4), 25–43. doi:10.4018/IJEPR.2018100102

ArcGIS Hub web. (n.d.). Available online: https://hub.arcgis.com/

Armitage, D. F., Berkes, F., & Doubleday, N. (2007). *Adaptive Co-management: Collaboration, Learning and Multi-level Governance*. University of British Columbia.

Arnold, H. F. (1980). *Trees in Urban Design* (1st ed.). Van Nostrand Reinhold Company.

Arnstein, S. (1969). A ladder of citizen participation. *Journal of the American Institute of Planners*, *35*, 216-224.

Arnstein, S. R. (1969). A ladder of Citizen Participation. *Journal of the American Planning Association*, *35*(4), 216–224.

Artz, M., Batty, M., Dangermond, J., Fisher, T., McElvaney, S., Richardson, K., & Steinitz, C. (2013). Geodesign: Past, Present, and Future. Esri Press.

Artz, M., Dangermond, J., Ball, M., & Abukhater, A. (2010). *Changing Geography by Design. Selected readings in GeoDesign*. Esri Press.

Ascher, F. (2004). Metapolis. A Third Modern Urban Revolution. Change in Urban Scale and Shape in France. In Mitten am Rand. Auf dem Weg von der Vorstadt über die Zwischenstadt zur Regionalen Stadtlandschaft (Zwischenstadt Band 1). Wuppertal: Verlag Müller + Busmann KG.

Associated Press. (2017). *Rolling Stone slams defamation claim in UVa rape story.* The Daily Progress. https://www.dailyprogress.com/news/local/rolling-stone-slams-defamation-claim-in-uva-rape-story/article_e6bc4080-cbff-11e7-bed8-0b3c378ba1bc.html

Atzmanstorfer, K., & Blaschke, T. (2013). The Geospatial Web: A Tool to Support the Empowerment of Citizens through E-Participation? In C. N. Silva (Ed.), *Citizen E-Participation in Urban Governance: Crowdsourcing and Collaborative Creativity* (pp. 144–171). doi:10.4018/978-1-4666-4169-3.ch009

Auerbach, R. (2009). *The seed pod: facilitating self organizing robust settlement patterns in developing countries* (Master's thesis). University of Cape Town.

Ayache, M. & Dumez, H. (2011). Le codage dans la recherche qualitative : une nouvelle perspective ? *Le libellio d'Aegis, 7*(2), 33–46.

Bagloee, S. A., Heshmati, M., Dia, H., Ghaderi, H., Pettit, Ch., & Asadi, M. (2021). Blockchain: The operating system of smart cities. *Cities, 112.* doi:10.1016/j.cities.2021.103104

Bakogiannis, E., Siti, M., Kyriakidis, C., Christodoulopoulou, G., & Vassi, A. (2019). Tools and Technologies for Enhancing Public Engagement in Sustainable Urban Mobility Planning—The Case of Rethymno, Crete. In *Mediterranean Cities and Island Communities* (pp. 237–255). Springer. doi:10.1007/978-3-319-99444-4_10

Bąkowska, E., Kaczmarek, T., Jankowski, P., Zwoliński, Z., Mikuła, Ł., Czepkiewicz, M., & Brudka, C. (2016). Geoquestionnaire in urban planning – preliminary results of the experimental application in Poland. *Rozwój Regionalny i Polityka Regionalna, 35*, 9–26.

Bąkowska, E., Kaczmarek, T., & Mikuła, Ł. (2017). Wykorzystanie geoankiety jako narzędzia konsultacji społecznych w procesie planowania przestrzennego w aglomeracji poznańskiej. *Roczniki Geomatyki, XV, 2*(77), 147–158.

Bak, P. (1990). Self-Organized Criticality. *Physica A, 163*(1), 403–409. doi:10.1016/0378-4371(90)90348-V

Ballal, H. (2015). *Collaborative Planning with Digital Design Synthesis* (Ph.D. Thesis). UCL (University College London), London, UK.

Bannister, R., & Connolly, R. (2012). Forward to the past: Lessons for the future of e-government from the story so far. *Information Polity, 17*(3,4), 211–226. doi:10.3233/IP-2012-000282

Bashshur, R., Doarn, C. R., Frenk, J. M., Kvedar, J. C., & Woolliscroft, J. O. (2020). Telemedicine and the COVID-19 pandemic, lessons for the future. *Telemedicine Journal and e-Health, 26*(5), 571–573. doi:10.1089/tmj.2020.29040.rb PMID:32275485

Batty, M. (2016). Urban Resilience: How Cities Need to Adapt to Unanticipated and Sudden Change. *Perspectives on Complex Global Challenges: Education, Energy, Healthcare, Security and Resilience: Education, Energy, Healthcare, Security and Resilience*, 169–171.

Batty, M. (2007). *Cities and Complexity.* MIT Press.

Batty, M. (2013). Defining Geodesign (=GIS + design?). *Environment and Planning. B, Planning & Design, 40*(1), 1–2. doi:10.1068/b4001ed

Batty, M. (2013a). *The new science of cities.* MIT Press. doi:10.7551/mitpress/9399.001.0001

Batty, M. (2013b). Big data, smart cities and city planning. *Dialogues in Human Geography, 3*(3), 274–279. doi:10.1177/2043820613513390 PMID:29472982

Batty, M. (2014). Can it happen again? Planning support, Lee's Requiem and the rise of the smart cities movement. *Environment and Planning. B, Planning & Design, 41*(3), 388–391. doi:10.1068/b4103c2

Batty, M. (2016). Big data and the city. *Built Environment, 42*(3), 321–337. doi:10.2148/benv.42.3.321

Batty, M. (2018). *Inventing future cities*. MIT Press.

Batty, M., & Longley, P. A. (1994). *Fractal cities: a geometry of form and function*. Academic press.

Batty, M., & Marshall, S. (2009). Centenary paper: The evolution of cities: Geddes, Abercrombie and the new physicalism. *The Town Planning Review, 80*(6), 551–575. doi:10.3828/tpr.2009.12

Beatley, T. (2011). *Biophilic Cities: Integrating Nature Into Urban Design and Planning*. Island Press. doi:10.5822/978-1-59726-986-5

Becker, H. (2006). *Sur le concept d'engagement. Sociologies Découverte.*

Bedê, M. C. (2015). Habitação de interesse social e reabilitação de assentamentos precarious. In *Entre o urbano, o social e o ambiental, a Praxis em perspectiva* (pp. 259–277). Gaia Cultural.

Belanger, F., & Lemuria, C. (2008). Trust and risk in e-government adoption. *The Journal of Strategic Information Systems, 17*(2), 165–176. doi:10.1016/j.jsis.2007.12.002

Benes, B., Deussen, O., Pirk, S., Chen, B., Mech, R., & Ijiri, T. (2016). Modeling plant life in computer graphics. *ACM SIGGRAPH 2016 Courses on - SIGGRAPH '16*, 1–180. doi:10.1145/2897826.2927332

Bernardo, M. (2019). Smart city governance: from e-government to smart governance. In *Smart cities and smart spaces: Concepts, methodologies, tools, and applications* (pp. 196–232). IGI Global. doi:10.4018/978-1-5225-7030-1.ch009

Bharosa, N., Lips, S., & Draheim, D. (2020). Making e-government work: learning from the Netherlands and Estonia. In *Proceedings of the international conference on electronic participation* (pp. 41-53). Cham: Springer. 10.1007/978-3-030-58141-1_4

Bickerstaff, K., Tolley, R., & Walker, G. (2002). Transport planning and participation: the rhetoric and realities of public involvement. *Journal of Transport Geography, 10*(1), 61-73. doi:10.1016/S0966-6923(01)00027-8

Birch, E. L. (2012). Cities, People, and Processes as Planning Case Studies. In R. Crane, R. Weber, & E. L. Birch (Eds.), *The Oxford Handbook of Urban Planning*. Oxford University Press. https://www.oxfordhandbooks.com/view/10.1093/oxfordhb/9780195374995.001.0001/oxfordhb-9780195374995-e-14

Björklund, F. (2016). E-government and moral citizenship: The case of Estonia. *Citizenship Studies, 20*(6-7), 914–931. doi:10.1080/13621025.2016.1213222

Blondiaux, L., & Sintomer, Y. (2002). L'impératif délibératif. *Plitix Revue des Sciences Sociales du Politique, 57*, 17–35.

Blum, C., & Zuber, C. I. (2016). Liquid democracy: Potentials, problems, and perspectives. *Journal of Political Philosophy, 24*(2), 162–182. doi:10.1111/jopp.12065

Bobo, L. D. (2018). Race as a complex adaptive system. *Du Bois Review, 15*(2), 211–215. doi:10.1017/S1742058X19000043

BOM. (2013). *2013 shaping up to be one of Australia's hottest years on record* [Climate Update]. Bureau of Meteorology. http://www.bom.gov.au/climate/updates/articles/a003-2013-temperature.shtml#july-update

Bonney, R. (1995). Citizen Science: A Lab Tradition. *Living Bird, 15*(4), 7–15.

Bonsón, E., Perea, D., & Bednárová, M. (2019). Twitter as a tool for citizen engagement: An empirical study of the Andalusian municipalities. *Government Information Quarterly, 36*(3), 480-489. doi:10.1016/j.giq.2019.03.001

Botrić, V., & Božić, L. (2020). The digital divide and E-government in European economies. *Economic Research-Ekonomska Istraživanja, 33*, 1–20.

Bottazzi, R. (2018). *Digital Architecture Beyond Computers: Fragments of a Cultural History of Computational Design.* Bloomsbury Publishing. doi:10.5040/9781474258159

Boudia, S., & Jas, N. (2014). *Powerless Science? Science and Politics in a Toxic World.* Berghahn Books.

Boudjelida, A., Mellouli, S., & Lee, J. (2016). Electronic Citizens Participation: Systematic Review. In *Proceedings of the 9th International Conference on Theory and Practice of Electronic Governance (ICEGOV '15-16).* Association for Computing Machinery. 10.1145/2910019.2910097

Bourdillon, F. (2016). *Santé Publique France - Une année au service de la santé des populations.* Academic Press.

Bournaris, T. (2020). Evaluation of e-government web portals: The case of agricultural e-government services in Greece. *Agronomy (Basel), 10*(7), 932–951. doi:10.3390/agronomy10070932

Boutaric, F. (2014). *Pollution atmosphérique et action publique*. Editions Rue d'Ulm.

Boutaric, F. (2021). *L'art de gouverner la qualité de l'air: l'action publique en question.* Presses Universitaires de Rennes.

Bovaird, T. (2017). Improving public services through e-government and citizen engagement: Comparing emerging lessons in east and west European countries. In E. Loffler & M. Vintar (Eds.), *Improving the quality of East and West European public services* (pp. 252–256). Routledge. doi:10.4324/9781351155922-20

Bremen transport plan. (n.d.). Available online: bauumwelt.bremen.de

Broadband. (n.d.). Available online:https://en.wikipedia.org/wiki/Broadband

Broday, D.M. & Citi-Sense Project Collaborators. (2017). Wireless Distributed Environmental Sensor Networks for Air Pollution Measurement - The Promise and the Current Reality. *Sensors (Basel)*, 17.

Brown, A. L., & Thérivel, R. (2000). Principles to guide the development of strategic environmental assessment methodology. *Impact Assessment and Project Appraisal, 18*(3), 183–189. doi:10.3152/147154600781767385

Brown, G., & Raymond, Ch. M. (2014). Methods for identifying land use conflict potential using participatory mapping. *Landscape and Urban Planning, 122*, 196–208. doi:10.1016/j.landurbplan.2013.11.007

BRUDA. (2020). *Cool walks*. https://www2.bcnregional.com/

Brussels Mobility. (n.d.). *Brussels regional public service*. Available online: https://mobilite-mobiliteit.brussels/en/good-move

Bryan, C. (2017). *Facebook's AI accidentally created its own language*. TNW News.

Buckley, W. (1968). *Modern systems research for the behavioral scientist*. Aldine.

Buckley, W. F. (1998). *Society—a complex adaptive system: essays in social theory* (Vol. 9). Taylor & Francis.

Bugs, G., Granell, C., Fonts, O., Huerta, J., & Painho, M. (2010). An assessment of Public Participation GIS and Web 2.0 technologies in urban planning practice in Canela, Brazil. *Cities (London, England)*, *27*(3), 172–181. doi:10.1016/j.cities.2009.11.008

Bulkeley, H., Coenen, L., Frantzeskaki, N., Hartmann, C., Kronsell, A., Mai, L., Marvin, S., McCormick, K., van Steenbergen, F., & Voytenko Palgan, Y. (2016). Urban living labs: Governing urban sustainability transitions. *Current Opinion in Environmental Sustainability*, *22*, 13–17. doi:10.1016/j.cosust.2017.02.003

Burkhard, H. D. (1998). Extending some concepts of CBR-Foundations of case retrieval nets. In *Case-Based Reasoning Technology* (pp. 17–50). Springer. doi:10.1007/3-540-69351-3_2

Bušelić, V., Slamić, M., & Kovačević, A. (2020). Complex digital citizen services in coronavirus age. In *Central European conference on information and intelligent systems 2020* (pp. 89-96). Faculty of Organization and Informatics Varazdin.

Callon, M., Lascoumes, P., & Barthe, Y. (2001). *Agir dans un monde incertain*. Editions le Seuil.

Campagna, M. (2014a). Geodesign from Theory to Practice: From Metaplanning to 2nd Generation of Planning Support Systems. *Tema. Journal of Land Use, Mobility and Environment*.

Campagna, M. (2014b). Geodesign: dai contenuti metodologici all'innovazione nelle pratiche. In Atti della XVII Conferenza Nazionale SIU – Società Italiana degli Urbanisti. Planum Publisher.

Campagna, M., & Di Cesare, E. A. (2016). Geodesign: Lost in Regulations (and in Practice). In Smart Energy in the Smart City (pp. 307-327). Springer International Publishing.

Campagna, M. (2016). Metaplanning: About designing the Geodesign process. *Landscape and Urban Planning*, *156*, 118–128. Advance online publication. doi:10.1016/j.landurbplan.2015.08.019

Campagna, M. (2016). Sistemas de Suporte ao Planejamento (Planning Support Systems): Retrospectivas e Prospectivas. In *Tecnologias de geoinformação para representar e planejar o território urbano* (pp. 217–248). Interciência.

Campagna, M., & Craglia, M. (2012). The socioeconomic impact of the spatial data infrastructure of Lombardy. *Environment and Planning. B, Planning & Design*, *39*(6), 1069–1083. doi:10.1068/b38006

Campagna, M., De Montis, A., & Deplano, G. (2006). PSS design: A general framework perspective. *International Journal of Environmental Technology and Management*, *6*(1-2), 163–179. doi:10.1504/IJETM.2006.008259

Campagna, M., Steinitz, C., Di Cesare, E. A., Cocco, C., Ballal, H., & Canfield, T. (2016). *Collaboration in Planning: the Geodesign approach*. Rozwój Regionalny I Polityka Regionalna.

Campbell, H., & Marshall, R. (2002). ` Utilitarianism's Bad Breath? A Re-evaluation of the Public Interest Justification for Planning '. *Planning Theory*, *1*(2), 163–187. doi:10.1177/147309520200100205

Caniggia, G., & Maffei, G. L. (2001). *Interpreting basic building: Architectural composition and building typology*. Alinea.

Capello, R. (2013). *Classical Contributions: Von Thünen, Weber, Christaller, Lösch*. Academic Press.

Caragliu, A., Del Bo, C., & Nijkamp, P. (2011). Smart Cities in Europe. *Journal of Urban Technology*, *18*(2), 65–82. doi:10.1080/10630732.2011.601117

Cardullo, P., & Kitchin, R. (2019). Smart urbanism and smart citizenship: The neoliberal logic of 'citizen-focused' smart cities in Europe. Environment and Planning C. *Politics and Space*, *37*(5), 813–830.

Carli, R., Dotoli, M., & Cianci, E. (2017). An optimization tool for energy efficiency of street lighting systems in smart cities. *IFAC-PapersOnLine*, *50*(1), 14460–14464. doi:10.1016/j.ifacol.2017.08.2292

Carlyle-Moses, D. E., Livesley, S., Baptista, M. D., Thom, J., & Szota, C. (2020). Urban Trees as Green Infrastructure for Stormwater Mitigation and Use. In D. F. Levia, D. E. Carlyle-Moses, S. Iida, B. Michalzik, K. Nanko, & A. Tischer (Eds.), *Forest-Water Interactions* (pp. 397–432). Springer International Publishing. doi:10.1007/978-3-030-26086-6_17

Carpo, M. (Ed.). (2013). *The digital turn in architecture 1992-2012*. John Wiley & Sons. doi:10.1002/9781118795811

Carton, L., & Ache, P. (2017). Citizen-sensor-networks to confront government decision-makers: Two lessons from the Netherlands. *Journal of Environmental Management*, *196*, 234–251.

Carver, S., Evans, A., Kingston, R., & Turton, I. (2001). Public participation, GIS, and cyberdemocracy: Evaluating on-line spatial decision support systems. *Environment and Planning. B, Planning & Design*, *28*(6), 907–921. doi:10.1068/b2751t

Cascetta, E., & Pagliara, F. (2013). Public engagement for planning and designing transportation systems. *Procedia-Social and Behavioral Sciences*, *87*, 103-116. doi:10.1016/j.sbspro.2013.10.597

Castells, M. (2011). *The rise of the network society* (Vol. 12). John Wiley & Sons.

Casti, J. L., & Berry, M. (1997). *Would-be worlds: How simulation is changing the frontiers of science*. J. Wiley.

Cegarra-Navarro, J. G., Pachón, J. R. C., & Cegarra, J. L. M. (2012). E-government and citizen's engagement with local affairs through e-websites: The case of Spanish municipalities. *International Journal of Information Management*, *32*(5), 469–478. doi:10.1016/j.ijinfomgt.2012.02.008

Cellini, N., Canale, N., Mioni, G., & Costa, S. (2020). Changes in sleep pattern, sense of time and digital media use during COVID-19 lockdown in Italy. *Journal of Sleep Research*, *29*(4). doi:10.1111/jsr.13074

Cerratto-Pargman, T., & Joshi, S. (2016). Reconfiguring civic participation: open source software in the political space. In *14th Participatory Design Conference, Aarhus, Denmark, 15-19 August 2016* (pp. 1-6). Retrieved from: http://urn.kb.se/resolve?urn=urn:nbn:se:su:diva-137488

Certomà, C., Corsini, F., & Rizzi, F. (2015). Crowdsourcing urban sustainability. Data, people and technologies in participatory governance. *Futures*, *74*, 93–106. doi:10.1016/j.futures.2014.11.006

Chadwick, A., & May, C. (2003). Interaction between states and citizens in the age of the internet: 'e-Government' in the United States, Britain, and the European Union. *Governance: An International Journal of Policy, Administration and Institutions*, *16*(2), 271–300. doi:10.1111/1468-0491.00216

Charlottesville Outdoor Lighting Ordinance. Chapter 34, Article IX, Division 3, Section 34 Chapter 34, Article IX, Division 3, Section 34. https://library.municode.com/va/charlottesville/codes/code_of_ordinances?nodeId=CO_CH34ZO_ARTIXGEAPRE_DIV3OULI_S34-1000PUIN

Charvolin, F. (2015). *Un air familier? Sociohistoire des pollutions atmosphériques*. Presses des Mines.

Chateauraynaud, F., & Debaz, J. (2013). *De la métrologie en démocratie. La nouvelle vage des capteurs citoyens*. https://socioargu.hypotheses.org/4505

Chateauraynaud, F., & Debaz, J. (2017). *Aux bords de l'irréversible : Sociologie pragmatique des transformations*. Editions P.

Chen, Q., Min, C., Zhang, W., Wang, G., Ma, X., & Evans, R. (2020). Unpacking the black box: How to promote citizen engagement through government social media during the COVID-19 crisis. *Computers in Human Behavior*, *110*. doi:10.1016/j.chb.2020.106380

Chobeaux, F. (2012). L'éducation par les pairs: Attentes et limites. *La Santé de l'Homme*, (421), 15–16.

Chrostowski, A., & Jemielniak, D. (2011). *Skuteczne doradztwo strategiczne: metoda Action Research w praktyce*. Poltext.

Cilliers, P. (2005). Knowing complex systems. In K. Richardson (Ed.), *Managing Organizational Complexity: philosophy, theory, and application* (pp. 7–20). Information Age Publishers.

Citizen Engagement and Participation. (n.d.). Available online: https://www.corkcity.ie/en/dbic/smart-city/citizen-engagement-and-participation/

Citizen Engagement and Stakeholder Consultations during COVID-19. (n.d.). Available online: https://pubdocs.worldbank.org/en/875401606251371090/Fact-Sheet-Citizen-Engagement-and-Stakeholder-Consultations-during-COVID-19.pdf

Citizenlab. (n.d.). *An e-democracy platform for stronger communities*. Available online: https://www.citizenlab.co/

City of Charlottesville. (2013). *2013 Charlottesville Comprehensive Plan*. City of Charlottesville, Virginia. http://www.charlottesville.org/home/showdocument?id=35059

City of Melbourne. (2011). *Urban Forest Diversity Guidelines: 2011 tree species selection strategy for the City of Melbourne*. Author.

Civicplus. (n.d.). *Website and Citizen Engagement Tools To Help You Build A Better Community*. Available online: https://www.civicplus.com/civicengage/local-government-website-design

Civitas Elan Final Evaluation Report. (n.d.). Available online: https://civitas.eu/sites/default/files/d_10_11_final_evaluation_report_civitas_elan.pdf

Clarke, G. P., & Wilson, A. G. (1994). A new geography of performance indicators for urban planning. In C.S. Bertuglia, G.P. Clarke G.P., & A.G. Wilson (Eds.), Modelling the city. Performance, policy and planning (pp. 55-81). Routledge.

Clark-Ginsberg, A. (2020). A Complexity Approach for Reducing Disaster Risks for Marginalized Urban Populations: Comparing DRR Interventions Across Four Cities. In *External Interventions for Disaster Risk Reduction* (pp. 171–192). Springer. doi:10.1007/978-981-15-4948-9_10

Coenen, F. (2009). Local Agenda 21: 'meaningful and effective' participation? In *Public participation and better environmental decisions* (pp. 165–182). Springer. doi:10.1007/978-1-4020-9325-8_10

COM. (2016). *Draft Arden Vision Framework*. City of Melbourne.

Commission, P.I. (2015). *Rapport fait au nom de la commission d'enquête sur le coût économique et financier de la pollution de l'air - remis le 8 juillet 2015*. Academic Press.

Considerit. (n.d.). *Civil, Organized, and Efficient online Dialofue*. Available online: https://consider.it/

Conti, A. (2004). A política de intervenção nos assentamentos informais em Belo Horizonte nas décadas de 1980 e 1990 e o "Plano Global Específico". *Cadernos de Arquitetura e Urbanismo*, *11*(12), 189–216.

Corbun, J. (2005). *Street Science: Community Knowledge and Environmental Health Justice*. The MIT Press.

Cork Smart Gateway. (n.d.). Available online: http://www.corksmartgateway.ie/citizenengagement/

Costa, G., Stroscia, M., Zengarini, N., & Demaria, M. (2017). *40 anni di salute a Torino. Spunti per leggere i bisogni e le politiche a Torino*. Inferenze.

Couclelis, H. (2005). "Where has the future gone?" Rethinking the role of integrated land-use models in spatial planning. *Environment & Planning A*, *37*(8), 1353–1371. doi:10.1068/a3785

Cournede, P.-H., Mathieu, A., Houllier, F., Barthelemy, D., & de Reffye, P. (2007). Computing Competition for Light in the GREENLAB Model of Plant Growth: A Contribution to the Study of the Effects of Density on Resource Acquisition and Architectural Development. *Annals of Botany*, *101*(8), 1207–1219. doi:10.1093/aob/mcm272 PMID:18037666

Coutts, Loughnan, Tapper, White, Thom, Broadbent, & Harris. (2014). *The Impacts of WSUD solutions on human thermal comfort.* Green Cities and Micro-Cilmate B3.1 Cooperatice Research Centre for Water Sensitive Cities.

COWI. (2009). *Study concerning the report on the application and effectiveness of the SEA Directive* (2001/42/EC). DG Environment European Commission, Report no. p-67683-a, Issue no. 2.

Cowley, R., Joss, S., & Dayot, Y. (2018). The smart city and its publics: Insights from across six UK cities. *Urban Research & Practice*, *11*(1), 53–77. doi:10.1080/17535069.2017.1293150

Craglia, M., & Annoni, A. (2007). INSPIRE: An innovative approach to the development of spatial data infrastructures in Europe. *Research and Theory in Advancing Spatial Data Infrastructure Concepts*, 93-105.

Craglia, M., & Campagna, M. (2009). *Advanced regional spatial data infrastructures in Europe.* Luxembourg: Office for official publications of the European Communities.

Cravo, T., & Hashiguchi, O. (2020). *Bringing health care to the patient: An overview of the use of telemedicine in OECD countries*, OECD Health Working Papers, N. 116. doi:10.1787/18152015

Craw, S., Wiratunga, N., & Rowe, R. C. (2006). Learning adaptation knowledge to improve case-based reasoning. *Artificial Intelligence*, *170*(16-17), 1175–1192. doi:10.1016/j.artint.2006.09.001

Cugurullo, F. (2013). How to build a sandcastle: An analysis of the genesis and development of Masdar City. *Journal of Urban Technology*, *20*(1), 23–37. doi:10.1080/10630732.2012.735105

Czepkiewicz, M., & Brudka, C. (2016b). *Raport z badania preferencji mieszkańców na temat kierunków zmian zagospodarowania przestrzennego osiedla Andrzejów w Łodzi.* Centrum Badań Metropolitalnych UAM.

Czepkiewicz, M., Brudka, C., Jankowski, P., Kaczmarek, T., Zwoliński, Z., Mikuła, Ł., Bąkowska, E., Młodkowski, M., & Wójcicki, M. (2016a). Public Participation GIS for sustainable urban mobility planning: Methods, applications and challenges. *Rozwój Regionalny i Polityka Regionalna*, *35*, 9–35.

Czepkiewicz, M., Jankowski, P., & Młodkowski, M. (2017). Geo-questionnaires in urban planning: Recruitment methods, participant engagement, and data quality. *Cartography and Geographic Information Science*, *44*(6), 551–567. doi:10.1080/15230406.2016.1230520

Dahl, A. J., Milne, G. R., & Peltier, J. W. (2021). Digital health information seeking in an omni-channel environment: A shared decision-making and service-dominant logic perspective. *Journal of Business Research*, *125*, 840–850. doi:10.1016/j.jbusres.2019.02.025

Damurski, L. (2021). How to Include Omnichannel Services in Land-Use Policy? E-Planning Holds the Key. *International Journal of E-Planning Research*, *10*(3), 70–85. doi:10.4018/IJEPR.20210701.oa5

Dangermond, J. (2010). *GeoDesign and GIS–Designing our Futures. Peer Reviewed Proceedings of Digital Landscape Architecture.* Anhalt University of Applied Science.

Davoudi, S., Brooks, E., & Mehmood, A. (2013). Evolutionary resilience and strategies for climate adaptation. *Planning Practice and Research*, *28*(3), 307–322. doi:10.1080/02697459.2013.787695

De Certeau, M. (1990). *L'invention du quotidien* (Vols. 1-2). Editions Gallimard.

De Filippi, F., Coscia, C., & Guido, R. (2017). How Technologies Can Enhance Open Policy Making and Citizen-Responsive Urban Planning: MiraMap - A Governing Tool for the Mirafiori Sud District in Turin (Italy). *International Journal of E-Planning Research, 6*(1), 23–42. doi:10.4018/IJEPR.2017010102

De los Rios, I. C., & Charnley, F. J. S. (2017). Skills and capabilities for a sustainable and circular economy: The changing role of design. *Journal of Cleaner Production, 160*, 109–122. doi:10.1016/j.jclepro.2016.10.130

De Mántaras, R. L., McSherry, D., Bridge, D., Leake, D., Smyth, B., Craw, S., & Keane, M. (2005). Retrieval, reuse, revision and retention in case-based reasoning. *The Knowledge Engineering Review, 20*(3), 215–240. doi:10.1017/S0269888906000646

De Montis, A., Ledda, A., Caschili, S., Ganciu, A., & Barra, M. (2014). a. SEA effectiveness for landscape and master planning: An investigation in Sardinia. *Environmental Impact Assessment Review, 47*, 1–13. doi:10.1016/j.eiar.2014.03.002

de Reffye, P., Edelin, C., Francon, J., Jaeger, M., & Puech, C. (1988). Plant models faithful to botanical structure and development p151-de_reffye 1988.pdf. *Computer Graphics, 22*(4), 151–158. doi:10.1145/378456.378505

de Roo, G., & Rauws, W. (2012). Positioning planning in the world of order, chaos and complexity: On perspectives, behaviour and interventions in a non-linear environment. In Complexity theories of cities have come of age: An overview with implications to urban planning and design (pp. 207-220). Springer.

De Roo, G., & Hillier, J. (2016). *Complexity and planning: Systems, assemblages and simulations*. Routledge. doi:10.4324/9781315573199

Decidim. (n.d.). *Free open-source participatory democracy for cities and organizations*. Available online: https://decidim.org/

Deleuze, G., & Guattari, F. (1987). A thousand plateaus: Capitalism and schizophrenia. University of Minnesota Press.

Desrosières, A. (2008). *Pour une sociologie historique de la quantification*. Presses des Mines.

Di Nunzio, A. (2019). *Envi_MET*. https://www.envi-met.com/learning-support/plugins/

Distel, B., & Becker, J. (2018). A long and winding road? Analyzing e-government website maturity in Germany. In *Tagungsband Multikonferenz Wirtschaftsinformatik* (pp. 621-632). https://mkwi2018.leuphana.de/wp-content/uploads/MKWI_103.pdf

Dobbs, C., Kendal, D., & Nitschke, C. R. (2014). Multiple ecosystem services and disservices of the urban forest establishing their connections with landscape structure and sociodemographics. *Ecological Indicators, 43*, 44–55. doi:10.1016/j.ecolind.2014.02.007

Douay, N. (2016). La numérisation des dispositifs de participation de la Mairie de Paris - Le cas du budget participatif et de la plateforme "Madame la Maire, j'ai une idée!" *NETCOM, 30*(3/4), 249–280.

Dovey, K. (2012). Informal settlement and complex adaptive assemblage. *International Development Planning Review, 34*(3), 371–390.

Drumond, W. (2017). *Entendendo o PRU como novo instrumento de planejamento para assentamentos precários de Belo Horizonte. Unpublished Final Undergraduate Research Project. Escola de Arquitetura, Universidade Federal de Minas Gerais*. UFMG.

Dubai Health Authority. (2021). *Guidelines for Reporting Telehealth Key Performance Indicators*. https://www.dha.gov.ae/en/HealthRegulation/Documents/Guidelines%20for%20Reporting%20Telehealth%20KPIs.pdf

Dugdale, A., Daly, A., Papandrea, F., & Maley, M. (2005). Accessing e-government: Challenges for citizens and organizations. *International Review of Administrative Sciences*, *71*(1), 109–118. doi:10.1177/0020852305051687

Duggan, P. (2018, October 2). Four alleged members of hate group charged in 2017 'Unite the Right' rally in Charlottesville. *Washington Post*. https://www.washingtonpost.com/local/public-safety/federal-officials-to-announce-additional-charges-in-2017-unite-the-right-rally-in-charlottesville/2018/10/02/60881262-c651-11e8-9b1c-a90f1daae309_story.html

Durán, A., Kutzin, J., Martin-Moreno, J.-M., & Travis, P. (2012). Understanding health systems: scope, functions and objectives. In J. Figueras & M. McKee (Eds.), *Health Systems, Health, Wealth and Societal Well-being. Assessing the case for investing in health systems* (pp. 19–36). Mc Maidenhead Graw Hill Open University Press.

Durdyev, S., Mbachu, J., Thurnell, D., Zhao, L., & Hosseini, M. R. (2021). BIM Adoption in the Cambodian Construction Industry: Key Drivers and Barriers. *ISPRS International Journal of Geo-Information*, *10*(4), 215. doi:10.3390/ijgi10040215

Dwivedi, Y., Rana, N., Tajvidi, M., Lal, B., Sahu, G., & Gupta, A. (2017). Exploring the role of social media in e-government: an analysis of emerging literature. In *Proceedings of the 10th international conference on theory and practice of electronic governance* (pp. 97-106). New York: Association for Computing Machinery. 10.1145/3047273.3047374

Dwivedi, Y., Weerakoddy, V., & Janssen, M. (2012). Moving towards maturity: Challenges to successful e-government implementation and diffusion. *ACM SIGMIS Database*, *42*(4), 11–22. doi:10.1145/2096140.2096142

Edelenbos, J. (1998). Article. In R. Monnikhof (Ed.), Spanning in interactie: een analyse van interactief beleid in de lokale democratie (pp. 11–48). Instituut voor Publiek en Politiek.

Edelenbos, J., & Monnikhof, R. (1977). Lokale interactieve beleidsvorming, Lemma, Utrecht. Cupps, D.S., 'Emerging problems of citizens participation'. *Public Administration Review*, *37*, 478–487.

Eigen, M. (1971). Molecular self-organization and the early stages of evolution. *Quarterly Reviews of Biophysics*, *4*(2-3), 149–212. doi:10.1017/S0033583500000627 PMID:5134461

Eitzel, M. V. (2017). Citizen Science Terminology Matters: Exploring Key Terms. *Citizen Science: Theory and Practice*, *2*(1), 1–20.

Ekeland, A. G., Bowesc, A., & Flottorpd, S. (2011). Methodologies for assessing telemedicine: A systematic review of reviews. *International Journal of Medical Informatics*, *81*(1), 1–11. doi:10.1016/j.ijmedinf.2011.10.009 PMID:22104370

El Khafif, M., Zhang, Z., & Mondschein, A. (2018). Mapping the Nocturnal City: Community-Centered Urban Sensing in Charlottesville, Virginia. *Proceedings of the CityStreet3 Conference*.

El Khafif, M., Mondschein, A., & Rosado Murillo, L. F. (2021). *Ostenda illuminata: A socio-ecological 1:1 prototype for networked public space*. Architectural Research Centers Consortium.

Elezaj, O., Tole, D., & Baci, N. (2018). Big data in e-government environments: Albania as a case study. *Academic Journal of Interdisciplinary Studies*, *7*(2), 117–124. doi:10.2478/ajis-2018-0052

Elwood, S. A. (2002). GIS Use in Community Planning: A Multidimensional Analysis of Empowerment. *Environment and Planning A. Economy and Space*, *34*(5), 905–922. doi:10.1068/a34117

Ely, M. (2009). Planning for Trees in Urban Environments. *TREENET Proceedings of the 10th National Street Tree Symposium 3rd and 4th September*, 87–98.

Ervin, S. (2011). A system for GeoDesign. *Proceedings of Digital Landscape Architecture, Anhalt University of Applied Science*, 145-154.

Escobedo, F. J., Kroeger, T., & Wagner, J. E. (2011). Urban Forests and Pollution Mitigation: Analyzing Ecosystem Services and Disservices. *Environmental Pollution, 159*(8), 2078–2087. doi:10.1016/j.envpol.2011.01.010 PMID:21316130

Esper, G. J., Sweeney, R. L., Winchell, E., Duffell, J. M., Kier, S. C., Lukens, H. W., & Krupinski, E. A. (2020). Rapid System Wide Implementation of Outpatient Telehealth in Response to the COVID-19 Pandemic. *Journal of Healthcare Management, 65*(6), 443–452. PMID:33074968

European Commission. (2002). *Guidance document on public participation in relation to the water framework directive. Active involvement, consultation, and public access to information.* Annex I Public participation techniques. Author.

European Commission. (2018). *Market study on telemedicine. Luxembourg: Publications Office of the European Union.* https://ec.europa.eu/health/sites/default/files/ehealth/docs/2018_ provision_marketstudy_telemedicine_en.pdf

European Commission. (2020). A European strategy for data. COM(2020) 66 final.

European Commission. (2021). *Digital Solutions During the Pandemic.* https://ec.europa.eu/info/live-work-travel-eu/coronavirus-response/digital-solutions-during-pandemic_en

Evans-Cowley, J. S., & Griffin, G. (2012). Microparticipation with social media for community engagement in transportation planning. *Transportation Research Record: Journal of the Transportation Research Board, 2307*(1), 90–98. doi:10.3141/2307-10

Evans, D. B., Hsua, J., & Boermaa, T. (2013). Universal health coverage and universal access. *Bulletin of the World Health Organization, 91*(8), 546–546a. doi:10.2471/BLT.13.125450 PMID:23940398

Ewing, R., & Handy, S. (2009). Measuring the Unmeasurable: Urban Design Qualities Related to Walkability. *Journal of Urban Design, 14*(1), 65–84. doi:10.1080/13574800802451155

Fainstein, S. S. (2000). ` New Directions in Planning Theory '. *Urban Affairs Review, 35*(4), 451–478. doi:10.1177/107808740003500401

Falanga, R. (2020). *Citizen participation during the COVID-19 pandemic: Insights from local practices in European cities.* http://library.fes.de/pdf-files/bueros/lissabon/17148.pdf

Falanga, R., Cebulla, A., Principi, A., & Socci, M. (2021). The participation of senior citizens in policy-making: Patterning initiatives in Europe. *International Journal of Environmental Research and Public Health, 18*(1), 34. doi:10.3390/ijerph18010034 PMID:33374566

Falco, E., & Reinout Kleinhans, R. (2018). Beyond technology: Identifying local government challenges for using digital platforms for citizen engagement. *International Journal of Information Management, 40*, 17-20. doi:10.1016/j.ijinfomgt.2018.01.007

Fard, A. (2018). Not Sharing: Urban Techno-Colonialism in the Age of Big Data. *PLAT.* https://www.platjournal.com/articles/not-sharing-urban-techno-colonialism-in-the-age-of-big-data

Faura, J. (2020). Longitudinal study of the development of e-government in Europe. In How can digital technologies build a more integrated Europe? (pp. 75-90). Nomos Verlagsgesellschaft mbH.

Fernandez-i-Marín, X. (2011). The impact of e-government promotion in Europe. Internet dependence and critical mass. *Policy and Internet, 3*(4), 1–29. doi:10.2202/1944-2866.1093

Ferrara, F. M. (2021). *Osservatorio sulla Telemedicina Operativa.* https://www.dati-sanita.it/wp/wp-content/uploads/2021/08/101-osservatorio-telemedicina.pdf

Feuillet, T., Charreire, H., Hercberg, S., & Oppert, J.-M. (2019). Pour une géographie régionale des spécificités loc ales. L'exemple des liens entre environnement urbain et comportements individuels de santé. In V. Chasles & L. De Biaggi (Eds.), *Santé et territoire, une réflexion géographique* (Vol. 3, pp. 13–13). Université Jean Moulin.

Fischer, T. B. (2010). Reviewing the quality of strategic environmental assessment reports for English spatial plan core strategies. *Environmental Impact Assessment Review, 30*(1), 62–69. doi:10.1016/j.eiar.2009.04.002

Fischer, T. B., & Gazzola, P. (2006). SEA effectiveness criteria—Equally valid in all countries? The case of Italy. *Environmental Impact Assessment Review, 26*(4), 396–409. doi:10.1016/j.eiar.2005.11.006

Flaxman, M. (2010). Fundamentals of geodesign. *Proceedings of Digital Landscape Architecture, Anhalt University of Applied Science.*

Flyvbjerg, B. (2002). Bringing power to planning research: One researcher's praxis story. *Journal of Planning Education and Research, 21*(4), 353–366. doi:10.1177/0739456X0202100401

Flyvbjerg, B. (2018). *Making social science matter: why social inquiry fails and how it can succeed again.* Cambridge University Press.

Fonseca, B. (2016). Geodesign: Conceitos e arcabouço Metodológico. In *Tecnologias de geoinformação para representar e planejar o território urbano* (pp. 217–248). Interciência.

Foucault, M. (1969). *Madness and Civilization– The Archaeology of knowledge and the Discourse on Language.* Academic Press.

Fountain, J. (2001). The virtual state: Transforming American government? *National Civic Review, 90*(3), 241–251. doi:10.1002/ncr.90305

Fox-Rogers, L., & Murphy, E. (2014). Informal strategies of power in the local planning system. *Planning Theory, 13*(3), 244–268. doi:10.1177/1473095213492512

Freitas, V. (2020). *Geodesign: uma alternativa para criação de propostas coletivas para assentamentos precários em Belo Horizonte* (Master's thesis). NPGAU, Universidade Federal de Minas Gerais (UFMG), Belo Horizonte.

Fujita, M., & Thisse, J. F. (1996). Economics of agglomeration. *Planning Theory, 14*(3), 248–267.

Fundingsland Tetlow, M., & Hanusch, M. (2012). Strategic environmental assessment: The state of the art. *Impact Assessment and Project Appraisal, 30*(1), 15–24. doi:10.1080/14615517.2012.666400

Gabryelczyk, R. (2020). Has COVID-19 accelerated digital transformation? Initial lessons learned for public administrations. *Information Systems Management, 37*(4), 303–309. doi:10.1080/10580530.2020.1820633

Gabrys, J. (2014). Programming environments: Environmentality and citizen sensing in the smart city. *Environment and Planning. D, Society & Space, 32*(1), 30–48. doi:10.1068/d16812

Ganapati, S. (2010). *Using Geographic Information Systems to Increase Citizen Engagement E-Government.* IBM Center for The Business of Government.

Gardner, M. (1976). Mathematical games. *Scientific American, 234*(6), 120–125. doi:10.1038cientificamerican0676-120

Garfinkel, H. (1967). *Studies in Ethomethodology.* Academic Press.

Gatrell, A. C., & Wood, D. J. (2012). Variation in geographic access to specialist inpatient hospices in England. *Health & Place, 18*(4), 812–840. doi:10.1016/j.healthplace.2012.03.009 PMID:22522100

Geddes, M. (2006). Partnership and the Limits to Local Governance in England: Institutionalist Analysis and Neoliberalism. *International Journal of Urban and Regional Research*, *30*(1), 76–97. doi:10.1111/j.1468-2427.2006.00645.x

Geddes, P. (1915). *Cities in evolution: An introduction to the town planning movement and to the study of civics*. Williams.

Geertman, S. (2008). Planning Support Systems (PSS) - a planner's perspective. In *Brail, R. K. Planning Support Systems for Cities and Regions* (pp. 213–274). Lincoln Institute.

Geertman, S., & Stillwell, J. (2004). Planning support systems: An inventory of current practice. *Computers, Environment and Urban Systems*, *28*(4), 291–310. doi:10.1016/S0198-9715(03)00024-3

Geertman, S., & Stillwell, J. (2009). Planning Support Systems: Content, Issues and Trends. In S. Geertman & J. Stillwell (Eds.), *Planning Support Systems Best Practice and New Methods* (Vol. 95, pp. 1–26). Springer Netherlands. doi:10.1007/978-1-4020-8952-7_1

Geertman, S., Toppen, F., & Stillwell, J. (Eds.), *Planning Support Systems for Sustainable Development* (pp. 451–467). Springer.

Giddens, A. (1987). *Social theory today*. Stanford University Press.

Gil, J., Montenegro, N., & Duarte, J. (2010). Assessing computational tools for urban design – towards a 'city information model'. *Proceedings of the 28th Conference (eCAADe)*, 361–361.

Gillaizeau, I. (2019). *Evaluation d'un programme de prévention par les pairs des consommations de substances psychoactives*. Santé Publique France.

Glaser, B., & Strauss, A. (1967). *Discovery of Grounded Theory: Strategies for Qualitative Research*. Aldine Publishing.

Gleick, J. (2008). *Chaos: Making a new science*. Penguin Books.

Goffman, E. (1974a). *Les cadres de l'expériences*. Editions de Minuit.

Goffman, E. (1974b). *Les rites d'interaction*. Editions de Minuit.

Golding, A. R., & Rosenbloom, P. S. (1996). Improving accuracy by combining rule-based and case-based reasoning. *Artificial Intelligence*, *87*(1-2), 215–254. doi:10.1016/0004-3702(95)00120-4

González, A., Thérivel, R., Fry, J., & Foley, W. (2015). Advancing practice relating to SEA alternatives. *Environmental Impact Assessment Review*, *53*, 52–63. doi:10.1016/j.eiar.2015.04.003

Goodchild, M. (2007). Citizens as sensors: The world of volunteered geography. *GeoJournal*, *69*(4), 211–221. doi:10.100710708-007-9111-y

Goodchild, M. F. (2007). Citizen as Voluntary Sensors: Spatial Data Infrastructure in the World of Web 2.0. *International Journal of Spatial Data Infrastructures Research*, *2*, 24–32.

Goodman, N., Zwick, A., Spicer, Z., & Carlsen, N. (2020). Public engagement in smart city development: Lessons from communities in Canada's Smart City Challenge: Smart city public engagement. *The Canadian Geographer. Geographe Canadien*, *64*(3), 416–432. Advance online publication. doi:10.1111/cag.12607

Government of the United Kingdom. (2017). *Digital Government – Maintaining the UK Government as a World Leader in Serving its Citizens Online*. https://www.gov.uk/government/publications/uk-digital-strategy/6-digital-government-maintaining-the-uk-government-as-a-world-leader-in-serving-its-citizens-online

Graffigna, G., Barello, S., Savarese, M., Palamenghi, L., Castellini, G., Bonanomi, A., & Lozza, E. (2020). Measuring Italian citizens' engagement in the first wave of the COVID-19 pandemic containment measures: A cross-sectional study. *PLoS One*, *15*(9), e0238613. doi:10.1371/journal.pone.0238613 PMID:32915822

Gramaglia, C., & Dauphin, C.E. (2017). Toucher la pollution industrielle du doigt grâce au lichen. *Technique et Culture*, *68*.

Granier, B., & Kudo, H. (2016). How are citizens involved in smart cities? Analysing citizen participation in Japanese "Smart Communities". *Information Polity*, *21*(1), 61–76. doi:10.3233/IP-150367

Green, B. (2019). *The Smart Enough City: Putting Technology in Its Place to Reclaim Our Urban Future*. MIT Press. doi:10.7551/mitpress/11555.001.0001

Greenwood, D. J., & Levin, M. (2007). *Introduction to action research. Social research for social change*. Sage Publications, Inc. doi:10.4135/9781412984614

Grêt-Regamey, A., Sirén, E., Brunner, S. H., & Weibel, B. (2017). Review of decision support tools to operationalize the ecosystem services concept. *Ecosystem Services*, *26*, 306–315. doi:10.1016/j.ecoser.2016.10.012

Grey, T., Dyer, M., & Gleeson, D. (2017). Using Big and Small Urban Data for Collaborative Urbanism. In C. Certomà, M. Dyer, L. Pocatilu, & F. Rizzi (Eds.), *Citizen Empowerment and Innovation in the Data-Rich City* (pp. 31–54). Springer International Publishing. doi:10.1007/978-3-319-47904-0_3

Gryl, I., Jekel, T., & Donert, K. (2010). Geoinformation and Spatial Citizenship. *Learning with GeoInformation*, *V*, 2–11.

Habitação e Cidades, G. T., & Tonuci, J. (2020). *Casa Verde e Amarela: pontos críticos do novo programa habitacional do Governo Federal*. https://www.observatoriodasmetropoles.net.br/casa-verde-e-amarela-pontos-criticos-do-novo-programa-habitacional-do-governo-federal/

Hache, E. (2007). Is responsibility a tool of neo-liberal governmentality? *Raisons politiques*, *28*, 49–65.

Haken, H. (1985). Synergetics—An interdisciplinary approach to phenomena of self-organization. *Geoforum*, *16*(2), 205–211. doi:10.1016/0016-7185(85)90029-6

Haken, H., & Portugali, J. (2003). The face of the city is its information. *Journal of Environmental Psychology*, *23*(4), 385–408. doi:10.1016/S0272-4944(03)00003-3

Haken, H., & Portugali, J. (2019). A synergetic perspective on urban scaling, urban regulatory focus and their interrelations. *Royal Society Open Science*, *6*(8), 191087. doi:10.1098/rsos.191087 PMID:31598267

Haklay, M. (2015). *Citizen Science and Policy : A European Perspective*. Woodrow Wilson International Center for Scholars.

Haklay, M., Singleton, A., & Parker, Ch. (2008). Web Mapping 2.0: The Neogeography of the GeoWeb. *Geography Compass*, *2*(6), 2011–2039. doi:10.1111/j.1749-8198.2008.00167.x

Haklay, M., & Tabón, C. (2003). Usability evaluation and PPGIS: Towards a user-centred design approach. *International Journal of Geographical Information Science*, *17*(6), 577–592. doi:10.1080/1365881031000114107

Haltofová, B. (2020). Critical success factors of geocrowdsourcing use in e-government: A case study from the Czech Republic. *Urban Research & Practice*, *13*(4), 434–451. doi:10.1080/17535069.2019.1586990

Hammond, A., Adriaanse, A., Rodenburg, E., Bryant, D., & Woodward, R. (1995). *Environmental Indicators: A Systematic Approach to Measuring and Reporting on Environmental Policy Performance in the Context of Sustainable Development*. World Resources Institute.

Hapanovych, Y. (2019). Prospects for the implementation of the European experience of e-democracy and e-government in Ukraine. *Rocznik Administracji Publicznej*, *5*(5), 115–124. doi:10.4467/24497800RAP.19.007.11470

Harris, B. (2001). Sketch planning: Systematic methods in planning and its support. *Planning Support Systems: Integrating Geographic Information Systems, Models, and Visualization Tools*, 59-80.

Harris, B., & Batty, M. (1993). Locational models, geographic information and Planning Support Systems. *Journal of Planning Education and Research*, *12*(3), 184–198. doi:10.1177/0739456X9301200302

Harvey, D. (1989). *The condition of postmodernity an enquiry into the origins of cultural change*. Academic Press.

Harvey, D. (2007). *A brief history of neoliberalism*. Oxford University Press.

Hatna, E., & Benenson, I. (2014). *Combining segregation and integration: Schelling model dynamics for heterogeneous population*. arXiv preprint arXiv:1406.5215.

Haughton, G., Allmendinger, P., Counsell, D., & Vigar, G. (2010). *The New Spatial Planning: Territorial Management with Soft Spaces and Fuzzy Boundaries*. Taylor & Francis.

Haynes, A., Rychetnik, L., Finegood, D., Irving, M., Freebairn, L., & Hawe, P. (2020). Applying systems thinking to knowledge mobilisation in public health. *Health Research Policy and Systems*, *18*(1), 134. doi:10.118612961-020-00600-1 PMID:33203438

Hays, E. (2018, August 7). City commission to bring community leaders into Comprehensive Plan process · Charlottesville Tomorrow. *Charlottesville Tomorrow*. https://www.cvilletomorrow.org/news/article/31273-city-to-bring-community-leaders-into-comp-plan/

He, S., Besuievsky, G., Tourre, V., Patow, G., & Moreau, G. (2012). *All range and heterogeneous multi-scale 3D city models*. Usage, Usability, and Utility of 3D City Models–European COST Action TU0801, 02006.

Heath, N. (2019). How Estonia became an e-government powerhouse. *TechRepublic*. https://www.techrepublic.com/article/how-estonia-became-an-e-government-powerhouse/

Heaton, L., Millerand, F., Dias Da Silva, P., & Proulx, S. (2018). *La reconfiguration du travail scientifique en biodiversité*. Les Presses de l'Université de Montréal.

Hensel, M., Menges, A., & Weinstock, M. (2013). *Emergent technologies and design: towards a biological paradigm for architecture*. Routledge. doi:10.4324/9781315881294

Heppenstall, A. J., Crooks, A. T., See, L. M., & Batty, M. (Eds.). (2011). *Agent-based models of geographical systems*. Springer Science & Business Media.

Hillier, B. (2012). The genetic code for cities: Is it simpler than we think? In Complexity theories of cities have come of age: An overview with implications to urban planning and design (pp. 129-152). Springer.

HIMSS. (2021) *Future Healthcare Report*. https://www.himss.org/sites/hde/files/handout-FutureOfHealthcareReport-EMC-0821-FINAL_156.pdf

Hodžić, S., Ravšelj, D., & Alibegović, D. (2021). e-Government effectiveness and efficiency in EU-28 and COVID-19. *Central European Public Administration Review*, *19*(1), 159–180. doi:10.17573/cepar.2021.1.07

Hoffmann, H., & Payton, D. W. (2018). Optimization by Self-Organized Criticality. *Scientific Reports*, *8*(1), 2358. doi:10.103841598-018-20275-7 PMID:29402956

Hofmann, S., Madsen, C., & Distel, D. (2020). Developing an analytical framework for analyzing and comparing national e-government strategies. In *International conference on electronic government* (pp. 15-28). Cham: Springer. 10.1007/978-3-030-57599-1_2

Hohmann, B., Havemann, S., Krispel, U., & Fellner, D. (2010). A GML shape grammar for semantically enriched 3D building models. *Computers & Graphics*, *34*(4), 322–334. doi:10.1016/j.cag.2010.05.007

Hollander, J., & Neinstein, A. (2020). *Maturation from Adoption-Based to Quality-Based Telehealth Metrics*. https://catalyst.nejm.org/doi/pdf/10.1056/CAT.20.0408

Horelli, L., & Wallin, S. (2013). Starting points. In L. Horelli (Ed.), *New Approaches to Urban Planning. Insights from Participatory Communities* (pp. 11–22). Aalto University.

Horizonte, B., & Municipal, C. (2020). Relatório Final do Grupo de Trabalho da Comissão de Direitos Humanos e Defesa do Consumidor sobre Direito à Moradia. Academic Press.

Houllier, F. (2016). *Sciences participative en France - État des lieux, bonnes pratiques et recommandation*. Mission Sciences Paricipatives.

Huck, J. J. (2017). Combining physiological, environmental and locational sensors for citizen-oriented health applications. *Environmental Monitoring and Assessment*, *189*(114).

Hussnain, M. Q., Waheed, A., Anjum, G. A., Naeem, M. A., Hussain, E., Wakil, K., & Pettit, C. J. (2020). A framework to bridge digital planning tools' utilization gap in peri-urban spatial planning; lessons from Pakistan. *Computers, Environment and Urban Systems*, *80*, 101451. doi:10.1016/j.compenvurbsys.2019.101451

Innes, J. E., & Booher, D. E. (2010). *Planning with complexity: An introduction to collaborative rationality for public policy*. Routledge. doi:10.4324/9780203864302

Internet World Stats. (2021). https://www.internetworldstats.com/

Iqbal, A., Khan, I. A., & Jan, S. (2019). *A Review and Comparison of the Traditional Collaborative and Online Collaborative Techniques for Software Requirement Elicitation. In 2019 ICACS*. IEEE.

Irwin, A. (1995). *Citiern Science: A Study of People, Expertise ans Suistainable Development*. Routledge.

Iveson, K. (2013). Cities within the City: Do-It-Yourself Urbanism and the Right to the City. *International Journal of Urban and Regional Research*, *37*(3), 941–956. doi:10.1111/1468-2427.12053

Jacobs, J. (2016). *The death and life of great American cities*. Vintage.

Janhäll, S. (2015). Review on urban vegetation and particle air pollution - Deposition and dispersion. *Atmospheric Environment*, *105*, 130–137.

Jankowski, P., Czepkiewicz, M., Młodkowski, M., & Zwoliński, Z. (2016). Geo-questionnaire: A Method and Tool for Public Preference Elicitation in Land Use Planning. *Transactions in GIS*, *20*(6), 903–924. doi:10.1111/tgis.12191

Jankowski, P., Czepkiewicz, M., Młodkowski, M., Zwoliński, Z., & Wójcicki, M. (2017a). Evaluating the scalability of public participation in urban land use planning: A comparison of Geoweb methods with face-to-face meetings. *Environment and Planning. B, Urban Analytics and City Science*. Advance online publication. doi:10.1177/2399808317719709

Jankowski, P., Kaczmarek, T., Zwoliński, Z., Mikuła, Ł., Wójcicki, M., Bąkowska, E., Czepkiewicz, M., Młodkowski, M., & Brudka, C. (2017b). *Narzędzia internetowe w konsultacjach społecznych. Idea, obszary zastosowań i wdrażanie*. Wydawnictwo Naukowe UAM.

Jannati, A., Sadeghi, V., Imani, A., & Saadati, M. (2018). Effective coverage as a new approach to health system performance assessment: A scoping review. *BMC Health Services Research, 18*(1), 886. doi:10.118612913-018-3692-7 PMID:30470214

Jantsch, E. (1980). *The self-organizing universe: Scientific and human implications of the emerging paradigm of evolution.* Pergamon Press.

Jencks & Kropf. (Eds.). (2006). Theories and Manifestoes of Contemporary Architecture. Wiley-Academy.

Jerett, M. (2017). Validating novel air pollution sensors to improve exposure estimates for epidemiological analyses and citizen science. *Environmental Research, 158*, 286–294.

Jiao, W. (2016). Community Air Sensor Network (CAIRSENSE) project: Evaluation of low-cost sensor performance in a suburban environment in the southearstern United States. *Atmospheric Measurement Techniques, 9*, 5281–5292.

Joint Action to support the eHealth Network. (2017). *Report on EU state of play on telemedicine services and uptake recommendations.* https://ec.europa.eu/health/sites/default/files/ ehealth/docs/ev_20171128_co09_en.pdf

Joliveau, T. (2004). *Géomatique et gestion environnementale du territoire. Recherche sur un usage géographique des SIG. HDR.* Université de Rouen.

Jones, B. I. (2020). A study of Building Information Modeling (BIM) uptake and proposed evaluation framework. *Journal of Information Technology in Construction, 25*, 452–468. doi:10.36680/j.itcon.2020.026

Jost, J. T., Barberá, P., Bonneau, R., Langer, M., Metzger, M., Nagler, J., Sterling, J., & Tucker, J. A. (2018). How Social Media Facilitates Political Protest: Information, Motivation, and Social Networks. *Political Psychology, 39*, 85–118. doi:10.1111/pops.12478

Jun, M. S. (2018). Blockchain government - a next form of infrastructure for the twenty-first century. *Journal of Open Innovation, 4*(1), 7. doi:10.118640852-018-0086-3

Kaczmarek, T., & Bąkowska, E. (2016a). *Raport z konsultacji społecznych na temat kierunków zagospodarowania przestrzennego centrum miejscowości Rokietnica.* Centrum Badań Metropolitalnych UAM.

Kaczmarek, T., & Bąkowska, E. (2016b). *Raport z konsultacji społecznych poświęconych nowym kierunkom zagospodarowania przestrzennego terenu lotniska w Kobylnicy.* Centrum Badań Metropolitalnych UAM.

Kaczmarek, T., & Mikuła, Ł. (2020). Konsultacje społeczne w planowaniu przestrzennym z użyciem narzędzi internetowych PPGIS. In A. Bochentyn & J. H. Szlachetko (Eds.), *Cyfrowa czy analogowa? Funkcjonowanie administracji publicznej w stanie kryzysu* (pp. 411–435). Wydawnictwo Uniwersytetu Gdańskiego.

Kahila, M., & Kytta, M. (2009). SoftGIS as a Bridge-Builder in Collaborative Urban Planning. In S. Geertman & J. Stillwell (Eds.), *Planning Support Systems Best Practice and New Methods* (pp. 389–411). Springer. doi:10.1007/978-1-4020-8952-7_19

Kahila-Tani, M., Broberg, A., Kyttä, M., & Tyger, T. (2015). Let the Citizens Map—Public Participation GIS as a Planning Support System in the Helsinki Master Plan Process. *Planning Practice and Research, 31*(2), 195–214. doi:10.1080/02697459.2015.1104203

Kalra, D., Kolitsi, Z., Zobell, O., Bujok, S., & Stroetmann, V. (2020). *Better Utilisation of Data Infrastructures to Support Secondary Uses of Health Data.* https://digitalhealtheurope.eu/wp-content/uploads/2020/10/Better-Utilisation-of-Data-Infrastructures.pdf

Kalvet, T., Toots, M., van Veenstra, A., & Krimmer, R. (2018). Cross-border e-government services in Europe: expected benefits, barriers and drivers of the once-only principle. In A. Kankanhalli, A. Ojo, & D. Soares (eds.) *Proceedings of the 11th International Conference on Theory and Practice of Electronic Governance* (pp. 69-72). New York: Association for Computing Machinery. 10.1145/3209415.3209458

Kapp, S. (2012). *Arquitetos nas favelas: três críticas e uma proposta de atuação.* [Paper presentation]. IV Congresso Brasileiro e III Congresso Ibero-Americano Habitação Social:ciência e tecnologia "Inovação e Responsabilidade". Florianópolis: UFSC.

Karagulian, F., Barbiere, M., Kotsev, A., Spinelle, L., Gerboles, M., Lagler, F., Redon, N., Crunaire, S., & Borowiak, A. (2019). Revew of the Performance of Low-Cost Sensors for Air Quality Monitoring. *Atmosphere*, *10*, 506.

Kassen, M. (2020). Open data and its peers: Understanding promising harbingers from Nordic Europe. *Aslib Journal of Information Management*, *72*(5), 765–785. doi:10.1108/AJIM-12-2019-0364

Kauffman, S. A., & Strohman, R. C. (1994). *The Origins of Order: self-organization and selection in evolution.* Oxford University Press.

Kendall, G. (2018). Is Evolutionary Computation evolving fast enough? *IEEE Computational Intelligence Magazine*, *13*(2), 42–51. doi:10.1109/MCI.2018.2807019

Kettl, D. F. (2002). *The Transformation of Governance: Public Administration for Twenty-first Century America.* John Hopkins University Press.

Key Factors for Effective Citizens Engagement in Smart City: The Case of Cork City. (n.d.). Available online: https://iot-smartcities.lero.ie/wp-content/uploads/2016/12/CorkCitiEngage-Identifying-key-factors-for-effective-citizens-engagement-in-Smart-City-initiatives.pdf

Khacef, L., Girau, B., Rougier, N., Upegui, A., & Miramond, B. (2018). *Neuromorphic hardware as a self-organizing computing system.* https://arxiv.org/abs/1810.12640

Kidholm, K., Ekeland, A. G., Jensen, L. K., Rasmussen, J., Pedersen, C. D., Bowes, A., Flottorp, S. A., & Bech, M. (2012). A model for assessment of telemedicine applications: Mast. *International Journal of Technology Assessment in Health Care*, *28*(1), 44–51. doi:10.1017/S0266462311000638 PMID:22617736

King, C. (2017). *Armidale Regional Council.* Armidale CBD Revitalisation Project. https://yoursay.armidale.nsw.gov.au/32201/widgets/185587/documents/68266

Kingston, R. (2011). Online Public Participation GIS for Spatial Planning. In T. L. Nyerges, H. Couclelis, & R. McMaster (Eds.), *The SAGE Handbook of GIS and Society* (pp. 361–380). SAGE. doi:10.4135/9781446201046.n19

Kleinhans, R., Van Ham, M., & Evans-Cowley, J. (2015). Using social media and mobile technologies to foster engagement and self-organization in participatory urban planning and neighbourhood governance. *Planning Practice and Research*, *30*(3), 237–247. doi:10.1080/02697459.2015.1051320

Klosterman, R. E. (1999). The what if? Collaborative planning support system. *Environment and Planning. B, Planning & Design*, *26*(3), 393–408. doi:10.1068/b260393

Kohonen, T. (1989). Self-organizing feature maps. In *Self-organization and associative memory* (pp. 119–157). Springer. doi:10.1007/978-3-642-88163-3_5

Kolb, D. A., & Fry, R. E. (1974). *Toward an applied theory of experiential learning. MIT Alfred P.* Sloan School of Management.

Kolodner, J. L. (1992). An introduction to case-based reasoning. *Artificial Intelligence Review*, *6*(1), 3–34. doi:10.1007/BF00155578

Kota, S., Haberl, J. S., Clayton, M. J., & Yan, W. (2014). Building Information Modeling (BIM)-based daylighting simulation and analysis. *Energy & Buildings. Complete*, *81*, 391–403. doi:10.1016/j.enbuild.2014.06.043

Krishnan, S., & Balasubramanian, T. (2016). Traffic flow optimization and vehicle safety in smart cities. *Traffic (Copenhagen, Denmark)*, *5*(5), 7814–7820.

Krivý, M. (2018). Towards a critique of cybernetic urbanism: The smart city and the society of control. *Planning Theory*, *17*(1), 8–30. doi:10.1177/1473095216645631

Kropf, K. S. (2001). Conceptions of change in the built environment. *Urban Morphology*, *5*(1), 29–46.

Krugman, P. (1996). *The self-organizing economy*. Blackwel Publishers.

Kugler, P. N., Kelso, J. S., & Turvey, M. T. (1980). On the concept of coordinative structures as dissipative structures. Advances in Psychology, (1), 3-47.

Kumar, P. (2015). The rise of low-cost sensing for managing air pollution in cities. *Environment International*, *75*, 199–205.

Kurasz, G. (2006). *Fifeville Neighborhood Lighting Assessment*. Kurasz & Associates.

Kuusela & Partanen. (2022). *Plenty of Planning, Scanty Guidance: The case of Tampere*. In Progress.

Kwon, S., Kim, B. C., Mun, D., & Han, S. (2018). User-assisted integrated method for controlling level of detail of large-scale B-rep assembly models. *International Journal of Computer Integrated Manufacturing*, *31*(9), 881–892. doi:10.1080/0951192X.2018.1448114

Lafia, S., Turner, A., & Kuhn, W. (2018). Improving discovery of open civic data. In *10th International Conference on Geographic Information Science (GIScience 2018)*. Schloss Dagstuhl--Leibniz-Zentrum fuer Informatik. UC Santa Barbara. Retrieved from https://escholarship.org/uc/item/6rn1b22w doi:10.4230/LIPIcs.GISCIENCE.2018.9

Landini, S., Occelli, S., Dalmasso, M., Durando, M., & Montano, G. (2019). *Elementi di analisi per una mobilità sostenibile*. http://ires.Piemonte.it/relazione2019/ Capitolo_5_allegato01.pdf

Lane, N. D., Eisenman, S. B., Musolesi, M., Miluzzo, E., & Campbell, A. T. (2008). Urban Sensing Systems: Opportunistic or Participatory? *Proceedings of the 9th Workshop on Mobile Computing Systems and Applications*, 11–16. 10.1145/1411759.1411763

Langenheim, N., White, M., Tapper, N., Livesley, S. J., & Ramirez-Lovering, D. (2020). Right tree, right place, right time: A visual-functional design approach to select and place trees for optimal shade benefit to commuting pedestrians. *Sustainable Cities and Society*, *52*, 101816. doi:10.1016/j.scs.2019.101816

Lascoumes, P. (2007). Les instruments d'action publique, traceurs de changement : L'exemple des transformations de la politique française de lutte contre la pollution artmosphérique (1961-2006). *Politique et Sociétés*, *26*(2–3), 73–89.

Lascoumes, P., & Le Galès, P. (2005). *Gouverner par les instruments*. Presses de Sciences Po.

Latour, B. (1985). Les "vues" de l'esprit - Une introduction à l'anthropologie des sciences et des techniques. *Culture Technique*, *14*, 4–30.

Latour, B. (2007). *Reassembling the social – An introduction to Actor-Network Theory*. Oxford University Press.

Lau, C. (2006). *Problem-based Medical Case Management*. Hong Kong University Press.

Lawrence, H. W. (2008). *City trees: A historical geography from the Renaissance through the nineteenth century*. University of Virginia Press.

Lei Federal 10.257 (O Estatuto das Cidades) (2001), Regulamenta os arts. 182 e 183 da Constituição Federal, que estabelece diretrizes gerais da política urbana e dá outras providências. Brasília.

Lei Federal nº 13.465 (Regularização Fundiária) (2017). Dispõe sobre a regularização fundiária rural e urbana, sobre a liquidação de créditos concedidos aos assentados da reforma agrária e sobre a regularização fundiária no âmbito da Amazônia Legal; institui mecanismos para aprimorar a eficiência dos procedimentos de alienação de imóveis da União e dá outras providências. Brasília.

Lei Municipal 7.166 (Lei de parcelamento, ocupação e uso do solo) (1996). Estabelece normas e condições para parcelamento, ocupação e uso do solo urbano no município. Belo Horizonte.

Lei Municipal 9.959 (Altera a Lei n° 7.165, de 27 de agosto de 1996, a Lei n° 7.166, de 27 de agosto de 1996, estabelece normas e condições para a urbanização e a regularização fundiária da Zona de Especial Interesse Social - ZEIS, dispõe sobre parcelamento, ocupação e uso do solo nas Áreas de Especial Interesse Social - AEIS, e dá outras providências.) (2010). Belo Horizonte.

Lenton, T. M. (2011). Early warning of climate tipping points. *Nature Climate Change*, *1*(4), 201–209. doi:10.1038/nclimate1143

Levesque, J.-F., Harris, M. F., & Russell, G. (2013). Patient-centred access to health care: Conceptualising access at the interface of health systems and populations. *International Journal for Equity in Health*, *12*(18), 1–9. doi:10.1186/1475-9276-12-18 PMID:23496984

Li, S. (2018). Application of Blockchain Technology in Smart City Infrastructure. *2018 IEEE International Conference on Smart Internet of Things (SmartIoT)*, 276-2766. 10.1109/SmartIoT.2018.00056

Liao, C., Goff, L., Chaparro, A., Chaparro, B., & Stumpfhauser, L. (2000). A comparison of website usage between young adults and the elderly. In *Proceedings of the Human Factors and Ergonomics Society... Annual Meeting* (Vol. 4, p. 101). Sage Publications Ltd. https://www.proquest.com/scholarly-journals/comparison-website-usage-between-young-adults/docview/235449180/se-2?accountid=49401

Lin Y, Benneker K. (2021). *Assessing collaborative planning and the added value of planning support apps in The Netherlands*. doi:10.1177/23998083211009239

Lindberg, F., & Grimmond, C. (2011). The influence of vegetation and building morphology on shadow patterns and mean radiant temperatures in urban areas: Model development and evaluation. *Theoretical and Applied Climatology*, *105*(3–4), 311–323. doi:10.100700704-010-0382-8

Lindenau, M., & Böhler-Baedeker, S. (2014). Citizen and stakeholder involvement: A precondition for sustainable urban mobility. *Transportation Research Procedia*, *4*, 347–360. doi:10.1016/j.trpro.2014.11.026

Lom, M., & Pribyl, O. (2021). Smart city model based on systems theory. *International Journal of Information Management*, *56*, 102092. doi:10.1016/j.ijinfomgt.2020.102092

López-Lópeza, V., Iglesias-Antelo, S., Vázquez-Sanmartín, A., Connolly, R., & Bannister, F. (2018). e-Government, transparency & reputation: An empirical study of Spanish local government. *Information Systems Management*, *35*(4), 1–18. doi:10.1080/10580530.2018.1503792

Loukis, E., Charalabidis, Y., & Androutsopoulou, A. (2015, August). Evaluating a passive social media citizensourcing innovation. In *International Conference on Electronic Government* (pp. 305-320). Springer. 10.1007/978-3-319-22479-4_23

Lydon, M., & Garcia, A. (2015). *Tactical Urbanism: Short-term Action for Long-term Change*. Island Press. doi:10.5822/978-1-61091-567-0

Lynn, G. (1998). *Folds, Bodies and Blobs–Collected Essays*. La Lettre Volee.

Mabi, C. (2021). La "civic tech" et "la démocratie numérique" pour "ouvrir" la démocratie? *Réseaux. La Découverte*, *1*(225), 215–248.

MacEachren, A. M. (1995). *How Maps Work: Representation, Visualization, and Design*. Guilford Press.

Magalhães, D., & Moura, A. (2020). Aerial Images and Three-Dimensional Models Generated by RPA to Support Geovisualization in Geodesign Workshops. Lecture Notes in Computer Science, 12252, 296-309.

Magro, M. (2012). A review of social media use in e-government. *Administrative Sciences*, *2*(2), 148–161. doi:10.3390/admsci2020148

Maguire, D. J., Goodchild, M. F., & Rhind, D. W. (1991). *Geographical Information Systems: Principles and Applications* (Vols. 1–2). Longman Scientific and Technical.

Maguire, D. J., & Longley, P. A. (2005). The emergence of geoportals and their role in spatial data infrastructures. *Computers, Environment and Urban Systems*, *29*(1), 3–14. doi:10.1016/S0198-9715(04)00045-6

Maher, B. A., Ahmed, I. A., Davison, B., Karloukovski, V., & Clarke, R. (2013). Impact of Roadside Tree Lines on Indoor Concentrations of Traffic-Derived Particulate Matter. *Environmental Science & Technology*, *47*(23), 13737–13744. doi:10.1021/es404363m PMID:24215538

Majerova, I. (2019). e-Government as a smart solution for public administration: A case of Visegrad group. *Balkans Journal of Emerging Trends in Social Sciences*, *2*(2), 115–123. doi:10.31410/Balkans.JETSS.2019.2.2.115-123

Ma, L., & Zheng, Y. (2018). Does e-government performance actually boost citizen use? Evidence from European countries. *Public Management Review*, *20*(10), 1513–1532. doi:10.1080/14719037.2017.1412117

Ma, L., & Zheng, Y. (2019). National e-government performance and citizen satisfaction: A multilevel analysis across European countries. *International Review of Administrative Sciences*, *85*(3), 506–526. doi:10.1177/0020852317703691

Malczewski, J. (2004). GIS-based land-use suitability analysis: A critical overview. *Progress in Planning*, *62*(1), 3–65. doi:10.1016/j.progress.2003.09.002

Mann, M., Mitchell, P., Foth, M., & Anastasiu, I. (2020). #BlockSidewalk to Barcelona: Technological sovereignty and the social license to operate smart cities. *Journal of the Association for Information Science and Technology*, *71*(9), 1103–1115. doi:10.1002/asi.24387

Manzo, C. L. (2003). Beyond house and haven: Toward a revisioning of emotional relationships with places. *Journal of Environmental Psychology*, *23*(1), 47–61. doi:10.1016/S0272-4944(02)00074-9

Marino, A., & Pariso, P. (2019). E-government and its impact on national economic development: a case study concerning Southern Italy. In *Proceedings of the 2019 3rd International Conference on E-commerce, E-Business and E-Government* (pp. 1-4). New York: ACM. 10.1145/3340017.3342242

Marino, V., & Presti, L. L. (2018). From citizens to partners: the role of social media content in fostering citizen engagement. Transforming Government: People. *Process and Policy.*, *12*(1), 39–60. doi:10.1108/TG-07-2017-0041

Marshall, S. (2012). Planning, design and the complexity of cities. In *Complexity theories of cities have come of age: An overview with implications to urban planning and design* (pp. 191–205). Springer., doi:10.7551/mitpress/11923.001.0001.

Masala, E., & Pensa, S. (2016). O papel da visualização no Planejamento Urbano: uma abordagem a partir dos conceitos por trás da imagem espacial. In *Tecnologias de geoinformação para representar e planejar o território urbano* (pp. 34–47). Interciência.

May, R. J., Maier, H. R., & Dandy, G. C. (2010). Data splitting for artificial neural networks using SOM-based stratified sampling. *Neural Networks*, *23*(2), 283–294. doi:10.1016/j.neunet.2009.11.009 PMID:19959327

McCall, M. K., & Dunn, Ch. E. (2012). Geo-information tools for participatory spatial planning: Fulfilling the criteria for 'good' governance? *Geoforum*, *43*(1), 81–94. doi:10.1016/j.geoforum.2011.07.007

McPherson, E. G., & Peper, P. J. (2012). Urban Tree Growth Modeling. *Arboriculture & Urban Forestry*, *38*, 175–183.

Měch, R., & Prusinkiewicz, P. (1996). Visual models of plants interacting with their environment. *Proceedings of the 23rd Annual Conference on Computer Graphics and Interactive Techniques - SIGGRAPH '96*, 397–410. 10.1145/237170.237279

Medaglia, R., Hedman, J., & Eaton, B. (2017). It takes two to tango: Power dependence in the governance of public-private e-government infrastructures. *Proceedings of the 38th international conference on information systems*. https://core.ac.uk/download/pdf/132283312.pdf

Meijer, A. (2016). Smart City Governance: A Local Emergent Perspective. In J. R. Gil-Garcia, T. A. Pardo, & T. Nam (Eds.), *Smarter as the New Urban Agenda: A Comprehensive View of the 21st Century City* (pp. 73–85). Springer International Publishing., doi:10.1007/978-3-319-17620-8_4

Melbourne Water Corporation. (2012). *Planning for sea level rise guidelines*. Author.

Mello, I. (2009). *O espaço da política e as políticas do espaço: Tensões entre o programa de urbanização de favelas "Vila Viva" e as práticas cotidianas no Aglomerado da Serra em Belo Horizonte* (Master's thesis). IGC, Universidade Federal de Minas Gerais (UFMG), Belo Horizonte.

Mettler, T. (2018). The road to digital and smart government in Switzerland. In A. Ladner, N. Soguel, Y. Emery, S. Weerts, & S. Nahrath (Eds.), *Swiss public administration: Making the state work successfully* (pp. 175–186). Springer.

Mikuła, Ł. (2019). *Zarządzanie rozwojem przestrzennym obszarów metropolitalnych w świetle koncepcji miękkich przestrzeni planowania. Studia i Prace z Geografii, 73*. Bogucki Wydawnictwo Naukowe.

Miller, W. R. (2012). *Introducing Geodesign: The concept Director of Geodesign services*. ESRI Press.

Ministério das Cidades. (2004). *Plano Diretor Participativo: guia para a elaboração pelos municípios e cidadãos*. Ministério das Cidades & Confea.

Ministério das Cidades. (2010). *Guia para o Mapeamento e Caracterização de Assentamentos Precários*.

Ministério do Desenvolvimento Regional. (2020). *Regularização Fundiária e Melhoria Habitacional*. https://www.gov.br/mdr/pt-br/assuntos/habitacao/casa-verde-e-amarela/regularizacao-fundiaria-e-melhoria-habitacional

Ministero della Salute. (2018). *Mappatura delle esperienze di Telemedicina sul territorio nell'anno 2018*. https://www.salute.gov.it/imgs/C_17_pagineAree_2515_2_file.pdf

Miskell, G., Salmond, J., & Williams, D. E. (2017). Low-cost sensors and crowd-sourced data: Observations of siting impacts on a network of air-quality instruments. *The Science of the Total Environment*, *575*, 1119–1129.

Mitchell, M. (2009). *Complexity: A guided tour*. Oxford University Press.

MMBW. (1979). *Elster Creek Drainage System Development* (No. D0022). Melbourne Meltropolitan Board of Works. https://www.portphillip.vic.gov.au/Elster%20Creek%20Drainage%20System%20Development_MMBW_Dec%20 1979%20(1).pdf

Mohd Sanusi, A. R. (2015). *The importance of street trees for human thermal comfort in a warming climate*. Academic Press.

Molnár, T., & Kő, A. (2020). Exploring usability and acceptance factors of m-government systems for elderly. In *Proceedings of the international conference on electronic government and the information systems perspective* (pp. 121-134). Springer. 10.1007/978-3-030-58957-8_9

Mone, G. (2015). The New Smart Cities. *Communications of the ACM, 58*(7), 20–21. doi:10.1145/2771297

Monteiro, L., Moura, A., Zyngier, C., Sena, I., & Paula, P. (2018). Geodesign Facing the Urgency of Reducing Poverty: The Cases of Belo Horizonte. *DisegnareCon, 11*(20), 6.1-6.25. Available in: http://disegnarecon.univaq.it/ojs/index.php/disegnarecon/article/view/362

Moore, J. (2011). *Exploring the moral underground: self-organizing teachers in the complex adaptive system of a school* (Doctoral dissertation). Pepperdine University.

Morawska, L. (2018). Applications of low-cost sensing technologies for air quality monitoring and exposure assessment: How far have they gone? *Environment International, 116*, 286–299.

Moreland. (2009). *Moreland Landscape Guidelines*. MCC.

Mori, M. (1970). The uncanny valley. *Energy, 7*(4), 33–35.

Moroni, S. (2014). Complexity and the inherent limits of explanation and prediction: Urban codes for self-Newman. *Landscape and Urban Planning, 44*(4), 219-226.

Moura, A. (2019). O Geodesign como processo de cocriação de acordos coletivos para a paisagem territorial e urbana. In Planejamento e gestão territorial: o papel e os instrumentos do Planejamento Territorial na Interface entre o Urbano e o Rural. Criciúma - SC: UNESC.

Moura, A., Marino, T., Ballal, H., Ribeiro, S., & Motta, S. (2016). Interoperability and visualization as a support for mental maps to face differences in scale in Brazilian Geodesign processes. *Rozwój Regionalny i Polityka Regionalna, 35*, 89–102.

Mulder, E., & Snijders, D. (2020). Playing the telephone game in a multilevel polity: On the implementation of e-government services for business in the EU. *Government Information Quarterly, 37*, 101526. doi:10.1016/j.giq.2020.101526

Murgante, B., & Borruso, G. (2015). Smart Cities in a Smart World. In S. Th. Rassia & P. M. Pardalos (Eds.), *Future City Architecture for Optimal Living* (pp. 13–35). Springer International Publishing., doi:10.1007/978-3-319-15030-7_2

Myrovali, G., Tsaples, G., Morfoulaki, M., Aifadopoulou, G., & Papathanasiou, J. (2018, May). An interactive learning environment based on system dynamics methodology for sustainable mobility challenges communication & citizens' engagement. In *International Conference on Decision Support System Technology* (pp. 88-99). Springer. 10.1007/978-3-319-90315-6_8

Nascimento, T. (2011) *Participação institucionalizada e resistência popular na intervenção em favelas: o caso Vila Antena. Belo Horizonte* (Master's thesis). NPGAU, Universidade Federal de Minas Gerais (UFMG), Belo Horizonte.

National Academies of Sciences, Engineering, and Medicine. (2021). *Population Health in Challenging Times: Insights from Key Domains. Proceedings of a Workshop*. The National Academies Press.

Niemeijer, D., & de Groot, R. S. (2008). Framing environmental indicators: Moving from causal chains to causal networks. *Environment, Development and Sustainability, 10*(1), 89–106. doi:10.100710668-006-9040-9

Nilssen, M. (2018). To the smart city and beyond? Developing a typology of smart urban innovation. *Technological Forecasting and Social Change*. Advance online publication. doi:10.1016/j.techfore.2018.07.060

Norton, B., Coutts, A., Livesley, S., Harris, R., Hunter, A., & Williams, N. (2015). Planning for cooler cities: A framework to prioritise green infrastructure to mitigate high temperatures in urban landscapes. *Landscape and Urban Planning, 134*, 127–138. doi:10.1016/j.landurbplan.2014.10.018

Novac, P. E., Upegui, A., Barrientos, D., Sousa, C., Rodriguez, L., & Miramond, B. (2020). Dynamic Structural and Computational Resource Allocation for Self-Organizing Architectures. In *2020 27th IEEE International Conference on Electronics, Circuits and Systems (ICECS)* (pp. 1-4) 10.1109/ICECS49266.2020.9294983

Nyerges, T., Ballal, H., Steinitz, C., Canfield, T., Roderick, M., Ritzman, J., & Thanatemaneerat, W. (2016). Geodesign dynamics for sustainable urban watershed development. *Sustainable Cities and Society, 25*(6), 13–24. doi:10.1016/j.scs.2016.04.016

O'Hare, E. (2021, June 10). Future Land Use Map conversations cause confusion, controversy. *Charlottesville Tomorrow*. https://www.cvilletomorrow.org/articles/future-land-use-map-conversations-cause-confusion-controversy/

O'Shea, J. (2020, September 2). *First phase of work to transform MacCurtain Street starts tomorrow*. Cork Beo. https://www.corkbeo.ie/news/local-news/work-transform-maccurtain-street-victorian-18865812

Occelli, S., & Dalmasso, M. (2019). Probing patient mobility in a regional setting: the case of Piedmont. In V. Chasles & L. De Biaggi (Eds.), Santé et territoire, une réflexion géographique (pp. 123-138). Université Jean Moulin, Lyon3.

Occelli, S., & Dalmasso, M. (2021). *Health mobility in the Piedmont region: an outlook in the middle of the tens* [Paper presentation]. MoHeaP 2021: Virtual Workshop on Mobility, Health, and Place.

Occelli, S., & Scelfo, B. (2020). Can We Enable Digital Transition in Healthcare Delivery?: Insights From a Survey of Telemedicine Services in the Piedmont Region. *International Journal of E-Planning Research, 9*(2), 43–65. doi:10.4018/IJEPR.2020040103

OECD (2019a). *Health in the 21st Century: Putting Data to Work for Stronger Health Systems*. OECD Health Policy Studies, OECD Publishing.

OECD. (2019b). Health for Everyone?: Social Inequalities in Health and Health Systems. OECD Health Policy Studies, OECD Publishing.

OECD. (2021). *State of Health in the EU. Italy Country Health Profile 2021*. https://ec.europa.eu/health/system/files/2021-12/2021_chp_it_english.pdf

Ogonek, N., & Becker, J. (2018). Can we Learn from Down Under How to Rise Up in E-Government? A Comparative Analysis of the Public Sector Competences in the German and Australian Higher Education Systems. *Proceedings of the 51st Hawaii international conference on system sciences*. https://scholarspace.manoa.hawaii.edu/bitstream/10125/50171/1/paper0284.pdf

Okesola, O. J., Okokpujie, K. O., Goddy-Worlu, R., Ogunbanwo, A., & Olamma, I. (2019). Qualitative comparisons of elicitation techniques in requirement engineering. *Qualitative Comparisons of Elicitation Techniques in Requirement Engineering, 14*(2), 565–570.

Oke, T. R. (1988). Street design and urban canopy layer climate. *Energy and Building, 11*(1), 103–113. doi:10.1016/0378-7788(88)90026-6

Olawoyin, R., Nieto, A., Grayson, R. L., Hardisty, F., & Oyewole, S. (2013). Application of artificial neural network (ANN)–self-organizing map (SOM) for the categorization of water, soil and sediment quality in petrochemical regions. *Expert Systems with Applications*, *40*(9), 3634–3648. doi:10.1016/j.eswa.2012.12.069

Oliveira, T. C. O., Barlow, J., Gonçalves, L., & Bayer, S. (2013). Teleconsultations reduce greenhouse gas emissions. *Journal of Health Services Research & Policy*, *18*(4), 209–214. doi:10.1177/1355819613492717 PMID:23945677

Ong, B. L. (2003). Green plot ratio: An ecological measure for architecture and urban planning. *Landscape and Urban Planning*, *63*(4), 197–211. doi:10.1016/S0169-2046(02)00191-3

Open Government Partnership. (n.d.). Available online: https://www.opengovpartnership.org/stories/strengthening-citizen-engagement-during-a-pandemic-a-look-at-sao-paulo-brazil/

Ortelium. (n.d.). *Smart cities & municipalities*. Available online: https://www.ortelium.com/client-solutions/smart-cities?gclid=EAIaIQobChMIy5nMmuWX8QIVEK-yCh1fgwGqEAAYAyAAEgLW0vD_BwE

Ortiz, J., Young, A., Myers, M. D., Bedeley, R. T., Carbaugh, D., Chughtai, H., ... Wigdor, A. (2019). Giving voice to the voiceless: The use of digital technologies by marginalized groups. *Communications of the Association for Information Systems*, *45*(1), 2. doi:10.17705/1CAIS.04502

Osservatorio demografico territoriale del Piemonte. (2022). https://www.demos.piemonte.it/previsioni-demografiche/indicatori-demografici

Osservatorio GIMBE. (2019). *La mobilità sanitaria interregionale nel 2017*. Fondazione GIMBE. https://www.gimbe.org/osservatorio/Report_Osservatorio_GIMBE_2019.06_Mobilita_sanitaria_2017.pdf

Ottaviano, N. (2015). Mécaniques et postures d'un dispositive numérique de projet urbain. *Revue Lieux Communs*, 17.

Otto, L., Whitehouse, D., & Schlieter, H. (2019). On the Road to Telemedicine Maturity: A Systematic Review and Classification of Telemedicine Maturity Models. In A. Pucihar, M.K. Borstnar, R.W.H. Bons, J. Seitz, H. Cripps, & D. Vidmar (Eds.), 32nd Bled eConference: Humanizing Technology for a Sustainable Society (pp. 485-506). University of Maribor Press. doi:10.18690/978-961-286-280-0.26

Otto, L., & Whitehouse, D. (2021). Empowering Communities to Support Telemedicine and its Business Model. *Health Management*, *21*(4), 219–222.

Päivärinta, T., Smolander, K., & Yli-Huumo, J. (2019). Towards Stakeholder Governance on Large E-Government Platforms-A Case of Suomi.fi. *10th Scandinavian Conference on Information Systems, 11,* 1-12. https://aisel.aisnet.org/scis2019/11

Pardal, I. (2020). E-government in Europe: On the backwardness of the Swiss government. *Futuribles*, *437*(4), 89–96. doi:10.3917/futur.437.0089

Parisi, A. V., Kimlin, M. G., Wong, J. C., & Wilson, M. (2001). Solar ultraviolet exposures at ground level in tree shade during summer in south east Queensland. *International Journal of Environmental Health Research*, *11*(2), 117–127. doi:10.1080/09603120020047500 PMID:11382345

Parisi, A. V., Kimlin, M. G., Wong, J., & Wilson, M. (2000). Personal Exposure Distribution of Solar Erythemal Ultraviolet Radiation in Tree Shade over Summer. *Physics in Medicine and Biology*, *45*(2), 349–356. doi:10.1088/0031-9155/45/2/307 PMID:10701508

Parker, J. (2007). *Strategic Environmental Assessment and SFs Operational Programmes: An assessment.* DG Environment European Commission presentation at the Aarhus Workshop on Public Participation in Strategic Decision- Making (PPSD), Sofia, Bulgaria.

Partanen, J. (2016). Liquid planning, wiki-design—Learning from the Case Pispala. *Environment and Planning. B, Planning & Design, 43*(6), 997–1018. doi:10.1177/0265813516647965

Partanen, J. (2018). *'Don't Fix It if It Ain't Broke': Encounters with Planning for Complex Self-Organizing Cities.* Tampere University of Technology.

Partanen, J. (2020). Guiding urban self-organization: Combining rule-based and case-based planning. *Environment and Planning. B, Urban Analytics and City Science, 47*(2), 304–320. doi:10.1177/2399808319893687

Partidário, M. R. (2012). *Strategic Environmental Assessment Better Practice Guide.* Agência Portuguesa do Ambiente e Redes Energéticas Nacionais.

Paschke, K., Austermann, M. I., Simon-Kutscher, K., & Thomasius, R. (2021). Adolescent gaming and social media usage before and during the COVID-19 pandemic. *Sucht: Zeitschrift für Wissenschaft und Praxis, 67*(1), 13–22. doi:10.1024/0939-5911/a000694

Paula, P., Moura, A., & Patata, S. (2018). *The application of Geodesign in a Brazilian illegal settlement: Participatory Planning in Dandara Occupation, Case Study* [Paper presentation]. Input. Available at https://www.researchgate.net/publication/327446994_The_application_of_Geodesign_in_a_Brazilian_illegal_settlement_PARTICIPATORY_PLANNING_IN_DANDARA_OCCUPATION_CASE_STUDY

Peets, J., & Leach, B. (2000). *Evaluating methods for public participation: Literature review. R&D technical Report: E135.* Environment Agency.

Penchansky, R., & Thomas, J. W. (1981). The concept of access: Definition and relationship to consumer satisfaction. *Medical Care, 19*(2), 127–140. doi:10.1097/00005650-198102000-00001 PMID:7206846

Perchoux, C., Kestens, Y., Thomas, F., Van Hulst, A., Thierry, B., & Chaix, B. (2014). Assessing patterns of spatial behavior in health studies: Their socio-demographic determinants and associations with transportation modes (the RECORD Cohort Study). *Social Science & Medicine, 119*, 64–73. doi:10.1016/j.socscimed.2014.07.026 PMID:25150652

Pérez-Morote, R., Pontones-Rosa, C., & Núñez-Chicharro, M. (2020). The effects of e-government evaluation, trust and the digital divide in the levels of e-government use in European countries. *Technological Forecasting and Social Change, 154*, 119973. doi:10.1016/j.techfore.2020.119973

Peters, D. H. (2014). The application of systems thinking in health: Why use systems thinking? *Health Research Policy and Systems, 12*(51), 1–6. doi:10.1186/1478-4505-12-51 PMID:25160707

Pettit, C., Hawken, S., Zarpelon, S., Ticzon, C., Afrooz, A., Steinitz, C., Ballal, H., Canfield, T., & Lieske, S. (2019). Breaking Down the Silos Through Geodesign—Envisioning Sydney's Urban future. *Environment and Planning. B, Urban Analytics and City Science, 46*(8), 1387–1404. doi:10.1177/2399808318812887

Pham, L., Mai, T. T., & Massey, B. (2016). Key factors for effective citizens engagement in smart city: the case of Cork City. *AIS Pre-ICIS Workshop on "IoT & Smart City Challenges and Applications.*

Pina, V., Torres, L., & Royo, S. (2017). Comparing online with offline citizen engagement for climate change: Findings from Austria, Germany and Spain. *Government Information Quarterly, 34*(1), 26–36. doi:10.1016/j.giq.2016.08.009

Pirk, S., Stava, O., Kratt, J., Said, M. A. M., Neubert, B., Měch, R., Benes, B., & Deussen, O. (2012). Plastic trees: Interactive self-adapting botanical tree models. *ACM Transactions on Graphics, 31*(4), 1–10. doi:10.1145/2185520.2185546

Polis web. (n.d.). Available online: https://pol.is/home

Polletta, F. (2016). Participatory enthusiasms: A recent history of citizen engagement initiatives. *Journal of Civil Society, 12*(3), 231–246. doi:10.1080/17448689.2016.1213505

Popoola, O. (2018). Use of networks of low cost air quality sensors to quantify air quality in urban settings. *Atmospheric Environment, 194*, 58–70.

Portugali, J. (1999). *Self-organization and the city*. Springer-Verlag.

Portugali, J. (2006). Complexity theory as a link between space and place. *Environment & Planning A, 38*(4), 647–664. doi:10.1068/a37260

Portugali, J., Meyer, H., Stolk, E., & Tan, E. (Eds.). (2012). *Complexity theories of cities have come of age: an overview with implications to urban planning and design*. Springer. doi:10.1007/978-3-642-24544-2

Prigogine, I., & Lefever, R. (1974). *Stability and self-organization in open systems*. Wiley.

Project Dynamo. (n.d.). Available online: https://civitas.eu/sites/default/files/participation_2.0_in_the_sump_process_dynamo_web.pdf

Protopappas, L., Sideridis, A., & Yialouris, C. (2020). Implementation issues of cross border e-government systems and services. In Proceedings of the 9th international conference on information and communications technologies in agriculture, food and the environment. Thessalonika. http://ceur-ws.org/Vol-2761/HAICTA_2020_paper25.pdf

Prusinkiewicz, P., Lindenmayer, A., & Hanan, J. (1990). *The Algorithmic Beauty of Plants*. The Virtual Laboratory. doi:10.1007/978-1-4613-8476-2

Purohit, A., Smith, J., & Hibble, A. (2021). Does telemedicine reduce the carbon footprint of healthcare? A systematic review. *Future Healthcare Journal, 8*(1), 85–91. doi:10.7861/fhj.2020-0080 PMID:33791483

Rantanen, H., & Kahila, M. (2008). The SoftGIS approach to local knowledge. *Journal of Environmental Management, 90*(6), 1981–1990. doi:10.1016/j.jenvman.2007.08.025 PMID:18556113

Rebollo, C. (2007). *An efficient continuous level of detail model for foliage*. Academic Press.

Rebollo, C., Remolar, I., Chover, M., & Ripollés, O. (2006). An Efficient Continuous Level of Detail Model for Foliage. *WSCG'2006*.

Rebollo, C., Remolar, I., Gumbau, J., & Chover, M. (2014). Three-dimensional trees for virtual globes. *International Journal of Digital Earth, 7*(10), 789–810. doi:10.1080/17538947.2013.783126

Reddel, T., & Woolcock, G. (2004). From consultation to participatory governance? A critical review of citizen engagement strategies in Queensland. *Australian Journal of Public Administration, 63*(3), 75–87. doi:10.1111/j.1467-8500.2004.00392.x

Reed, J., Cook, G., Bolter, V., & Douglas, B. (2006). *Older people 'getting things done 'Involvement in policy and planning initiatives*. Academic Press.

Rega, C., & Baldizzone, G. (2015). Public participation in Strategic Environmental Assessment: A practitioners' perspective. *Environmental Impact Assessment Review, 50*, 105–115. doi:10.1016/j.eiar.2014.09.007

Rega, C., & Bonifazi, A. (2014). Strategic Environmental Assessment and spatial planning in Italy: Sustainability, integration and democracy. *Journal of Environmental Planning and Management, 57*(9), 1333–1358. doi:10.1080/09640568.2013.804404

Rhodes, R. A. W. (2000). *Debating Governance*. Oxford University Press.

Richard, E., & David, L. F. (2018). The future of citizen engagement in cities—The council of citizen engagement in sustainable urban strategies (ConCensus). *Futures*, *101*, 80–91. doi:10.1016/j.futures.2018.06.012

Riddlesden, D., & Singleton, A. (2014). Broadband speed equity: A new digital divide? *Applied Geography*, *52*, 25–33. doi:10.1016/j.apgeog.2014.04.008

Ridell, S. (2019). Mediated bodily routines as infrastructure in the algorhythmic city. *Media Theory*, *3*(2), 27–62.

Riesbeck, C. K., & Schank, R. C. (2013). *Inside case-based reasoning*. Psychology Press. doi:10.4324/9780203781821

Rivero, R., Smith, A., Ballal, H., & Steinitz, C. (2015) Promoting Collaborative Geodesign in a Multidisciplinary and Multiscale Environment: Coastal Georgia 2050, USA. *Proceedings of the 2015 Digital Landscape Architecture Conference*.

Rizmal, Z. (2019, October 24). *Hate Melbourne's plane trees? Relief is coming, but don't hold your breath*. ABC News. https://www.abc.net.au/news/2019-10-24/plane-trees-to-be-culled-in-melbourne-city/11635846

RK&K Consultants. (2016). *City of Charlottesville Pedestrian Lighting Study Improvement Plan Report*. http://www.charlottesville.org/home/showdocument?id=48739

Robinson, Y., Paraskevopoulou, A., & Hollingworth, S. (2019). Developing 'active citizens': Arts Award, creativity and impact. *British Educational Research Journal*, *45*(6), 1203–1219. doi:10.1002/berj.3558

Robles, J., Torres-Albero, C., & Villarino, G. (2021). Inequalities in digital welfare take-up: Lessons from e-government in Spain. *Policy Studies*, *42*(3), 1–16. doi:10.1080/01442872.2021.1929916

Rocheleau, B. (2007). Whither e-government? *Public Administration Review*, *67*(3), 584–588. doi:10.1111/j.1540-6210.2007.00741_1.x

Rodriguez-Hevía, L., Navío-Marco, J., & Ruiz-Gómez, L. (2020). Citizens' involvement in e-government in the European Union: The rising importance of the digital skills. *Sustainability*, *12*(17), 6807. doi:10.3390u12176807

Rogers, B., Bertram, N., Gunn, A., Löwe, R., Murphy, C., Pasman, R., Radhakrishnan, M., Urich, C., Wong, T., & Arnbjerg-Nielsen, K. (2016). An interdisciplinary approach to identify adaptation strategies that enhance flood resilience and urban liveability. *IWA World Water Congress & Exhibition*.

Rouse, W. B., & Serban, N. (2014). *Understanding and Managing the Complexity of Healthcare*. The MIT Press.

Ruhlandt, R. W. S. (2018). The governance of smart cities: A systematic literature review. *Cities (London, England)*, *81*, 1–23. https://doi.org/10.1016/j.cities.2018.02.014

Runhaar, H. (2009). Putting SEA in context: A discourse perspective on how SEA contributes to decision-making. *Environmental Impact Assessment Review*, *29*(3), 200–209. doi:10.1016/j.eiar.2008.09.003

Salha, R., El-Hallaq, M., & Alastal, A. (2019). Blockchain in Smart Cities: Exploring Possibilities in Terms of Opportunities and Challenges. *Journal of Data Analysis and Information Processing*, *7*(03), 118–139. doi:10.4236/jdaip.2019.73008

Salingaros, N. A. (2000). Complexity and urban coherence. *Journal of Urban Design*, *5*(3), 291–31. doi:10.1080/713683969

Salingaros, N. A. (2012). Urbanism as Computation. In *Complexity Theories of Cities Have Come of Age* (pp. 245–268). Springer. doi:10.1007/978-3-642-24544-2_14

Salomão, T. (2016) *Linguagem Técnica e (Im)Possibilidades para a Produção Democrática do Espaço Urbano: uma análise a partir de duas experiências participativas em Belo Horizonte* (Master's thesis). UFMG. Belo Horizonte.

Santos, O., & Montandon, D. (2011). *Os Planos Diretores Municipais pós-estatuto da cidade: balanço crítico e perspectivas*. Letra Capital, Observatório das Cidades & Ippur/UFRJ.

Santoveña-Casal, S., Gil-Quintana, J., & Ramos, L. (2021). *Digital Citizens' Feelings in National #Covid19 Campaigns in Spain*. doi:10.20944/preprints202103.0734.v1

Sarriot, E., & Kouletio, M. (2015). Community health systems as complex adaptive systems: Ontology and praxis lessons from an urban health experience with demonstrated sustainability. *Systemic Practice and Action Research*, *28*(3), 255–272. doi:10.100711213-014-9329-9

Satorras, M., Ruiz-Mallén, I., Monterde, A., & March, H. (2020). Co-production of urban climate planning: Insights from the Barcelona Climate Plan. *Cities (London, England)*, *106*, 102887. doi:10.1016/j.cities.2020.102887

Saunders, J. R., & Shackelford, R. N. (2005). Urban Renewal and the End of Black Culture in Charlottesville, Virginia: An Oral History of Vinegar Hill. McFarland.

Sawicki, D. S., & Craig, W. J. (1996). The Democratization of Data: Bridging the Gap for Community Groups. *Journal of the American Planning Association*, *62*(4), 512–523. https://doi.org/10.1080/01944369608975715

Schank, R. C. (1983). *Dynamic memory: A theory of reminding and learning in computers and people*. Cambridge university press.

Schank, R., Riesbeck, C., & Kass, A. (1994). *Inside Case-Based Explanation*. Lawrence Erlbaum Associates.

Schauer, F. (1991). *Playing by the rules: A philosophical examination of rule-based decision-making in law and in life*. Clarendon Press.

Schelling, T. C. (1969). Models of segregation. *The American Economic Review*, *59*(2), 488–493.

Schroeder, P. (1996). Criteria for the design of a GIS/2. *Specialists' meeting for NCGIA Initiative 19: GIS and society*.

See, L., Mooney, P., Foody, G., Bastin, L., Comber, A., Estima, J., Fritz, S., Kerle, N., Jiang, B., Laakso, M., Liu, H.-Y., Milčinski, G., Nikšič, M., Painho, M., Pődör, A., Olteanu-Raimond, A.-M., Rutzinger, M., See, L., Mooney, P., … Rutzinger, M. (2016). Crowdsourcing, Citizen Science or Volunteered Geographic Information? The Current State of Crowdsourced Geographic Information. *ISPRS International Journal of Geo-Information*, *5*(5), 55. doi:10.3390/ijgi5050055

See, L. (2016). Crowdsourcing, Citizen Science or Volunteered Geographic Information? The Current State of Crowdsourced Geographic Information. *International Journal of Géo-Information*, *5*(55), 1–23.

Seljan, S., Miloloza, I., & Pejicbach, M. (2020). E-government in European countries: Gender and ageing digital divide. *Interdisciplinary Management Research*, *16*, 1563–1584.

Shane, D. G. (2011). *Urban Design since 1945. A Global Perspective*. Wiley.

Shashua-Bar, L., Tsiros, I. X., & Hoffman, M. E. (2010). A Modeling Study for Evaluating Passive Cooling Scenarios in Urban Streets with Trees. Case Study: Athens, Greece. *Building and Environment*, *45*(12), 2798–2807. doi:10.1016/j.buildenv.2010.06.008

Shehu, J., & Xhina, E. (2019). Modeling an ontology for public e-government services in Albania. *International Academic Conference on Research in Engineering and Technology*. https://www.dpublication.com/wp-content/uploads/2019/09/620-R.pdf

Shkarlet, S., Oliychenko, I., Dubyna, M., Ditkovska, M., & Zhovtok, V. (2020). Comparative analysis of best practices in e-Government implementation and use of this experience by developing countries. *Administratie si Management Public*, *34*, 118–136. doi:10.24818/amp/2020.34-07

Sideridis, A., Protopappas, L., Tsiafoulis, S., & Pimenidis, E. (2017). Smart cross-border e-gov systems: An application to refugee mobility. *International Journal of Electronic Governance*, *9*(3-4), 246–266. doi:10.1504/IJEG.2017.088219

Sieber, R. (2006). Public Participation Geographic Information Systems: A Literature Review and Framework. *Annals of the Association of American Geographers*, *96*(3), 491–507. doi:10.1111/j.1467-8306.2006.00702.x

Siemiński W. (2014). Partycypacja spontaniczna a kształtowanie przestrzeni. *Partycypacja społeczna w planowaniu przestrzennym, Konferencja Towarzystwa Urbanistów Polskich Oddziału w Warszawie oraz Biura Architektury i Planowania Przestrzennego Urzędu m.st. Warszawy.*

Silva, M. (2013) *Água em meio urbano, favelas nas cabeceiras* (Doctoral dissertation). NPGAU, Universidade Federal de Minas Gerais (UFMG), Belo Horizonte.

Slaev, A. D., Kovachev, A., Nozharova, B., Daskalova, D., Nikolov, P., & Petrov, P. (2019). Overcoming the failures of citizen participation: The relevance of the liberal approach in planning. *Planning Theory*, *18*(4), 448–469. doi:10.1177/1473095219848472

Slotterback, C. S., & Lauria, M. (2019). Building a Foundation for Public Engagement in Planning. *Journal of the American Planning Association*, *85*(3), 183–187. https://doi.org/10.1080/01944363.2019.1616985

Smeets, E., & Weterings, R. (1999). *Environmental indicators: Typology and overview*. Copenhagen: European Environment Agency, EEA Technical report No 25.

Spencer, H. (2018). *Summer of Hate*. University of Virginia Press. https://www.upress.virginia.edu/title/5323

Sprague, R. H. (1980). A Framework for the Development of Decision Support Systems. *Management Information Systems Quarterly*, *4*(4), 1. doi:10.2307/248957

Standard Deviation and Variance. (n.d.). Available online: https://www.investopedia.com/terms/

Star, S. L., & Griesemer, J. R. (1989). Institutional Ecology, "Translations" and Boundary Objects: Amateurs and Professionals in Berkeley's Museum of Vertebrate Zoology. *Social Studies of Science*, *19*(3), 387–420.

Steiniger, S., & Hunter, A. J. (2012). Free and open source GIS software for building a spatial data infrastructure. *Geospatial free and open source software in the 21st century*, 247-261.

Steinitz, C. (2012). *A Framework for Geodesign. Changing Geography by Design*. Esri Press.

Steinitz, C. (2012). *A Framework for Geodesign: Changing Geography By Design*. ESRI Press.

Steintz, C. (2012). *A Framework for Geodesign*. ESRI Press.

Stevens, R. G., Brainard, G. C., Blask, D. E., Lockley, S. W., & Motta, M. E. (2013). Adverse Health Effects of Nighttime Lighting: Comments on American Medical Association Policy Statement. *American Journal of Preventive Medicine*, *45*(3), 343–346. https://doi.org/10.1016/j.amepre.2013.04.011

Stojanovski, T., Partanen, J., Samuels, I., Sanders, P., & Peters, C. (2020). City Information Modelling (CIM) and Digitizing Urban Design Practices. *Built Environment*, *46*(4), 637–646. doi:10.2148/benv.46.4.637

Stratigea, A. (2012). The concept of 'smart cities'. Towards community development*? Netcom. Réseaux, communication et territoires, 26*(3/4), 375-388. https://journals.openedition.org/netcom/1105

Stratu-Strelet, D., Gil-Gómez, H., Oltra-Badenes, R., & Oltra-Gutierrez, J. (2021). Critical factors in the institutionalization of e-participation in e-government in Europe: Technology or leadership? *Technological Forecasting and Social Change*, *164*, 120489. doi:10.1016/j.techfore.2020.120489

Streich, B. (2004). Information, Wissen, Planungsprozesse und Macht, Die Wissengesllschaft als Herausforderung an die Planungsethik. In Ethik in der Raumplanung, Zagänge und Reflexionen Akademie für Raumforschung und Landesplanung, Forschungs - und Sitzungsberichte (pp. 31-49). Hannover: Band 221.

Strouhal, J., Horák, J., Bokšová, J., Bokša, M., & Pavlica, K. (2020). Corporate perspective on development of Czech e-government services. In *Proceedings of the 13th International Scientific Conference* (pp. 114-125). https://fes.upce.cz/sites/default/files/public/rast3126/public_administration_2020_155249.pdf#page=115

Suarez, C. (2017, August 23). Stop-and-frisk data show African-Americans disproportionately detained. *The Daily Progress*. https://www.dailyprogress.com/news/local/stop-and-frisk-data-show-african-americans-disproportionately-detained/article_1b606292-886e-11e7-960b-57d65b2b63d4.html

Sullivan, H., & Skelcher, C. K. (2002). *Managing across Boundaries*. Palgrave Macmillan. doi:10.1007/978-1-4039-4010-0

Sustainable Urban Mobility Plan Bremen 2025. (n.d.). Available online: https://www.bauumwelt.bremen.de/sixcms/media.php/13/SUMP_Bremen2025_web.pdf

Szopiński, T., & Staniewski, M. (2017). Manifestations of e-government usage in post-communist European countries. *Internet Research*, *27*(2), 199–210. doi:10.1108/IntR-01-2015-0011

Szota, C., McCarthy, M. J., Sanders, G. J., Farrell, C., Fletcher, T. D., Arndt, S. K., & Livesley, S. J. (2018). Tree water-use strategies to improve stormwater retention performance of biofiltration systems. *Water Research*, *144*, 285–295. doi:10.1016/j.watres.2018.07.044 PMID:30048867

Taipale, S. (2013). The use of e-government services and the Internet: The role of socio-demographic, economic and geographical predictors. *Telecommunications Policy*, *47*(4-5), 413–422. doi:10.1016/j.telpol.2012.05.005

Taleb, N. N. (2010). The Black Swan: Second Edition: The Impact of the Highly Improbable. Random House.

Tan, E., & Portugali, J. (2012). The responsive city design game. In *Complexity Theories of Cities have come of Age* (pp. 369–390). Springer. doi:10.1007/978-3-642-24544-2_20

Tanwar, S., Popat, A., Bhattacharya, P., Gupta, R., & Kumar, N. (2021). A taxonomy of energy optimization techniques for smart cities: Architecture and future directions. *Expert Systems: International Journal of Knowledge Engineering and Neural Networks*, 12703. doi:10.1111/exsy.12703

Taylor, M. S., Wheeler, B. W., White, M. P., Economou, T., & Osborne, N. J. (2015). Research note: Urban street tree density and antidepressant prescription rates—A cross-sectional study in London, UK. *Landscape and Urban Planning*, *136*, 174–179. doi:10.1016/j.landurbplan.2014.12.005

te Brömmelstroet, M. (2013). Performance of planning support systems: What is it, and how do we report on it? *Computers, Environment and Urban Systems*, *41*, 299–308. doi:10.1016/j.compenvurbsys.2012.07.004

te Brömmelstroet, M., & Schrijnen, P. M. (2010). From planning support systems to mediated planning support: A structured dialogue to overcome the implementation gap. *Environment and Planning. B, Planning & Design*, *37*(1), 3–20. doi:10.1068/b35019

Terzidis, K. (2006). *Algorithmic architecture*. Elsevier. doi:10.4324/9780080461298

The Johns Hopkins ACG® System. (2014). *Excerpt from Version 11.0 Technical Reference Guide*. https://www2.gov.bc.ca/assets/gov/health/conducting-health-research/data-access/johns-hopkins-acg-system-technical-reference-guide.pdf

The Urban Mobility Observatory. (n.d.). Available online: https://www.eltis.org/mobility-plans/sump-guidelines

Thévenot, L. (2011). Grand résumé de L'Action au pluriel. Sociologie des régimes d'engagement, Paris, Editions La Découverte, 2006. Sociologies, 1–13.

Thévenot, L. (2006). *L'action au pluriel : sociologie des régimes d'engagement*. Edititions la Découverte.

Thompson, S. C. G., & Barton, M. A. (1994). Ecocentric and anthropocentric attitudes toward the environment. *Journal of Environmental Psychology*, *14*(2), 149–157. doi:10.1016/S0272-4944(05)80168-9

Topçu, S. (2008). Le nucléaire à l'épreuve de l'expertise citoyenne. Quels savoirs? Quels enjeux. In S. Topçu, C. Cuny, & K. Serrano-Verlade (Eds.), *Savoir en débat. Perspectives franco-allemandes* (pp. 146–169). Editions l'Harmattan.

Turner, F. (2012). *From counterculture to cyberculture. Stewart Brand, the Whole Earth network, and the rise of digital utopianism*. University of Chicago Press.

Urbel. (2021). *PRU: Plano de Regularização Urbanística*. https://prefeitura.pbh.gov.br/urbel/pru-plano-de-regularizacao-urbanistica

Urich, C., Bach, P. M., Sitzenfrei, R., Kleidorfer, M., McCarthy, D. T., Deletic, A., & Rauch, W. (2013). Modelling cities and water infrastructure dynamics. *Proceedings of the Institution of Civil Engineers. Engineering Sustainability*, *166*(5), 301–308. doi:10.1680/ensu.12.00037

Urich, C., & Rauch, W. (2014). Exploring critical pathways for urban water management to identify robust strategies under deep uncertainties. *Water Research*, *66*, 374–389. doi:10.1016/j.watres.2014.08.020 PMID:25240118

Valverde, M., & Flynn, A. (2020). *Smart Cities in Canada: Digital Dreams, Corporate Designs: Independent experts analyze often-controversial schemes from Nunavut to Montreal to Toronto's failed Sidewalk Labs waterfront scheme*. James Lorimer & Company.

Van AekenK. (2016). Digital democracy in Belgium and the Netherlands. A socio-legal analysis of citizenlab. be and consultatie. nl. Koen Van Aeken, Digital Democracy in Belgium and the Netherlands. A Socio-Legal Analysis of Citizenlab. be and Consultatie. nl. In *Digital Democracy in a Globalized World*. https://ssrn.com/abstract=2932962 doi:10.2139/ssrn.2932962

Van Den Brink, A. (2007). *Imaging the future geovisualization for participatory spatial planning in Europe*. Wageningen Academic Publishers. doi:10.3920/978-90-8686-625-0

van Dyk, L. A. (2014). Review of telehealth service implementation frameworks. *International Journal of Environmental Research and Public Health*, *11*(2), 1279–1298. doi:10.3390/ijerph110201279 PMID:24464237

Vandenbroucke, D., & Biliouris, D. (2011). Spatial Data Infrastructures in Italy: State of play spring 2011. K.U. Leuven (SADL + ICRI).

Vanolo, A. (2014). Smartmentality: The smart city as disciplinary strategy. *Urban Studies (Edinburgh, Scotland)*, *51*(5), 883–898.

Velicogna, M. (2018). e-Justice in Europe: From national experiences to EU cross-border service provision. In International E-Government Development (pp. 39-72). Cham: Palgrave Macmillan.

Veliu, S., & Kocibellinj, M. (2018). The implementation of e-government in Albania. *Ditet e Studimeve Shqiptare*, 87-192. https://uet.edu.al/new/wp-content/uploads/2021/06/Proceedings-Book-1.pdf#page=187

Viale Pereira, G., Cunha, M. A., Lampoltshammer, T. J., Parycek, P., & Testa, M. G. (2017). Increasing collaboration and participation in smart city governance: A cross-case analysis of smart city initiatives. *Information Technology for Development*, *23*(3), 526–553. doi:10.1080/02681102.2017.1353946

Vonk, G. A. (2006). *Improving planning support: the use of planning support systems for spatial planning*. KNAG/ Netherlands Geographical Studies.

Vonk, G., & Geertman, S. (2008). Improving the adoption and use of planning support systems in practice. *Applied Spatial Analysis and Policy*, *1*(3), 153–173. doi:10.100712061-008-9011-7

Vonk, G., Geertman, S., & Schot, P. P. (2005). Bottlenecks blocking widespread usage of planning support systems. *Environment & Planning A*, *37*(5), 909–924. doi:10.1068/a3712

VoteIt web. (n.d.). Available online: https://www.voteit.se/in-english/

Waeterloos, C., De Meulenaere, J., Walrave, M., & Ponnet, K. (2021). Tackling COVID-19 from below: civic participation among online neighbourhood network users during the COVID-19 pandemic. *Online Information Review*. https://www.emerald.com/insight/content/doi/10.1108/OIR-08-2020-0379/full/html

Walker, B., & Salt, D. (2012). *Resilience practice: building capacity to absorb disturbance and maintain function*. Island Press. doi:10.5822/978-1-61091-231-0

Warf, B. (2018). Geographies of e-government in Europe. *International Journal of E-Planning Research*, *7*(4), 61–77. doi:10.4018/IJEPR.2018100104

Washington Post. (2014). Timeline: The search for Hannah Graham. *Washington Post*. https://www.washingtonpost.com/apps/g/page/local/timeline-the-search-for-hannah-graham/1332/

Weber, M. (1917). *Le Savant et le Politique*. 10x18.

Weber, J., & Penn, J. (1995). Creation and rendering of realistic trees. *Proceedings of the 22nd Annual Conference on Computer Graphics and Interactive Techniques*, 119–128.

Weinstock, M. (2004). Morphogenesis and the Mathematics of Emergence. *Architectural Design*, (3), 10-17.

White, M., & Kimm, G. (2015). Measuring sky view factor of urban canyons using hacked Gopro hemispheric video processing. In R. H. Crawford & A. Stephan (Eds.), *Research for a Better Built Environment: 49th International Conference of the Architectural Science Association 2015* (pp. 525–535). Faculty of Architecture, Building and Planning (UniMelb). http://apo.org.au/node/61762

White, M., & Langenheim, N. (2018). A Spatio-Temporal Decision Support System for Designing with Street Trees. *International Journal of E-Planning Research*, *7*(4), 1–24. doi:10.4018/IJEPR.2018100101

WHO/Europe. (2013). *Review of evidence on health aspect of air pollution - REVIHAAP Project*. Author.

Widholm, A., & Appelgren, E. (2020). A softer kind of hard news? Data journalism and the digital renewal of public service news in Sweden. *New Media & Society*.

Wiench J. (1991). *Metody i techniki konsultacji społecznych w planowaniu przestrzennym*. IGPiK.

Williams, I., Falch, M., & Tadayoni, R. (2018b). Internationalization of e-government services. In *2018 11th CMI international conference: Prospects and challenges towards developing a digital economy within the EU* (pp. 19-31). IEEE. 10.1109/PCTDDE.2018.8624828

Williams, I., Falch, M., & Tadayoni, R. (2018a). Institutional legitimacy and digital public cross-border service delivery between Denmark/Sweden and Denmark/Germany. *Journal of NBICT*, *1*, 1–30.

Wilson, B. B. (2018). *Resilience for All: Striving for Equity Through Community-Driven Design*. Island Press.

Wilson, B., & Chakraborty, A. (2019). Planning Smart(er) Cities: The Promise of Civic Technology. *Journal of Urban Technology*, *0*(0), 1–23. https://doi.org/10.1080/10630732.2019.1631097

Wolfram, S. (1984). Cellular automata as models of complexity. *Nature*, *311*(5985), 419–424. doi:10.1038/311419a0

World Health Organization. (2016a). *Defining evaluation indicators for telemedicine as a tool for reducing health inequities.* Pan American Health Organization. https://iris.paho.org/bitstream/handle/10665.2/28562/PAHOKBR16006eng.pdf?sequence=8

World Health Organization. (2016b). *Integrated care models: an overview.* https://www.euro.who.int/en/health-topics/Health-systems/health-services- delivery/publications/2016/integrated-care-models-an-overview-2016

World Justice Project. (n.d.). Available online: https://worldjusticeproject.org/our-work/programs/democracyos

Xia, T., Jatowt, A., Wang, Z., Si, R., Zhang, H., Liu, X., Shibasaki, R., Song, X., & Kim, K. (2020). CoolPath: An Application for Recommending Pedestrian Routes with Reduced Heatstroke Risk. In S. Di Martino, Z. Fang, & K.-J. Li (Eds.), *Web and Wireless Geographical Information Systems* (pp. 14–23). Springer International Publishing. doi:10.1007/978-3-030-60952-8_2

Yager, J. (2021). *Mapping Cville.* https://mappingcville.com/author/jordyyager/

Yang, B., Li, S., Elder, B. R., & Wang, Z. (2013). Community-planning approaches and residents' perceived safety: A landscape analysis of park design in the Woodlands, Texas. *Journal of Architectural and Planning Research*, *30*(3), 311–327.

Yang, Y., Gatto, E., Gao, Z., Buccolieri, R., Morakinyo, T. E., & Lan, H. (2019). The "plant evaluation model" for the assessment of the impact of vegetation on outdoor microclimate in the urban environment. *Building and Environment*, *159*, 106151. doi:10.1016/j.buildenv.2019.05.029

Yates, J. (2017). Saving the Soul of the Smart City. *The Hedgehog Review: Critical Reflections on Contemporary Culture*, *19*(2), 18–35.

Yeh, H. (2017). The effects of successful ICT- based smart city services: From citizens' perspectives. *Government Information Quarterly*, *34*(3), 556-565. doi:10.1016/j.giq.2017.05.001

Yera, A., Arbelaitz, O., Jauregui, O., & Muguerza, J. (2020). Characterization of e-government adoption in Europe. *PLoS One*, *15*(4). doi:10.1371/journal.pone.0231585 PMID:32302326

Yigitcanlar, T., & Baum, S. (2006). E-government and the digital divide. In M. Khosrow-Pour (Ed.), *Encyclopedia of e-commerce, e-government, and mobile commerce* (pp. 353–358). IGI Global. doi:10.4018/978-1-59140-799-7.ch058

Yigitcanlar, T., Kamruzzaman, Md., Buys, L., Ioppolo, G., Sabatini-Marques, J., da Costa, E. M., & Yun, J. J. (2018). Understanding 'smart cities': Intertwining development drivers with desired outcomes in a multidimensional framework. *Cities (London, England)*, *81*, 145–160. https://doi.org/10.1016/j.cities.2018.04.003

Yildiz, M. (2007). E-government research: Reviewing the literature, limitations and ways forward. *Government Information Quarterly*, *24*(3), 646–665. doi:10.1016/j.giq.2007.01.002

Yin, R. K. (2003). *Case study research. Design and Methods Applied Social Research Methods Series.* International Educational and Professional Publisher.

Yoon, A., & Copeland, A. (2020). Toward community-inclusive data ecosystems: Challenges and opportunities of open data for community-based organizations. *Journal of the Association for Information Science and Technology*, *71*(12), 1439–1454. https://doi.org/10.1002/asi.24346

Zellner, M., Lyons, L., Hoch, C. J., Weizeorick, J., Kunda, C., & Milz, D. C. (2012). Modeling, learning, and Planning together: An Application of Participatory Agent-based Modeling to Environmental Planning. *URISA Journal*, *1*, 77–92.

Zhang, Z., Mondschein, A., & El Khafif, M. (2018). Human-Tool Assemblage: Designers in the Big Data World. *Journal of Digital Landscape Architecture*, *3*, 397–405.

Zoppi, C. (2012). Valutazione come sostegno all'efficacia del piano. In Valutazione e pianificazione delle trasformazioni territoriali nei processi di governance ed e-governance–Sostenibilità ed e-governance nella pianificazione del territorio. Milano: FrancoAngeli.

Zoroja, J., Strugar, I., & Jaković, B. (2020). E-government in Croatia: Perceived quality and obstacles among younger users. *International Journal of E-Services and Mobile Applications*, *12*(3), 85–96. doi:10.4018/IJESMA.2020070105

Zwattendorfer, Stranacher, Tauber, & Reichstädter. (2013). Cloud computing in e-government across europe. In *International conference on electronic government and the information systems perspective* (pp. 181-195). Springer.

Zyngier, C., Moura, A., Araújo, R., & Carsalade, F. (2016). Geodesign in Pampulha cultural and heritage urban area: Visualization tools to orchestrate urban growth and dynamic transformations. *Rozwój Regionalny i Polityka Regionalna*, *35*, 103–117.

Related References

To continue our tradition of advancing information science and technology research, we have compiled a list of recommended IGI Global readings. These references will provide additional information and guidance to further enrich your knowledge and assist you with your own research and future publications.

Abdel-Hameid, S. O., & Wilson, E. (2018). Gender, Organization, and Change in Sudan. In N. Mahtab, T. Haque, I. Khan, M. Islam, & I. Wahid (Eds.), *Handbook of Research on Women's Issues and Rights in the Developing World* (pp. 107–120). Hershey, PA: IGI Global. doi:10.4018/978-1-5225-3018-3.ch007

Abdullahi, R. B. (2018). Volunteerism in Urban Development the Case of Non-Cash, Non-Digital Crowdfunding Growth in Nigeria. In U. Benna & A. Benna (Eds.), *Crowdfunding and Sustainable Urban Development in Emerging Economies* (pp. 188–210). Hershey, PA: IGI Global. doi:10.4018/978-1-5225-3952-0.ch010

Abioye, T. O., Oyesomi, K., Ajiboye, E., Omidiora, S., & Oyero, O. (2017). Education, Gender, and Child-Rights: Salient Issues in SDGS Years in ADO-ODO/OTA Local Government Area of Ogun State, Nigeria. In O. Nelson, B. Ojebuyi, & A. Salawu (Eds.), *Impacts of the Media on African Socio-Economic Development* (pp. 141–154). Hershey, PA: IGI Global. doi:10.4018/978-1-5225-1859-4.ch009

Adalı, G. K. (2022). Measuring the Attitudes of Governmental Policies and the Public Towards the COVID-19 Pandemic. In Ş. Omeraki Çekirdekci, Ö. İngün Karkış, & S. Gönültaş (Eds.), *Handbook of Research on Interdisciplinary Perspectives on the Threats and Impacts of Pandemics* (pp. 163–187). IGI Global. https://doi.org/10.4018/978-1-7998-8674-7.ch009

Adisa, W. B. (2018). Land Use Policy and Urban Sprawl in Nigeria: Land Use and the Emergence of Urban Sprawl. In A. Eneanya (Ed.), *Handbook of Research on Environmental Policies for Emergency Management and Public Safety* (pp. 256–274). Hershey, PA: IGI Global. doi:10.4018/978-1-5225-3194-4.ch014

Afolabi, O. S., Amao-Kolawole, T. G., Shittu, A. K., & Oguntokun, O. O. (2018). Rule of Law, Governance, and Sustainable Development: The Nigerian Perspective. In K. Teshager Alemu & M. Abebe Alebachew (Eds.), *Handbook of Research on Sustainable Development and Governance Strategies for Economic Growth in Africa* (pp. 273–290). Hershey, PA: IGI Global. doi:10.4018/978-1-5225-3247-7.ch015

Agyemang, O. S. (2018). Institutional Structures and the Prevalence of Foreign Ownership of Firms: Empirical Evidence From Africa. In K. Teshager Alemu & M. Abebe Alebachew (Eds.), *Handbook of Research on Sustainable Development and Governance Strategies for Economic Growth in Africa* (pp. 455–479). Hershey, PA: IGI Global. doi:10.4018/978-1-5225-3247-7.ch024

Aham-Anyanwu, N. M., & Li, H. (2017). E-State: Realistic or Utopian? *International Journal of Public Administration in the Digital Age*, *4*(2), 56–76. doi:10.4018/IJPADA.2017040105

Al Balushi, T., & Ali, S. (2020). Theoretical Approach for Instrument Development in Measuring User-Perceived E-Government Service Quality: A Case of Oman E-Government Services. *International Journal of Electronic Government Research*, *16*(1), 40–58. doi:10.4018/IJEGR.2020010103

Al-Jamal, M., & Abu-Shanab, E. (2018). Open Government: The Line between Privacy and Transparency. *International Journal of Public Administration in the Digital Age*, *5*(2), 64–75. doi:10.4018/IJPADA.2018040106

Alsaç, U. (2017). EKAP: Turkey's Centralized E-Procurement System. In R. Shakya (Ed.), *Digital Governance and E-Government Principles Applied to Public Procurement* (pp. 126–150). Hershey, PA: IGI Global. doi:10.4018/978-1-5225-2203-4.ch006

Amadi, L. A., & Igwe, P. (2018). Open Government and Bureaucratic Secrecy in the Developing Democracies: Africa in Perspective. In A. Kok (Ed.), *Proliferation of Open Government Initiatives and Systems* (pp. 1–28). Hershey, PA: IGI Global. doi:10.4018/978-1-5225-4987-1.ch001

Arble, E., & Arnetz, B. B. (2021). Four Fundamental Principles to Enhance Police Performance and Community Safety. In E. Arble & B. Arnetz (Eds.), *Interventions, Training, and Technologies for Improved Police Well-Being and Performance* (pp. 231–245). IGI Global. https://doi.org/10.4018/978-1-7998-6820-0.ch015

Arora, G. C. (2020). The Problem of Climate-Induced Displacement: Analyzing the International Framework on Protection of Rights of Climate Migrants. In R. Das & N. Mandal (Eds.), *Interdisciplinary Approaches to Public Policy and Sustainability* (pp. 67–82). IGI Global. https://doi.org/10.4018/978-1-7998-0315-7.ch004

Arora, T., & Mehra, N. (2021). Administration of Civil Justice in India: Ancient and Modern Perspectives. In E. Yin & N. Kofie (Eds.), *Advancing Civil Justice Reform and Conflict Resolution in Africa and Asia: Comparative Analyses and Case Studies* (pp. 17–45). IGI Global. https://doi.org/10.4018/978-1-7998-7898-8.ch002

Arteta, A. (2019). *Democracy and Government: Making Decisions*. IGI Global. doi:10.4018/978-1-5225-7558-0.ch001

Asante, M., & Botchway, T. P. (2021). Shielding Members of Parliament Against Court Summons: Interrogating the Question of Parliamentary Immunity. In E. Yin & N. Kofie (Eds.), *Advancing Civil Justice Reform and Conflict Resolution in Africa and Asia: Comparative Analyses and Case Studies* (pp. 210–229). IGI Global. https://doi.org/10.4018/978-1-7998-7898-8.ch012

Ateş, V. (2021). Critical Antecedents of Trust in E-Government Services in Turkey. In C. Babaoğlu, E. Akman, & O. Kulaç (Eds.), *Handbook of Research on Global Challenges for Improving Public Services and Government Operations* (pp. 94–116). IGI Global. https://doi.org/10.4018/978-1-7998-4978-0.ch006

Ayeni, A. O. (2018). Environmental Policies for Emergency Management and Public Safety: Implementing Green Policy and Community Participation. In A. Eneanya (Ed.), *Handbook of Research on Environmental Policies for Emergency Management and Public Safety* (pp. 40–59). Hershey, PA: IGI Global. doi:10.4018/978-1-5225-3194-4.ch003

Ayodele, J. O. (2017). The Influence of Migration and Crime on Development in Lagos, Nigeria. In G. Afolayan & A. Akinwale (Eds.), *Global Perspectives on Development Administration and Cultural Change* (pp. 192–230). Hershey, PA: IGI Global. doi:10.4018/978-1-5225-0629-4.ch009

Baarda, R. (2017). Digital Democracy in Authoritarian Russia: Opportunity for Participation, or Site of Kremlin Control? In R. Luppicini & R. Baarda (Eds.), *Digital Media Integration for Participatory Democracy* (pp. 87–100). Hershey, PA: IGI Global. doi:10.4018/978-1-5225-2463-2.ch005

Balakrishnan, K. (2017). The Rationale for Offsets in Defence Acquisition from a Theoretical Perspective. In K. Burgess & P. Antill (Eds.), *Emerging Strategies in Defense Acquisitions and Military Procurement* (pp. 263–276). Hershey, PA: IGI Global. doi:10.4018/978-1-5225-0599-0.ch015

Banerjee, S. (2017). Globalization and Human Rights: How Globalization Can Be a Tool to Protect the Human Rights. In C. Akrivopoulou (Ed.), *Defending Human Rights and Democracy in the Era of Globalization* (pp. 1–16). Hershey, PA: IGI Global. doi:10.4018/978-1-5225-0723-9.ch001

Bessant, J. (2017). Digital Humour, Gag Laws, and the Liberal Security State. In R. Luppicini & R. Baarda (Eds.), *Digital Media Integration for Participatory Democracy* (pp. 204–221). Hershey, PA: IGI Global. doi:10.4018/978-1-5225-2463-2.ch010

Bhat, M. Y. (2020). Environmental Problems of Delhi and Governmental Concern. In M. Merviö (Ed.), *Global Issues and Innovative Solutions in Healthcare, Culture, and the Environment* (pp. 133–167). IGI Global. https://doi.org/10.4018/978-1-7998-3576-9.ch008

Bhat, R. A. (2021). Identification of Various Dimensions and Indicators of Immigrant Integration Into Global Scenarios. In M. Mafukata (Ed.), *Impact of Immigration and Xenophobia on Development in Africa* (pp. 247–260). IGI Global. https://doi.org/10.4018/978-1-7998-7099-9.ch014

Boachie, C. (2017). Public Financial Management and Systems of Accountability in Sub-National Governance in Developing Economies. In E. Schoburgh & R. Ryan (Eds.), *Handbook of Research on Sub-National Governance and Development* (pp. 193–217). Hershey, PA: IGI Global. doi:10.4018/978-1-5225-1645-3.ch009

Boachie, C., & Adu-Darko, E. (2018). Socio-Economic Impact of Foreign Direct Investment in Developing Countries. In V. Malepati & C. Gowri (Eds.), *Foreign Direct Investments (FDIs) and Opportunities for Developing Economies in the World Market* (pp. 66–81). Hershey, PA: IGI Global. doi:10.4018/978-1-5225-3026-8.ch004

Bogdanoski, M., Stoilkovski, M., & Risteski, A. (2020). Novel First Responder Digital Forensics Tool as a Support to Law Enforcement. In I. Management Association (Ed.), *Improving the Safety and Efficiency of Emergency Services: Emerging Tools and Technologies for First Responders* (pp. 239-270). IGI Global. https://doi.org/10.4018/978-1-7998-2535-7.ch011

Borràs, S. (2017). Rights of Nature to Protect Human Rights in Times of Environmental Crisis. In C. Akrivopoulou (Ed.), *Defending Human Rights and Democracy in the Era of Globalization* (pp. 225–261). Hershey, PA: IGI Global. doi:10.4018/978-1-5225-0723-9.ch010

Bouaziz, F. (2021). E-Government and Digital Transformation: A Conceptual Framework for Risk Factors Identification. In K. Sandhu (Ed.), *Disruptive Technology and Digital Transformation for Business and Government* (pp. 67–90). IGI Global. https://doi.org/10.4018/978-1-7998-8583-2.ch004

Bradford, A. C., McElroy, H. K., & Rosenblatt, R. (2019). Social Climate Change and the Modern Police Department: Millennials, Marijuana, and Mass Media. In I. Management Association (Ed.), *Police Science: Breakthroughs in Research and Practice* (pp. 34-51). IGI Global. https://doi.org/10.4018/978-1-5225-7672-3.ch003

Brusca, I., Olmo, J., & Labrador, M. (2018). Characterizing the Risk Factors for Financial Sustainability in Spanish Local Governments. In M. Rodríguez Bolívar & M. López Subires (Eds.), *Financial Sustainability and Intergenerational Equity in Local Governments* (pp. 206–223). Hershey, PA: IGI Global. doi:10.4018/978-1-5225-3713-7.ch009

Budd, J. R., & Littrell, M. W. (2021). Law Enforcement Challenges to Gathering Intelligence in the Street: The Fourth Amendment. In E. de Silva & A. Abeyagoonesekera (Eds.), *Intelligence and Law Enforcement in the 21st Century* (pp. 18–40). IGI Global. https://doi.org/10.4018/978-1-7998-7904-6.ch002

Bush, C. L. (2021). Policing Strategies and Approaches to Improving Community Relations: Black Citizens' Perceptions of Law Enforcement Efforts to Intentionally Strengthen Relationships. In M. Pittaro (Ed.), *Global Perspectives on Reforming the Criminal Justice System* (pp. 56–75). IGI Global. https://doi.org/10.4018/978-1-7998-6884-2.ch004

Canto Moniz, G. (2020). Is There Anything Left of the Portuguese Law Implementing the GDPR?: The Decision of the Portuguese Supervisory Authority. In M. Tzanou (Ed.), *Personal Data Protection and Legal Developments in the European Union* (pp. 125–139). IGI Global. https://doi.org/10.4018/978-1-5225-9489-5.ch007

Casoria, M., & AlSarraf, E. M. (2020). The Impact of the GDPR on Extra-EU Legal Systems: The Case of the Kingdom of Bahrain. In M. Tzanou (Ed.), *Personal Data Protection and Legal Developments in the European Union* (pp. 224–237). IGI Global. https://doi.org/10.4018/978-1-5225-9489-5.ch011

Chaves, A. (2022). Government Response Capacity to the COVID-19 Pandemic: Estimating the Impact of Lockdown Measures in Colombia. In G. Antošová (Ed.), *Innovative Strategic Planning and International Collaboration for the Mitigation of Global Crises* (pp. 114–137). IGI Global. https://doi.org/10.4018/978-1-7998-8339-5.ch008

Chen, M., & Su, F. (2017). Global Civic Engagement as an Empowering Device for Cross-Ethnic and Cross-Cultural Understanding in Taiwan. In R. Shin (Ed.), *Convergence of Contemporary Art, Visual Culture, and Global Civic Engagement* (pp. 24–45). Hershey, PA: IGI Global. doi:10.4018/978-1-5225-1665-1.ch002

Chigwata, T. C. (2017). Fiscal Decentralization: Constraints to Revenue-Raising by Local Government in Zimbabwe. In E. Schoburgh & R. Ryan (Eds.), *Handbook of Research on Sub-National Governance and Development* (pp. 218–240). Hershey, PA: IGI Global. doi:10.4018/978-1-5225-1645-3.ch010

Chowdhury, M. A. (2017). The Nexus Between Institutional Quality and Foreign Direct Investments (FDI) in South Asia: Dynamic Heterogeneous Panel Approach. In T. Dorożyński & A. Kuna-Marszałek (Eds.), *Outward Foreign Direct Investment (FDI) in Emerging Market Economies* (pp. 293–310). Hershey, PA: IGI Global. doi:10.4018/978-1-5225-2345-1.ch015

Christopher, M. E., & Tsushima, V. G. (2017). Police Interactions with Persons-in-Crisis: Emergency Psychological Services and Jail Diversion. In C. Mitchell & E. Dorian (Eds.), *Police Psychology and Its Growing Impact on Modern Law Enforcement* (pp. 274–294). Hershey, PA: IGI Global. doi:10.4018/978-1-5225-0813-7.ch014

Ciftci, D. (2022). Digital Citizenship and E-Government Integration: The Case of North Cyprus. In E. Öngün, N. Pembecioğlu, & U. Gündüz (Eds.), *Handbook of Research on Digital Citizenship and Management During Crises* (pp. 34–55). IGI Global. https://doi.org/10.4018/978-1-7998-8421-7.ch003

Citro, F., Lucianelli, G., & Santis, S. (2018). Financial Conditions, Financial Sustainability, and Intergenerational Equity in Local Governments: A Literature Review. In M. Rodríguez Bolívar & M. López Subires (Eds.), *Financial Sustainability and Intergenerational Equity in Local Governments* (pp. 101–124). Hershey, PA: IGI Global. doi:10.4018/978-1-5225-3713-7.ch005

Correa de Mello, M. B. (2019). The Information Access Law as a Full Constitutional Citizenship Instrument. In A. Melro & L. Oliveira (Eds.), *Constitutional Knowledge and Its Impact on Citizenship Exercise in a Networked Society* (pp. 55–72). IGI Global. https://doi.org/10.4018/978-1-5225-8350-9.ch003

Covell, C. E. (2018). Theoretical Application of Public Sector Planning and Budgeting. In M. Rodríguez Bolívar & M. López Subires (Eds.), *Financial Sustainability and Intergenerational Equity in Local Governments* (pp. 248–279). Hershey, PA: IGI Global. doi:10.4018/978-1-5225-3713-7.ch011

Cunha, A. M., Ferreira, A. D., & Fernandes, M. J. (2018). The Impact of Accounting Information and Socioeconomic Factors in the Re-Election of Portuguese Mayors. In G. Azevedo, J. da Silva Oliveira, R. Marques, & A. Ferreira (Eds.), *Handbook of Research on Modernization and Accountability in Public Sector Management* (pp. 406–432). Hershey, PA: IGI Global. doi:10.4018/978-1-5225-3731-1.ch019

da Rosa, I., & de Almeida, J. (2017). Digital Transformation in the Public Sector: Electronic Procurement in Portugal. In R. Shakya (Ed.), *Digital Governance and E-Government Principles Applied to Public Procurement* (pp. 99–125). Hershey, PA: IGI Global. doi:10.4018/978-1-5225-2203-4.ch005

Daramola, O. (2018). Revisiting the Legal Framework of Urban Planning in the Global South: An Explanatory Example of Nigeria. In K. Teshager Alemu & M. Abebe Alebachew (Eds.), *Handbook of Research on Sustainable Development and Governance Strategies for Economic Growth in Africa* (pp. 258–271). Hershey, PA: IGI Global. doi:10.4018/978-1-5225-3247-7.ch014

Dau, L. A., Moore, E. M., Soto, M. A., & LeBlanc, C. R. (2017). How Globalization Sparked Entrepreneurship in the Developing World: The Impact of Formal Economic and Political Linkages. In B. Christiansen & F. Kasarcı (Eds.), *Corporate Espionage, Geopolitics, and Diplomacy Issues in International Business* (pp. 72–91). Hershey, PA: IGI Global. doi:10.4018/978-1-5225-1031-4.ch005

De Man, P. (2019). In-Situ Resource Utilization: Legal Aspects. In A. Nakarada Pecujlic & M. Tugnoli (Eds.), *Promoting Productive Cooperation Between Space Lawyers and Engineers* (pp. 211–224). IGI Global. https://doi.org/10.4018/978-1-5225-7256-5.ch013

Drenner, K. (2017). Introduction to Faith in State Legislatures: Land of the Brave and the Home of the Free – The Star-Spangled Banner. In *Impacts of Faith-Based Decision Making on the Individual-Level Legislative Process: Emerging Research and Opportunities* (pp. 1–25). Hershey, PA: IGI Global. doi:10.4018/978-1-5225-2388-8.ch001

Drenner, K. (2017). The Holy Wars of Marriage. In *Impacts of Faith-Based Decision Making on the Individual-Level Legislative Process: Emerging Research and Opportunities* (pp. 92–116). Hershey, PA: IGI Global. doi:10.4018/978-1-5225-2388-8.ch004

Drenner, K. (2017). The Implications of Religious Liberty. In *Impacts of Faith-Based Decision Making on the Individual-Level Legislative Process: Emerging Research and Opportunities* (pp. 143–162). Hershey, PA: IGI Global. doi:10.4018/978-1-5225-2388-8.ch006

Duncan, S. T., & Geczi, H. (2021). Body-Worn Cameras: Panacea or Distraction for Increased Police Use of Force Accountability? In M. Pittaro (Ed.), *Global Perspectives on Reforming the Criminal Justice System* (pp. 1–25). IGI Global. https://doi.org/10.4018/978-1-7998-6884-2.ch001

Duran, D. Ş. (2019). Reflections of the Multi-Level Governance Approach on the Turkish Metropolitan Municipality System: Evaluations on the Metropolitan Law No 6360. In T. Uysal & C. Aldemir (Eds.), *Multi-Level Governance in Developing Economies* (pp. 52–91). IGI Global. https://doi.org/10.4018/978-1-5225-5547-6.ch003

Durnalı, M., & Eriçok, B. (2019). Legal Challenges on Developing Education Policy for Immigrants in Turkey. In I. Management Association (Ed.), *Immigration and Refugee Policy: Breakthroughs in Research and Practice* (pp. 223-237). IGI Global. https://doi.org/10.4018/978-1-5225-8909-9.ch013

Ehalaiye, D., Redmayne, N. B., & Laswad, F. (2020). The Case of Accounting Information for Infrastructural Assets Reporting: Local Government Borrowings and Investment Choices in the Context of Moral Hazard and Local Government Politics. In A. Cunha, A. Ferreira, M. Fernandes, & P. Gomes (Eds.), *Financial Determinants in Local Re-Election Rates: Emerging Research and Opportunities* (pp. 176–201). IGI Global. https://doi.org/10.4018/978-1-5225-7820-8.ch007

Elena, S., & van Schalkwyk, F. (2017). Open Data for Open Justice in Seven Latin American Countries. In C. Jiménez-Gómez & M. Gascó-Hernández (Eds.), *Achieving Open Justice through Citizen Participation and Transparency* (pp. 210–231). Hershey, PA: IGI Global. doi:10.4018/978-1-5225-0717-8.ch011

Eneanya, A. N. (2018). Integrating Ecosystem Management and Environmental Media for Public Policy on Public Health and Safety. In A. Eneanya (Ed.), *Handbook of Research on Environmental Policies for Emergency Management and Public Safety* (pp. 321–338). Hershey, PA: IGI Global. doi:10.4018/978-1-5225-3194-4.ch017

Erickson, G. S. (2019). Sharing Knowledge With the Government: Implications of FOIA Requests. In Y. Albastaki, A. Al-Alawi, & S. Abdulrahman Al-Bassam (Eds.), *Handbook of Research on Implementing Knowledge Management Strategy in the Public Sector* (pp. 143–158). IGI Global. https://doi.org/10.4018/978-1-5225-9639-4.ch007

Essien, E. D. (2018). Strengthening Performance of Civil Society Through Dialogue and Critical Thinking in Nigeria: Its Ethical Implications. In S. Chhabra (Ed.), *Handbook of Research on Civic Engagement and Social Change in Contemporary Society* (pp. 82–102). Hershey, PA: IGI Global. doi:10.4018/978-1-5225-4197-4.ch005

Eweida, A. M. (2021). Urban Laws in Harmony or at Odds With Knowledge-Based Urban Policies: A Case Study on Egypt. In A. Galaby & A. Abdrabo (Eds.), *Handbook of Research on Creative Cities and Advanced Models for Knowledge-Based Urban Development* (pp. 23–40). IGI Global. https://doi.org/10.4018/978-1-7998-4948-3.ch002

Fanaian, T. (2017). The Theocratic Deception Trap: Khomeini's Persuasion Techniques and Communication Patterns in His Books, Guardianship of the Jurist 1979 and Testament 1989. In E. Lewin, E. Bick, & D. Naor (Eds.), *Comparative Perspectives on Civil Religion, Nationalism, and Political Influence* (pp. 62–105). Hershey, PA: IGI Global. doi:10.4018/978-1-5225-0516-7.ch003

Faura-Martínez, U., & Cifuentes-Faura, J. (2020). Does E-Government Promote Transparency and the Fight Against Corruption in the European Union? *International Journal of Electronic Government Research, 16*(4), 42–57. https://doi.org/10.4018/IJEGR.2020100103

Fawsitt, J. (2020). Government Policy and the Disintegration of Village Community Life and Individual Identity in Urbanising Japan. In M. Merviö (Ed.), *Recent Social, Environmental, and Cultural Issues in East Asian Societies* (pp. 76–94). IGI Global. https://doi.org/10.4018/978-1-7998-1807-6.ch005

Fidanoski, F., Sergi, B. S., Simeonovski, K., Naumovski, V., & Sazdovski, I. (2018). Effects of Foreign Capital Entry on the Macedonian Banking Industry: Two-Edged Sword. In B. Sergi, F. Fidanoski, M. Ziolo, & V. Naumovski (Eds.), *Regaining Global Stability After the Financial Crisis* (pp. 308–338). Hershey, PA: IGI Global. doi:10.4018/978-1-5225-4026-7.ch015

Franconi, A. I. (2018). Economic Variations and Their Impact on Labor Legislation Throughout History in Argentina. In S. Amine (Ed.), *Employment Protection Legislation in Emerging Economies* (pp. 77–98). Hershey, PA: IGI Global. doi:10.4018/978-1-5225-4134-9.ch004

Friedrich, P., & Chebotareva, M. (2017). Options for Applying Functional Overlapping Competing Jurisdictions (FOCJs) for Municipal Cooperation in Russia. In M. Lewandowski & B. Kożuch (Eds.), *Public Sector Entrepreneurship and the Integration of Innovative Business Models* (pp. 73–107). Hershey, PA: IGI Global. doi:10.4018/978-1-5225-2215-7.ch004

Garad, A., & Qamari, I. N. (2021). Determining Factors Influencing Establishing E-Service Quality in Developing Countries: A Case Study of Yemen E-Government. *International Journal of Electronic Government Research, 17*(1), 15–30. https://doi.org/10.4018/IJEGR.2021010102

Garita, M. (2018). The Negotiation and Effects of Fiscal Privileges in Guatemala. In M. Garita & C. Bregni (Eds.), *Economic Growth in Latin America and the Impact of the Global Financial Crisis* (pp. 119–137). Hershey, PA: IGI Global. doi:10.4018/978-1-5225-4981-9.ch008

Gascó-Hernández, M. (2017). Digitalizing Police Requirements: Opening up Justice through Collaborative Initiatives. In C. Jiménez-Gómez & M. Gascó-Hernández (Eds.), *Achieving Open Justice through Citizen Participation and Transparency* (pp. 157–172). Hershey, PA: IGI Global. doi:10.4018/978-1-5225-0717-8.ch008

Gáspár-Szilágyi, S. (2017). Human Rights Conditionality in the EU's Newly Concluded Association Agreements with the Eastern Partners. In C. Akrivopoulou (Ed.), *Defending Human Rights and Democracy in the Era of Globalization* (pp. 50–79). Hershey, PA: IGI Global. doi:10.4018/978-1-5225-0723-9.ch003

Gavrielides, T. (2017). Reconciling Restorative Justice with the Law for Violence Against Women in Europe: A Scheme of Structured and Unstructured Models. In D. Halder & K. Jaishankar (Eds.), *Therapeutic Jurisprudence and Overcoming Violence Against Women* (pp. 106–120). Hershey, PA: IGI Global. doi:10.4018/978-1-5225-2472-4.ch007

Gechlik, M., Dai, D., & Beck, J. C. (2017). Open Judiciary in a Closed Society: A Paradox in China? In C. Jiménez-Gómez & M. Gascó-Hernández (Eds.), *Achieving Open Justice through Citizen Participation and Transparency* (pp. 56–92). Hershey, PA: IGI Global. doi:10.4018/978-1-5225-0717-8.ch004

Gerger, A. (2021). Technologies for Connected Government Implementation: Success Factors and Best Practices. In Z. Mahmood (Ed.), *Web 2.0 and Cloud Technologies for Implementing Connected Government* (pp. 36–66). IGI Global. https://doi.org/10.4018/978-1-7998-4570-6.ch003

Germann, S. K. (2019). The International Space Station: Legal Aspects. In A. Nakarada Pecujlic & M. Tugnoli (Eds.), *Promoting Productive Cooperation Between Space Lawyers and Engineers* (pp. 96–113). IGI Global. doi:10.4018/978-1-5225-7256-5.ch007

Gillath, N. (2017). Avoiding Conscription in Israel: Were Women Pawns in the Political Game? In E. Lewin, E. Bick, & D. Naor (Eds.), *Comparative Perspectives on Civil Religion, Nationalism, and Political Influence* (pp. 226–256). Hershey, PA: IGI Global. doi:10.4018/978-1-5225-0516-7.ch009

Górski, J. (2019). Northern Sea Route: International Law Perspectives. In V. Erokhin, T. Gao, & X. Zhang (Eds.), *Handbook of Research on International Collaboration, Economic Development, and Sustainability in the Arctic* (pp. 292–313). IGI Global. https://doi.org/10.4018/978-1-5225-6954-1.ch014

Grant, B., Woods, R., & Tan, S. F. (2017). Subnational Finance in Australia and China: The Case for Municipal Bond Banks. In E. Schoburgh & R. Ryan (Eds.), *Handbook of Research on Sub-National Governance and Development* (pp. 150–166). Hershey, PA: IGI Global. doi:10.4018/978-1-5225-1645-3.ch007

Gurpinar, B. (2018). Supporter, Activist, Rebel, Terrorist: Children in Syria. In C. Akrivopoulou (Ed.), *Global Perspectives on Human Migration, Asylum, and Security* (pp. 97–114). Hershey, PA: IGI Global. doi:10.4018/978-1-5225-2817-3.ch005

Gussen, B. F. (2018). The United States. In *Ranking Economic Performance and Efficiency in the Global Market: Emerging Research and Opportunities* (pp. 109–136). Hershey, PA: IGI Global. doi:10.4018/978-1-5225-2756-5.ch005

Halder, D., & Bhati, D. (2021). Culture, Ethnicity, and Hate Crimes: A Comparative Analysis of Preventive Laws Between India and the USA. In M. Pittaro (Ed.), *Global Perspectives on Reforming the Criminal Justice System* (pp. 236–245). IGI Global. https://doi.org/10.4018/978-1-7998-6884-2.ch013

Haque, T. (2018). Women-Friendly Working Environment in Bangladesh: Critical Analysis. In N. Mahtab, T. Haque, I. Khan, M. Islam, & I. Wahid (Eds.), *Handbook of Research on Women's Issues and Rights in the Developing World* (pp. 52–68). Hershey, PA: IGI Global. doi:10.4018/978-1-5225-3018-3.ch004

Heuva, W. E. (2017). Deferring Citizens' "Right to Know" in an Information Age: The Information Deficit in Namibia. In N. Mhiripiri & T. Chari (Eds.), *Media Law, Ethics, and Policy in the Digital Age* (pp. 245–267). Hershey, PA: IGI Global. doi:10.4018/978-1-5225-2095-5.ch014

Hwangbo, Y. (2022). Government Challenges Over Global Electronic Commerce Using FinTech: Design of Consumer Payment Tax (CPT) System. In M. Anshari, M. Almunawar, & M. Masri (Eds.), *FinTech Development for Financial Inclusiveness* (pp. 197–213). IGI Global. https://doi.org/10.4018/978-1-7998-8447-7.ch011

Ibrahim, F., Suhip, H., Kura, K. M., & Noor, H. M. L. (2022). Exploratory Study on User Satisfaction of E-HRM: Evidence From Brunei Government Employee Management System (GEMS). In P. Ordóñez de Pablos (Eds.), *Handbook of Research on Developing Circular, Digital, and Green Economies in Asia* (pp. 243-271). IGI Global. https://doi.org/10.4018/978-1-7998-8678-5.ch014

Islam, M. R. (2018). Abuse Among Child Domestic Workers in Bangladesh. In I. Tshabangu (Ed.), *Global Ideologies Surrounding Children's Rights and Social Justice* (pp. 1–21). Hershey, PA: IGI Global. doi:10.4018/978-1-5225-2578-3.ch001

Jankovic-Milic, V., & Džunić, M. (2017). Measuring Governance: The Application of Grey Relational Analysis on World Governance Indicators. In J. Stanković, P. Delias, S. Marinković, & S. Rochhia (Eds.), *Tools and Techniques for Economic Decision Analysis* (pp. 104–128). Hershey, PA: IGI Global. doi:10.4018/978-1-5225-0959-2.ch005

Jenkins, B., Semple, T., Quail, J., & Bennell, C. (2021). Optimizing Scenario-Based Training for Law Enforcement. In E. Arble & B. Arnetz (Eds.), *Interventions, Training, and Technologies for Improved Police Well-Being and Performance* (pp. 18–37). IGI Global. https://doi.org/10.4018/978-1-7998-6820-0.ch002

Jiménez-Gómez, C. E. (2017). Open Judiciary Worldwide: Best Practices and Lessons Learnt. In C. Jiménez-Gómez & M. Gascó-Hernández (Eds.), *Achieving Open Justice through Citizen Participation and Transparency* (pp. 1–15). Hershey, PA: IGI Global. doi:10.4018/978-1-5225-0717-8.ch001

Karatzimas, S., & Miquela, C. G. (2018). Two Approaches on Local Governments' Financial Sustainability: Law vs. Practice in Catalan Municipalities. In M. Rodríguez Bolívar & M. López Subires (Eds.), *Financial Sustainability and Intergenerational Equity in Local Governments* (pp. 58–81). Hershey, PA: IGI Global. doi:10.4018/978-1-5225-3713-7.ch003

Khachaturyan, M., & Klicheva, E. (2021). Risks of Introducing E-Governance Into Strategic Management Systems of Russian Companies in the Context of the Pandemic. *International Journal of Electronic Government Research, 17*(4), 84–102. https://doi.org/10.4018/IJEGR.2021100105

Khan, S. A., & Alag, A. (2020). Arbitration in Patent Disputes: To What Extent Is It Possible? In N. Dewani & A. Gurtu (Eds.), *Intellectual Property Rights and the Protection of Traditional Knowledge* (pp. 76–102). IGI Global. https://doi.org/10.4018/978-1-7998-1835-9.ch004

Kirsanov, S., Safonov, E., Tuzcuoglu, F., & Mammadov, Z. (2022). Modern Problems of Staff Training for State and Municipal Services in Russia. In O. Kulaç, C. Babaoğlu, & E. Akman (Eds.), *Public Affairs Education and Training in the 21st Century* (pp. 325–337). IGI Global. https://doi.org/10.4018/978-1-7998-8243-5.ch021

Kita, Y. (2017). An Analysis of a Lay Adjudication System and Open Judiciary: The New Japanese Lay Adjudication System. In C. Jiménez-Gómez & M. Gascó-Hernández (Eds.), *Achieving Open Justice through Citizen Participation and Transparency* (pp. 93–109). Hershey, PA: IGI Global. doi:10.4018/978-1-5225-0717-8.ch005

Kuatova, A., Bekbasarova, T., & Abdrashev, R. (2020). Introducing E-Government in Kazakhstan: The Concept of E-Democracy for the State-Public Interaction. In G. Tazhina & J. Parker (Eds.), *Toward Sustainability Through Digital Technologies and Practices in the Eurasian Region* (pp. 1–16). IGI Global. https://doi.org/10.4018/978-1-7998-2551-7.ch001

Kumari, S., Patil, Y., & Rao, P. (2017). An Approach to Sustainable Watershed Management: Case Studies on Enhancing Sustainability with Challenges of Water in Western Maharashtra. In P. Rao & Y. Patil (Eds.), *Reconsidering the Impact of Climate Change on Global Water Supply, Use, and Management* (pp. 252–271). Hershey, PA: IGI Global. doi:10.4018/978-1-5225-1046-8.ch014

Kumburu, N. P., & Pande, V. S. (2018). Decentralization and Local Governance in Tanzania: Theories and Practice on Sustainable Development. In K. Teshager Alemu & M. Abebe Alebachew (Eds.), *Handbook of Research on Sustainable Development and Governance Strategies for Economic Growth in Africa* (pp. 131–148). Hershey, PA: IGI Global. doi:10.4018/978-1-5225-3247-7.ch007

Kunock, A. I. (2017). Boko Haram Insurgency in Cameroon: Role of Mass Media in Conflict Management. In N. Mhiripiri & T. Chari (Eds.), *Media Law, Ethics, and Policy in the Digital Age* (pp. 226–244). Hershey, PA: IGI Global. doi:10.4018/978-1-5225-2095-5.ch013

Kurebwa, J. (2019). Young People-Sensitive and Participatory Governance Approaches: Lessons for the Zimbabwean Government. In J. Kurebwa & O. Dodo (Eds.), *Participation of Young People in Governance Processes in Africa* (pp. 80–99). IGI Global. https://doi.org/10.4018/978-1-5225-9388-1.ch005

Lawrie, A. (2017). The Subnational Region: A Utopia? The Challenge of Governing Through Soft Power. In E. Schoburgh & R. Ryan (Eds.), *Handbook of Research on Sub-National Governance and Development* (pp. 96–115). Hershey, PA: IGI Global. doi:10.4018/978-1-5225-1645-3.ch005

Lentzis, D. (2020). Revisiting the Basics of EU Data Protection Law: On the Material and Territorial Scope of the GDPR. In M. Tzanou (Ed.), *Personal Data Protection and Legal Developments in the European Union* (pp. 19–33). IGI Global. https://doi.org/10.4018/978-1-5225-9489-5.ch002

Leote, F. J., Teixeira, N. M., & Galvão, R. (2021). Entrepreneurship and Unemployment: Government Grants and Challenges in a Pandemic Context. In N. Teixeira, & I. Lisboa (Eds.), *Handbook of Research on Financial Management During Economic Downturn and Recovery* (pp. 20-40). IGI Global. https://doi.org/10.4018/978-1-7998-6643-5.ch002

Lewin, E., & Bick, E. (2017). Introduction: Civil Religion and Nationalism on a Godly-Civil Continuum. In E. Lewin, E. Bick, & D. Naor (Eds.), *Comparative Perspectives on Civil Religion, Nationalism, and Political Influence* (pp. 1–31). Hershey, PA: IGI Global. doi:10.4018/978-1-5225-0516-7.ch001

Luyombya, D. (2018). Management of Records and Archives in Uganda's Public Sector. In P. Ngulube (Ed.), *Handbook of Research on Heritage Management and Preservation* (pp. 275–297). Hershey, PA: IGI Global. doi:10.4018/978-1-5225-3137-1.ch014

Mabe, M., & Ashley, E. A. (2017). The Local Command Structure and How the Library Fits. In *In The Developing Role of Public Libraries in Emergency Management: Emerging Research and Opportunities* (pp. 44–60). Hershey, PA: IGI Global. doi:10.4018/978-1-5225-2196-9.ch004

Macilotti, G. (2020). Online Child Pornography: Conceptual Issues and Law Enforcement Challenges. In A. Balloni & R. Sette (Eds.), *Handbook of Research on Trends and Issues in Crime Prevention, Rehabilitation, and Victim Support* (pp. 226–247). IGI Global. https://doi.org/10.4018/978-1-7998-1286-9.ch013

Maher, C. (2018). Legal Framework, Funding, and Procurement Polices to Accelerate the Growth of the Social Enterprise Ecosystem. In *Influence of Public Policy on Small Social Enterprises: Emerging Research and Opportunities* (pp. 52–83). Hershey, PA: IGI Global. doi:10.4018/978-1-5225-2770-1.ch003

Mahmood, M. (2019). Transformation of Government and Citizen Trust in Government: A Conceptual Model. In A. Molnar (Ed.), *Strategic Management and Innovative Applications of E-Government* (pp. 107–122). IGI Global. https://doi.org/10.4018/978-1-5225-6204-7.ch005

Mahmood, Z. (2021). Cloud Computing Technologies for Connected Digital Government. In Z. Mahmood (Ed.), *Web 2.0 and Cloud Technologies for Implementing Connected Government* (pp. 19–35). IGI Global. https://doi.org/10.4018/978-1-7998-4570-6.ch002

Malik, I., Putera, V. S., & Putra, I. E. (2018). Traditional Leaders in the Reconciliation of Muslim-Christian Conflicts in Moluccas. In A. Campbell (Ed.), *Global Leadership Initiatives for Conflict Resolution and Peacebuilding* (pp. 235–248). Hershey, PA: IGI Global. doi:10.4018/978-1-5225-4993-2.ch011

Martin, S. M. (2017). Transnational Crime and the American Policing System. In M. Dawson, D. Kisku, P. Gupta, J. Sing, & W. Li (Eds.), Developing Next-Generation Countermeasures for Homeland Security Threat Prevention (pp. 72-92). Hershey, PA: IGI Global. https://doi.org/ doi:10.4018/978-1-5225-0703-1.ch004

Mekonnen, G. A., & Kassie, W. A. (2020). Fiscal Decentralization at Local Government of Ethiopia: Theory vs Practices. In S. Chhabra & M. Kumar (Eds.), *Civic Engagement Frameworks and Strategic Leadership Practices for Organization Development* (pp. 171–197). IGI Global. https://doi.org/10.4018/978-1-7998-2372-8.ch008

Mensah, I. K., Luo, C., & Abu-Shanab, E. (2021). Citizen Use of E-Government Services Websites: A Proposed E-Government Adoption Recommendation Model (EGARM). *International Journal of Electronic Government Research, 17*(2), 19–42. https://doi.org/10.4018/IJEGR.2021040102

Mhiripiri, N. A., & Chikakano, J. (2017). Criminal Defamation, the Criminalisation of Expression, Media and Information Dissemination in the Digital Age: A Legal and Ethical Perspective. In N. Mhiripiri & T. Chari (Eds.), *Media Law, Ethics, and Policy in the Digital Age* (pp. 1–24). Hershey, PA: IGI Global. doi:10.4018/978-1-5225-2095-5.ch001

Mishaal, D. A., & Abu-Shanab, E. A. (2017). Utilizing Facebook by the Arab World Governments: The Communication Success Factor. *International Journal of Public Administration in the Digital Age, 4*(3), 53–78. doi:10.4018/IJPADA.2017070105

Morim, A. C., Inácio, H., & Vieira, E. (2018). Internal Control in a Public Hospital: The Case of Financial Services Expenditure Department. In G. Azevedo, J. da Silva Oliveira, R. Marques, & A. Ferreira (Eds.), *Handbook of Research on Modernization and Accountability in Public Sector Management* (pp. 77–102). Hershey, PA: IGI Global. doi:10.4018/978-1-5225-3731-1.ch005

Mupepi, M. G. (2017). Developing Democratic Paradigms to Effectively Manage Business, Government, and Civil Society: The African Spring. In E. Schoburgh & R. Ryan (Eds.), *Handbook of Research on Sub-National Governance and Development* (pp. 432–462). Hershey, PA: IGI Global. doi:10.4018/978-1-5225-1645-3.ch020

Mwakisisya, H. J., Rugeiyamu, R., & Cyprian, S. (2021). Blending Local Government Authorities and Grassroots for Industrial Economy Through Participatory Development Communication in Tanzania. In F. Nafukho & A. Boniface Makulilo (Eds.), *Handbook of Research on Nurturing Industrial Economy for Africa's Development* (pp. 74–98). IGI Global. https://doi.org/10.4018/978-1-7998-6471-4.ch005

Naidoo, V., & Nzimakwe, T. I. (2019). M-Government and Its Application on Public Service Delivery. In R. Abassi & A. Ben Chehida Douss (Eds.), *Security Frameworks in Contemporary Electronic Government* (pp. 1–14). IGI Global. https://doi.org/10.4018/978-1-5225-5984-9.ch001

Navaratnam, R., & Lee, I. Y. (2017). Globalization as a New Framework for Human Rights Protection. In C. Akrivopoulou (Ed.), *Defending Human Rights and Democracy in the Era of Globalization* (pp. 17–49). Hershey, PA: IGI Global. doi:10.4018/978-1-5225-0723-9.ch002

Nemec, J., Meričková, B. M., Svidroňová, M. M., & Klimovský, D. (2017). Co-Creation as a Social Innovation in Delivery of Public Services at Local Government Level: The Slovak Experience. In E. Schoburgh & R. Ryan (Eds.), *Handbook of Research on Sub-National Governance and Development* (pp. 281–303). Hershey, PA: IGI Global. doi:10.4018/978-1-5225-1645-3.ch013

Neupane, A., Soar, J., Vaidya, K., & Aryal, S. (2017). Application of E-Government Principles in Anti-Corruption Framework. In R. Shakya (Ed.), *Digital Governance and E-Government Principles Applied to Public Procurement* (pp. 56–74). Hershey, PA: IGI Global. doi:10.4018/978-1-5225-2203-4.ch003

Njie, S. N., Wogu, I. A., Ogbuehi, U. K., Misra, S., & Udoh, O. D. (2021). Rising Global Challenges in Energy Demand and the Politics of Climate Change in Government Operations: Policy and Economic Development Implications. In C. Babaoğlu, E. Akman, & O. Kulaç (Eds.), *Handbook of Research on Global Challenges for Improving Public Services and Government Operations* (pp. 242–263). IGI Global. https://doi.org/10.4018/978-1-7998-4978-0.ch013

Ogunde, O. (2017). Democracy and Child Rights Protection: The Problem of the Nigerian Constitution. In C. Akrivopoulou (Ed.), *Defending Human Rights and Democracy in the Era of Globalization* (pp. 123–144). Hershey, PA: IGI Global. doi:10.4018/978-1-5225-0723-9.ch006

Ojedokun, U. A. (2017). Crime Witnesses' Non-Cooperation in Police Investigations: Causes and Consequences in Nigeria. In S. Egharevba (Ed.), *Police Brutality, Racial Profiling, and Discrimination in the Criminal Justice System* (pp. 89–99). Hershey, PA: IGI Global. doi:10.4018/978-1-5225-1088-8.ch005

Okeke, G. S. (2018). The Politics of Environmental Pollution in Nigeria: Emerging Trends, Issues, and Challenges. In A. Eneanya (Ed.), *Handbook of Research on Environmental Policies for Emergency Management and Public Safety* (pp. 300–320). Hershey, PA: IGI Global. doi:10.4018/978-1-5225-3194-4.ch016

Ökten, S., Akman, E., & Akman, Ç. (2018). Modernization and Accountability in Public-Sector Administration: Turkey Example. In G. Azevedo, J. da Silva Oliveira, R. Marques, & A. Ferreira (Eds.), *Handbook of Research on Modernization and Accountability in Public Sector Management* (pp. 18–39). Hershey, PA: IGI Global. doi:10.4018/978-1-5225-3731-1.ch002

Oladapo, O. A., & Ojebuyi, B. R. (2017). Nature and Outcome of Nigeria's #NoToSocialMediaBill Twitter Protest against the Frivolous Petitions Bill 2015. In O. Nelson, B. Ojebuyi, & A. Salawu (Eds.), *Impacts of the Media on African Socio-Economic Development* (pp. 106–124). Hershey, PA: IGI Global. doi:10.4018/978-1-5225-1859-4.ch007

Olmos, S., & Nares, J. J. (2020). Applying Fault Trees to the Analysis at the Minimum Age for Sexual Consent in the Criminal Law of México. In A. Balloni & R. Sette (Eds.), *Handbook of Research on Trends and Issues in Crime Prevention, Rehabilitation, and Victim Support* (pp. 60–78). IGI Global. https://doi.org/10.4018/978-1-7998-1286-9.ch005

Olukolu, Y. R. (2017). Harmful Traditional Practices, Laws, and Reproductive Rights of Women in Nigeria: A Therapeutic Jurisprudence Approach. In D. Halder & K. Jaishankar (Eds.), *Therapeutic Jurisprudence and Overcoming Violence Against Women* (pp. 1–14). Hershey, PA: IGI Global. doi:10.4018/978-1-5225-2472-4.ch001

Onyebadi, U., & Mbunyuza-Memani, L. (2017). Women and South Africa's Anti-Apartheid Struggle: Evaluating the Political Messages in the Music of Miriam Makeba. In U. Onyebadi (Ed.), *Music as a Platform for Political Communication* (pp. 31–51). Hershey, PA: IGI Global. doi:10.4018/978-1-5225-1986-7.ch002

Osmani, A. R. (2017). Tipaimukh Multipurpose Hydroelectric Project: A Policy Perspective – Indo-Bangla Priorities, Indigenous Peoples' Rights, and Environmental Concerns. In P. Rao & Y. Patil (Eds.), *Reconsidering the Impact of Climate Change on Global Water Supply, Use, and Management* (pp. 227–251). Hershey, PA: IGI Global. doi:10.4018/978-1-5225-1046-8.ch013

Owolabi, T. O. (2018). Free Media and Bank Reforms in West Africa: Implications for Sustainable Development. In A. Salawu & T. Owolabi (Eds.), *Exploring Journalism Practice and Perception in Developing Countries* (pp. 18–39). Hershey, PA: IGI Global. doi:10.4018/978-1-5225-3376-4.ch002

Öztürk, N. K. (2021). Government Systems and Control of Bureaucracy. In C. Babaoğlu, E. Akman, & O. Kulaç (Eds.), *Handbook of Research on Global Challenges for Improving Public Services and Government Operations* (pp. 133–150). IGI Global. https://doi.org/10.4018/978-1-7998-4978-0.ch008

Pacheco, F. M., & Alves, D. R. (2019). The New Paths of Fundamental Rights in the 21st Century: Globalization and Knowledge in a Digital Age as a Proposal. In A. Melro & L. Oliveira (Eds.), *Constitutional Knowledge and Its Impact on Citizenship Exercise in a Networked Society* (pp. 1–26). IGI Global. https://doi.org/10.4018/978-1-5225-8350-9.ch001

Panda, P., & Sahu, G. P. (2017). Public Procurement Framework in India: An Overview. In R. Shakya (Ed.), *Digital Governance and E-Government Principles Applied to Public Procurement* (pp. 229–248). Hershey, PA: IGI Global. doi:10.4018/978-1-5225-2203-4.ch010

Pande, V. S., & Kumburu, N. P. (2018). An Overview of Population Growth and Sustainable Development in Sub-Saharan Africa. In K. Teshager Alemu & M. Abebe Alebachew (Eds.), *Handbook of Research on Sustainable Development and Governance Strategies for Economic Growth in Africa* (pp. 480–499). Hershey, PA: IGI Global. doi:10.4018/978-1-5225-3247-7.ch025

Parikh, M., & Krishna, V. S. (2021). Recent Trends and Repercussions in Civil and Criminal Justice Systems: A Comparative Analysis of England, Singapore, and India. In E. Yin, & N. Kofie (Ed.), *Advancing Civil Justice Reform and Conflict Resolution in Africa and Asia: Comparative Analyses and Case Studies* (pp. 230-242). IGI Global. https://doi.org/10.4018/978-1-7998-7898-8.ch013

Pattnaik, P. N., & Shukla, M. K. (2020). Understanding Indian Political Parties Through the Lens of Marketing Management: Towards a Conceptual Political Marketing Model. In S. Kavoğlu & M. Salar (Eds.), *Political Propaganda, Advertising, and Public Relations: Emerging Research and Opportunities* (pp. 170–190). IGI Global. https://doi.org/10.4018/978-1-7998-1734-5.ch008

Paulin, A. A. (2017). Informating Public Governance: Towards a Basis for a Digital Ecosystem. *International Journal of Public Administration in the Digital Age*, *4*(2), 14–32. doi:10.4018/IJPADA.2017040102

Perelló-Sobrepere, M. (2017). Building a New State from Outrage: The Case of Catalonia. In M. Adria & Y. Mao (Eds.), *Handbook of Research on Citizen Engagement and Public Participation in the Era of New Media* (pp. 344–359). Hershey, PA: IGI Global. doi:10.4018/978-1-5225-1081-9.ch019

Pohl, G. M. (2017). The Role of Social Media in Enforcing Environmental Justice around the World. In K. Demirhan & D. Çakır-Demirhan (Eds.), *Political Scandal, Corruption, and Legitimacy in the Age of Social Media* (pp. 123–156). Hershey, PA: IGI Global. doi:10.4018/978-1-5225-2019-1.ch006

Popescu, C. R. (2022). Environmental, Social, and Corporate Governance by Avoiding Management Bias and Tax Minimization: Reaching a General Consensus Regarding a Minimum Global Tax Rate. In C. Popescu (Ed.), *COVID-19 Pandemic Impact on New Economy Development and Societal Change* (pp. 94–132). IGI Global. https://doi.org/10.4018/978-1-6684-3374-4.ch006

Popoola, T. (2017). Ethical and Legal Challenges of Election Reporting in Nigeria: A Study of Four General Elections, 1999-2011. In N. Mhiripiri & T. Chari (Eds.), *Media Law, Ethics, and Policy in the Digital Age* (pp. 78–100). Hershey, PA: IGI Global. doi:10.4018/978-1-5225-2095-5.ch005

Prakash, O. (2020). History, Policy Making, and Sustainability. In R. Das & N. Mandal (Eds.), *Interdisciplinary Approaches to Public Policy and Sustainability* (pp. 1–17). IGI Global. https://doi.org/10.4018/978-1-7998-0315-7.ch001

Provazníková, R., Sobotková, L., & Sobotka, M. (2021). Local Government Development in the Czech Republic: Dilemmas and Challenges. In C. Babaoğlu, E. Akman, & O. Kulaç (Eds.), *Handbook of Research on Global Challenges for Improving Public Services and Government Operations* (pp. 151–171). IGI Global. https://doi.org/10.4018/978-1-7998-4978-0.ch009

Qawasmeh, F. A. (2022). Understanding the Field of Public Policy in the Context of Public Administration Evolution. In O. Kulaç, C. Babaoğlu, & E. Akman (Eds.), *Public Affairs Education and Training in the 21st Century* (pp. 22–44). IGI Global. https://doi.org/10.4018/978-1-7998-8243-5.ch002

Rahman, M. S. (2017). Politics-Administration Relations and the Effect on Local Governance and Development: The Case of Bangladesh. In E. Schoburgh & R. Ryan (Eds.), *Handbook of Research on Sub-National Governance and Development* (pp. 256–279). Hershey, PA: IGI Global. doi:10.4018/978-1-5225-1645-3.ch012

Ramachandran, M., Chelliah, P. R., & Soundarabai, P. B. (2021). Towards Connected Government Services: A Cloud Software Engineering Framework. In Z. Mahmood (Ed.), *Web 2.0 and Cloud Technologies for Implementing Connected Government* (pp. 113–135). IGI Global. https://doi.org/10.4018/978-1-7998-4570-6.ch006

Ravotti, N. (2020). In the [Source] C\ode of the Conquerors: On the Need for Culturally-Minded Tribal Law Research Databases. In S. Edwards III, & D. Santos (Eds.), *Digital Transformation and Its Role in Progressing the Relationship Between States and Their Citizens* (pp. 124-137). IGI Global. https://doi.org/10.4018/978-1-7998-3152-5.ch006

Razzante, R. (2020). The Fight Against Corruption. In A. Balloni & R. Sette (Eds.), *Handbook of Research on Trends and Issues in Crime Prevention, Rehabilitation, and Victim Support* (pp. 167–186). IGI Global. doi:10.4018/978-1-7998-1286-9.ch010

Reddy, P. S. (2017). Political-Administrative Interface at the Local Sphere of Government with Particular Reference to South Africa. In E. Schoburgh & R. Ryan (Eds.), *Handbook of Research on Sub-National Governance and Development* (pp. 242–255). Hershey, PA: IGI Global. doi:10.4018/978-1-5225-1645-3.ch011

Rombo, D. O., & Lutomia, A. N. (2018). Tracing the Rights of Domestic and International Kenyan House Helps: Profiles, Policy, and Consequences. In N. Mahtab, T. Haque, I. Khan, M. Islam, & I. Wahid (Eds.), *Handbook of Research on Women's Issues and Rights in the Developing World* (pp. 1–18). Hershey, PA: IGI Global. doi:10.4018/978-1-5225-3018-3.ch001

Rouzbehani, K. (2017). Health Policy Implementation: Moving Beyond Its Barriers in United States. In N. Wickramasinghe (Ed.), *Handbook of Research on Healthcare Administration and Management* (pp. 541–552). Hershey, PA: IGI Global. doi:10.4018/978-1-5225-0920-2.ch032

Ryan, R., & Woods, R. (2017). Decentralization and Subnational Governance: Theory and Praxis. In E. Schoburgh & R. Ryan (Eds.), *Handbook of Research on Sub-National Governance and Development* (pp. 1–33). Hershey, PA: IGI Global. doi:10.4018/978-1-5225-1645-3.ch001

Sabao, C., & Chingwaramusee, V. R. (2017). Citizen Journalism on Facebook and the Challenges of Media Regulation in Zimbabwe: Baba Jukwa. In N. Mhiripiri & T. Chari (Eds.), *Media Law, Ethics, and Policy in the Digital Age* (pp. 193–206). Hershey, PA: IGI Global. doi:10.4018/978-1-5225-2095-5.ch011

Saidane, A., & Al-Sharieh, S. (2019). A Compliance-Driven Framework for Privacy and Security in Highly Regulated Socio-Technical Environments: An E-Government Case Study. In R. Abassi & A. Ben Chehida Douss (Eds.), *Security Frameworks in Contemporary Electronic Government* (pp. 15–50). IGI Global. https://doi.org/10.4018/978-1-5225-5984-9.ch002

Sandill, S. (2020). Law, Equality, and Entrepreneurship Through a Gendered Lens: Bridging the Gap Between Academia, Legislature, and Politics. In T. Moeke-Pickering, S. Cote-Meek, & A. Pegoraro (Eds.), *Critical Reflections and Politics on Advancing Women in the Academy* (pp. 124–149). IGI Global. https://doi.org/10.4018/978-1-7998-3618-6.ch008

Santos, H. R., & Tonelli, D. F. (2019). Smart Government and the Maturity Levels of Sociopolitical Digital Interactions: Analysing Temporal Changes in Brazilian E-Government Portals. In A. Molnar (Ed.), *Strategic Management and Innovative Applications of E-Government* (pp. 176–199). IGI Global. https://doi.org/10.4018/978-1-5225-6204-7.ch008

Saponaro, A. (2021). "Visible" and "Invisible" Victims in the Criminal Justice System: Victim-Oriented Paradigms and Models. In R. Blasdell, L. Krieger-Sample, & M. Kilburn (Eds.), *Invisible Victims and the Pursuit of Justice: Analyzing Frequently Victimized Yet Rarely Discussed Populations* (pp. 1–23). IGI Global. doi:10.4018/978-1-7998-7348-8.ch001

Shaikh, A. K., Ahmad, N., Khan, I., & Ali, S. (2021). E-Participation Within E-Government: A Bibliometric-Based Systematic Literature Review. *International Journal of Electronic Government Research, 17*(4), 15–39. https://doi.org/10.4018/IJEGR.2021100102

Shakya, R. K., & Schapper, P. R. (2017). Digital Governance and E-Government Principles: E-Procurement as Transformative. In R. Shakya (Ed.), *Digital Governance and E-Government Principles Applied to Public Procurement* (pp. 1–28). Hershey, PA: IGI Global. doi:10.4018/978-1-5225-2203-4.ch001

Shambare, R. (2019). Facilitating Consumers' Adoption of E-Government in South Africa: Supply Side-Driven Virtuous Cycles. In A. Gbadamosi (Ed.), *Exploring the Dynamics of Consumerism in Developing Nations* (pp. 243–265). IGI Global. https://doi.org/10.4018/978-1-5225-7906-9.ch011

Siphambe, H., Kolobe, M., & Oageng, I. P. (2018). Employment Protection Legislation and Unemployment in Botswana. In S. Amine (Ed.), *Employment Protection Legislation in Emerging Economies* (pp. 157–191). Hershey, PA: IGI Global. doi:10.4018/978-1-5225-4134-9.ch008

Song, M. Y., & Abelson, J. (2017). Public Engagement and Policy Entrepreneurship on Social Media in the Time of Anti-Vaccination Movements. In M. Adria & Y. Mao (Eds.), *Handbook of Research on Citizen Engagement and Public Participation in the Era of New Media* (pp. 38–56). Hershey, PA: IGI Global. doi:10.4018/978-1-5225-1081-9.ch003

Sood, P., Malhotra, M., & Nijjer, S. (2021). Government Policies During Lockdown: Indian vs. International Perspective. In V. Kumar & G. Malhotra (Eds.), *Stakeholder Strategies for Reducing the Impact of Global Health Crises* (pp. 18–39). IGI Global. https://doi.org/10.4018/978-1-7998-7495-9.ch002

Stacey, E. (2018). Networked Protests: A Review of Social Movement Literature and the Hong Kong Umbrella Movement (2017). In S. Chhabra (Ed.), *Handbook of Research on Civic Engagement and Social Change in Contemporary Society* (pp. 347–363). Hershey, PA: IGI Global. doi:10.4018/978-1-5225-4197-4.ch020

Stamatakis, N. (2017). Authority and Legitimacy: A Quantitative Study of Youth's Perceptions on the Brazilian Police. In S. Egharevba (Ed.), *Police Brutality, Racial Profiling, and Discrimination in the Criminal Justice System* (pp. 151–213). Hershey, PA: IGI Global. doi:10.4018/978-1-5225-1088-8.ch009

Steinert, S. W. (2021). An Interview With Chief Sargent of the Worcester, MA Police Department. In E. Arble & B. Arnetz (Eds.), *Interventions, Training, and Technologies for Improved Police Well-Being and Performance* (pp. 185–197). IGI Global. https://doi.org/10.4018/978-1-7998-6820-0.ch011

Sugars, J. M. (2017). Refoulement and Refugees. In C. Akrivopoulou (Ed.), *Defending Human Rights and Democracy in the Era of Globalization* (pp. 181–197). Hershey, PA: IGI Global. doi:10.4018/978-1-5225-0723-9.ch008

Takaya-Umehara, Y. (2019). Suborbital Spaceflight: Legal Aspects. In A. Nakarada Pecujlic & M. Tugnoli (Eds.), *Promoting Productive Cooperation Between Space Lawyers and Engineers* (pp. 64–78). IGI Global. https://doi.org/10.4018/978-1-5225-7256-5.ch005

Tan, S. F. (2017). Local Representation in Australia: Preliminary Findings of a National Survey. In E. Schoburgh & R. Ryan (Eds.), *Handbook of Research on Sub-National Governance and Development* (pp. 368–384). Hershey, PA: IGI Global. doi:10.4018/978-1-5225-1645-3.ch017

Tavares, M. D., & Rodrigues, L. L. (2018). Strategic Responses of Public Sector Entities to GRI Sustainability Reports. In G. Azevedo, J. da Silva Oliveira, R. Marques, & A. Ferreira (Eds.), *Handbook of Research on Modernization and Accountability in Public Sector Management* (pp. 159–188). Hershey, PA: IGI Global. doi:10.4018/978-1-5225-3731-1.ch008

Thakre, A. G. (2017). Sexual Harassment of Women in Workplace in India: An Assessment of Implementation of Preventive Laws and Practicing of Therapeutic Jurisprudence in New Delhi. In D. Halder & K. Jaishankar (Eds.), *Therapeutic Jurisprudence and Overcoming Violence Against Women* (pp. 135–146). Hershey, PA: IGI Global. doi:10.4018/978-1-5225-2472-4.ch009

Tiwary, A. (2017). Key Elements of CEAF. In *Driving Efficiency in Local Government Using a Collaborative Enterprise Architecture Framework: Emerging Research and Opportunities* (pp. 25–61). Hershey, PA: IGI Global. doi:10.4018/978-1-5225-2407-6.ch002

Toscano, J. P. (2017). Social Media and Public Participation: Opportunities, Barriers, and a New Framework. In M. Adria & Y. Mao (Eds.), *Handbook of Research on Citizen Engagement and Public Participation in the Era of New Media* (pp. 73–89). Hershey, PA: IGI Global. doi:10.4018/978-1-5225-1081-9.ch005

Trompetter, P. S. (2019). A History of Police Psychology. In I. Management Association (Eds.), *Police Science: Breakthroughs in Research and Practice* (pp. 377-402). IGI Global. https://doi.org/10.4018/978-1-5225-7672-3.ch019

Tryma, K., & Salnikova, N. (2021). The Influence of Religion on Political Parties of the European Union. In E. Alaverdov & M. Bari (Eds.), *Global Development of Religious Tourism* (pp. 98-112). IGI Global. https://doi.org/10.4018/978-1-7998-5792-1.ch007

Tsabedze, V. W. (2020). Strategies for Managing E-Records for Good Governance: Reflection on E-Government in the Kingdom of Eswatini. In M. Rodríguez Bolívar & M. Cortés Cediel (Eds.), *Digital Government and Achieving E-Public Participation: Emerging Research and Opportunities* (pp. 63–86). IGI Global. https://doi.org/10.4018/978-1-7998-1526-6.ch004

Tshishonga, N. (2017). Operation Sukuma-Sakhe: A New Social Contract for Decentralized Service Delivery and Responsive Governance in KwaZulu-Natal. In E. Schoburgh & R. Ryan (Eds.), *Handbook of Research on Sub-National Governance and Development* (pp. 304–323). Hershey, PA: IGI Global. doi:10.4018/978-1-5225-1645-3.ch014

Tsygankov, S., & Gasanova, E. (2017). Electronification of the Public Procurement System: A Comparative Analysis of the Experience of the Russian Federation and Ukraine. In R. Shakya (Ed.), *Digital Governance and E-Government Principles Applied to Public Procurement* (pp. 267–277). Hershey, PA: IGI Global. doi:10.4018/978-1-5225-2203-4.ch013

Tüzünkan, D. (2018). The International Migration Movements and Immigrant Policies From the Ottoman Empire 1299 to Republican Turkey 2016. In Ş. Erçetin (Ed.), *Social Considerations of Migration Movements and Immigration Policies* (pp. 13–45). Hershey, PA: IGI Global. doi:10.4018/978-1-5225-3322-1.ch002

Ullah, Z. (2022). Violent Extremism and the Politics of Education in Pakistan: An Analysis of the Links Between Anti-Terror Laws, the Curriculum, and the Islamised Public Sphere. In O. Kulaç, C. Babaoğlu, & E. Akman (Eds.), *Public Affairs Education and Training in the 21st Century* (pp. 256–275). IGI Global. https://doi.org/10.4018/978-1-7998-8243-5.ch017

Valenzuela, R., & Ochoa, A. (2018). Open Mexico Network in the Implementation of National Open Data Policy. In A. Kok (Ed.), *Proliferation of Open Government Initiatives and Systems* (pp. 50–67). Hershey, PA: IGI Global. doi:10.4018/978-1-5225-4987-1.ch003

Valera-Ordaz, L., & Humanes, M. L. (2022). What Drives Selective Exposure to Political Information in Spain? Comparing Political Interest and Ideology. In D. Palau-Sampio, G. López García, & L. Iannelli (Eds.), *Contemporary Politics, Communication, and the Impact on Democracy* (pp. 93–112). IGI Global. https://doi.org/10.4018/978-1-7998-8057-8.ch006

van der Vliet-Bakker, J. M. (2019). Environmentally Forced Migration and Human Rights. In I. Management Association (Eds.), *Immigration and Refugee Policy: Breakthroughs in Research and Practice* (pp. 336-362). IGI Global. https://doi.org/10.4018/978-1-5225-8909-9.ch019

Waller, P. (2017). Co-Production and Co-Creation in Public Services: Resolving Confusion and Contradictions. *International Journal of Electronic Government Research, 13*(2), 1–17. doi:10.4018/IJEGR.2017040101

Wodecka-Hyjek, A. (2017). Co-Operation between the Public Administration and Non-Profit Organisations as a Condition of the Development of Public Entrepreneurship: On the Example of the Selected World Solutions. In V. Potocan, M. Ünğan, & Z. Nedelko (Eds.), *Handbook of Research on Managerial Solutions in Non-Profit Organizations* (pp. 253–275). Hershey, PA: IGI Global. doi:10.4018/978-1-5225-0731-4.ch012

Yang, K. C., & Kang, Y. (2017). Social Media, Political Mobilization, and Citizen Engagement: A Case Study of the March 18, 2014, Sunflower Student Movement in Taiwan. In M. Adria & Y. Mao (Eds.), *Handbook of Research on Citizen Engagement and Public Participation in the Era of New Media* (pp. 360–388). Hershey, PA: IGI Global. doi:10.4018/978-1-5225-1081-9.ch020

Yang, K. C., & Kang, Y. (2020). Will Microblogs Shape China's Civil Society Under President's Xi's Surveillance State?: The Case of Anti-Extradition Law Protests in Hong Kong. In V. Kumar & G. Malhotra (Eds.), *Examining the Roles of IT and Social Media in Democratic Development and Social Change* (pp. 156–184). IGI Global. https://doi.org/10.4018/978-1-7998-1791-8.ch007

Yasmeen, H., Wang, Y., Zameer, H., & Ismail, H. (2020). Modeling the Role of Government, Firm, and Civil Society for Environmental Sustainability. In I. Management Association (Ed.), *Developing Eco-Cities Through Policy, Planning, and Innovation: Can It Really Work?* (pp. 62-83). IGI Global. https://doi.org/10.4018/978-1-7998-0441-3.ch003

Yavuz, N., Karkın, N., & Sevinç Çubuk, E. B. (2020). Explaining Government Crowdsourcing Decisions: A Theoretical Model. In M. Rodríguez Bolívar & M. Cortés Cediel (Eds.), *Digital Government and Achieving E-Public Participation: Emerging Research and Opportunities* (pp. 159–183). IGI Global. https://doi.org/10.4018/978-1-7998-1526-6.ch008

Zhaleleva, S., Zhaleleva, R., & Pasternak, A. (2020). Evolution of Business-Government Interaction Models: Their Use and Management. In G. Tazhina & J. Parker (Eds.), *Toward Sustainability Through Digital Technologies and Practices in the Eurasian Region* (pp. 17–35). IGI Global. https://doi.org/10.4018/978-1-7998-2551-7.ch002

Zhao, B. (2018). A Privacy Perspective of Open Government: Sex, Wealth, and Transparency in China. In A. Kok (Ed.), *Proliferation of Open Government Initiatives and Systems* (pp. 29–48). Hershey, PA: IGI Global. doi:10.4018/978-1-5225-4987-1.ch002

About the Contributors

Carlos Nunes Silva, Geographer, PhD, Professor Auxiliar at the Institute of Geography and Spatial Planning, University of Lisbon, Portugal, holds a degree in Geography (University of Coimbra), a post-graduation in European Studies (University of Coimbra - Faculty of Law), a Master degree in Human Geography: Regional and Local Planning (University of Lisbon), and a PhD in Geography: Regional and Local Planning (University of Lisbon). His research interests focus mainly on urban and metropolitan governance, the history and theory of urban planning, urban planning in Africa, urban e-planning, urban planning ethics, local government policies, local e-government, and research methods. He is the Chair of the International Geographical Union Commission on 'Geography of Governance', and the founding editor-in-chief of the 'International Journal of E-Planning Research' (IJEPR).

* * *

Rogério Palhares Zschaber de Araújo received his PhD in Geography from the Universidade Federal de Minas Gerais - UFMG, Brazil in 2009, his Master's Degree in Community Planning from the University of Rhode Island, USA in 1985 and his Bachelor Degree in Architecture from UFMG in 1979. He was a partner and planning consultant at Praxis Projetos e Consultoria from 1989 until 2014 and lectured at the Architecture Department of the Pontifícia Universidade Católica de Minas Gerais from1997 until 2009, when he joined the UFMG Urban Planning Department as an Adjunct Professor. Having a full-time position at UFMG since 2014, he lectures and develops extension and research projects on urban and environmental planning, environmental regulation, urban design, cultural heritage, land use and water resource management. In 2016-17, he was a Post-Doctoral Research Fellow at the Faculdade de Arquitetura da Universidade de Lisboa, where he collaborates with the CIAUD Research Center on Architecture, Planning and Design.

Edyta Bąkowska-Waldmann is a socio-economic geographer and urban planner holding a position of assistant professor at the Adam Mickiewicz University in Poznań. She specializes in urban planning and local governance, with particular interest in public participation, PPGIS and decision-making processes. Winner of prestigious awards including The Prime Minister of Poland Award for Scientific and Technical Achievements and the Polish Academy of Science Award for doctoral thesis in the field of spatial management. Author of numerous scientific publications in the field of spatial management and public participation. Academic teacher and certified Trainer of Science.

Michele Campagna is Professor of Spatial Planning at the Department of Civil and Environmental Engineering and Architecture of the University of Cagliari, where he teaches geodesign. Author of over 130 publications, his research deals with the use of Information and Communication Technology in Spatial planning, Planning Support Systems, metaplanning, Volunteered and Social Media Geographic information, and geodesign.

Mikulas Cerny is a PhD student at the Department of Road and Urban Transport at the University of Zilina, and his research topic is focused on Smart technologies in Transport.

Marco Dalmasso is a computer scientist at the Regional Epidemiology Unit - ASL TO3. He is responsible of the data management and development of the health information system of this Unit. His research interests lie in the area of data mining and innovative use of tools and methods for information treatment. He collaborates with the other departments of the regional epidemiology network to support health surveillance activities in the Piedmont region.

Mona El Khafif, Dr. techn., Dipl. Ing., is an Associate Professor at UVA School of Architecture and Graduate Urban Design Program Director. She is co-author of the award winning publication 'URBAN-build: Local/Global' (with Ila Berman) and author of 'Staged Urbanism: Urban Spaces for Art, Culture and Consumption in the Age of Leisure Society' (German edition). Her research operates at multiple scales, examining the interdisciplinary aspects of urban design, temporary urbanism, urban prototyping, and strategies for the smart city. At UVA El Khafif serves as the RCN director of the recently funded NSF Grant entitled "MainStreet21" supporting a network of small and midscale cities in Virginia and co-directs - together with Andrew Mondschein - the Networked Public Space project.

Marian Gogola is an associate professor at the Department of Road and Urban Transport, Faculty of Operations and Economics of Transport and Communications at the University of Žilina. He obtained his PhD from the University of Zilina. His research focuses on transport planning and modelling, travel behaviour, sustainable mobility, and public transport and non-motorised transport, mainly cycling. During his work, he had research stays and visits at the University of Genoa (Italy), Kielce University of Technology (Poland), Kazimierz Pułaski University of Technology and Humanities in Radom (Poland), ETH Zurich (Switzerland), University of Texas at El Paso (USA), Vienna University of Technology (Austria). Marián took part in several European projects such as Central MeetBike, Benefit, Advance, RECOAUDIT, YOUMOBIL, RUMOBIL, MoTiV.

Tomasz Kaczmarek is a professor at the University of Adam Mickiewicz in Poznań. Director of the Metropolitan Research Center, Head of the Department of Settlement Systems and Territorial Management. Member of the Steerings Committee of the Commission on Geography of Governance - International Geographical Union. A member of the national scientific bodies: the Committee for the Spatial Development of the Country of the Polish Academy of Sciences, the Committee on Geographical Sciences of the Polish Academy of Sciences, the Council of Scientific Excellence. Research specialization: local governance in Europe, strategic management and spatial planning in metropolitan areas, social participation in decision making processes in the public sector. Contractor of international grants, among others Accommodating Creative Knowledge - Competitiveness of European Metropolitan Regions within the Enlarged Union (Framework Program of the EU). Grant Manager National Center for Research and Development "Geoportal supporting social participation in spatial planning." Author of 180 publications, 8 books, including "Territorial and administrative structures and their reforms in European countries"," Territorial and administrative systems of metropolitan areas in Europe". Laureate, with a research team, of the Prime Minister's Award for scientific and technical achievements (2017). Other publications; Top-Down and Bottom-Up Metropolitan Integration in Poland, Models of Governance in the Urban Functional Areas, Soft Planning for Soft Spaces, Concept of Poznań Metropolitan Area Development, Geoweb Methods for Public Participation in Urban Planning, The Use of Geo-Questionnaire in Spatial Planning.

Stanislav Kubal'ák is a PhD student at the Department of Road and Urban Transport at the University of Zilina, and his research topic is focused on sustainable mobility.

Gwendoline l'Her is a researcher in social sciences, member of the CRENAU team of the Ambiances Architectures Urbanités laboratory of Nantes University. Her research focuses on the relationship between sensitivity and metrology. During her PhD she studied participatory metrology in the context of transformations in environmental public action. She is currently working on the processes of production of knowledge on the urban environment by experts.

Nano Langenheim is a landscape architect, horticulturist, arborist, a James Cox Scholar with the CRC for Water Sensitive Cities (MADA, Monash University), and lecturer in landscape architecture and urban design at the Melbourne School of Design at the University of Melbourne. Dr Langenheim specialises in the use of procedural modelling and geospatial data visualisation, and knowledge transfer techniques to support multidisciplinary decision-making in design education, research and practice. She is also an Atlas of Living Australia & Biodiversity and Climate Change Virtual Laboratory E-infrastructure Champion (a National Built Environment Champion of EcoED partnership program).

Andrew Mondschein, PhD AICP, is an associate professor of urban and environmental planning at the University of Virginia School of Architecture. He studies transportation systems and travel behavior, seeking to foster equitable, sustainable accessibility in cities and regions. He addresses a rapidly changing terrain of transportation and information technologies, identifying means to assert social imperatives during a period of urban transformation. His research emphasizes the role of information and knowledge in fostering individual- and community-level capability and control over mobility. With Mona El Khafif, he co-directs the Networked Public Space project.

Ana Clara Mourão Moura received her Bachelor's Degree in Architecture and Planning from Federal University of Minas Gerais – UFMG in 1988, and a Post Graduate Degree in Territorial and Urban Planning from PUC-MG and University of Bologna in 1990, a Master's Degree in Geography from UFMG in 1993 and a PhD in Geography (GIS) from Federal University of Rio de Janeiro in 2002. She is currently Professor at UFMG, Department of Urban Planning, where she coordinates the geoprocessing Laboratory in the School of Architecture. Her main interests and experience mainly are related to the following themes: GIS, Landscape, Cultural Heritage, Environmental Analysis and Urban Analysis. She is also the coordinator of the CNPq research group: "Geoprocessing in the management of the urban landscape and environment analysis." She was awarded the Cartographic Medal of Merit by the Brazilian Cartography Society in 2010 and was nominated one of the 5 personalities of the decade in Geoprocessing by MundoGeo.

Thaisa Daniele Apóstolo Nogueira is an Architect and Urbanist graduated from UFMG in 2014, with a period of international exchange at the Escuela Tecnica Superior de Arquitetectura de Madrid - ETSAM, at the Polytechnic University of Madrid. She has graduated in Specialization in Geoprocessing and Spatial Analysis at the Pontifical Catholic University of Minas Gerais in 2018 and is currently a master's candidate in the Program for Analysis and Modeling of Environmental Systems, at the Institute of Geosciences at UFMG (IGC/UFMG). Professionally, she works at Práxis Projetos e Consultoria, where she participates in the elaboration of environmental impact studies, Neighborhood Impact Studies (EIV), master plans and urban projects.

Sylvie Occelli is an urban and regional planner. She served as a senior research fellow at the Socio-Economic Research Institute of Piedmont (IRES) and led the Socio-Technical Research Unit until 2019. She has worked in various fields of regional analysis such as housing, metropolitan systems, socio-environmental indicators, transportation and mobility, urban modelling and spatial analysis. Main research interests include: transport, mobility and road safety policy, ICT and broadband impact on regional development, e-health and telemedicine, and the role of model-based activity as a way to support modernization in policy practices.

Jenni Partanen, DSc (Arch), works as a professor of urban studies at the Tallinn University of Technology, Academy of Architecture and Urban Studies, Estonia, and as a visiting scholar in Tampere University, Finland. She has formerly worked at Aalto University and Tampere University as a Research Fellow. Partanen leads a research team at TalTech exploring ongoing transition towards technology mediated 'smart' urbanity; new mobility; and planning of sustainable urban futures. Her personal research interests include urban self-organizing processes and their simulation, spatial data analytics, problematic of ubiquitous 'programmable city', and the possibilities of planning in controlling the complex urbanity.

Myriam Servières is an associate professor in computer science at Ecole Centrale Nantes (ECN) since 2006. She obtained her MSc and Ph.D. in Applied Computer Science in ECN in 2002 and in the University of Nantes in 2005. Her research activity takes place in the CNRS JRU « Ambiences, Architecture and Urbanity » where she studies Computer Vision, on-site geolocalization, GIS, Augmented Reality and the use of mobile tools for urban design. She is assistant Director of CNRS' UMR Ambiances Architectures Urbanités.

Daniel Siret holds a PhD in Architecture. He is a Research Fellow at the School of Architecture of Nantes.

Barney Warf is a professor of Geography at the University of Kansas. His research and teaching interests lie within the broad domain of human geography. Much of his research concerns producer services and telecommunications, particularly the geographies of the internet, including the digital divide, e-government, and internet censorship. He views these topics through the lens of political economy and social theory. One interest concerns time-space compression, or how societies fold time and space in different ways. He maintains an active interest in political geography, including elections, voting technologies, and the U.S. electoral college. He has also studied a range of topics that fall under the umbrella of globalization, such as international producer services, fiber optics, the satellite industry, offshore banking, electronic funds transfer, military spending, and religious diversity. More recently he has studied the geographies of corruption. He has authored, co-authored, or co-edited nine books, three encyclopedias, 65 book chapters, and roughly 110 refereed journal articles. Currently, he serves as editor of Geojournal, editor-in-chief for geography for Oxford Bibliographies On-Line, and edits a series of geography texts for Springer, de Gruyter, and Rowman and Littlefield.

Marcus White is an award-winning architect and urban designer, director of Harrison and White, Professor of Architecture and Urban Design at Swinburne University of Technology, and the Deputy Director of the Centre for Design Innovation. He has been awarded the RAIA Haddon Travelling Scholarship, the inaugural AIA National Emerging Architect Award for his contribution to practice, teaching and research, the Graham Treloar Fellowship, AIA awards for Residential Alt/Add, Interior Architecture, Sustainable Architecture, Joseph Reed Award for Urban Design, Henry Bastow Award for Educational Architecture, the Melbourne Prize, and the Victorian Architecture Medal. His design research into urban modelling and new design methods has been widely published and exhibited throughout Australia, North America, Asia and Europe.

Index

www.ingramcontent.com/pod-product-compliance
Lightning Source LLC
LaVergne TN
LVHW062145300926
849161LV00010B/2626